# Interaction of Mechanics and Mathematics

Marcelo Epstein · Marek Elżanowski

# Material Inhomogeneities and their Evolution

A Geometric Approach

With 29 Figures

# IMM Advisory Board

D. Colton (USA) . R. Knops (UK) . G. DelPiero (Italy) . Z. Mroz (Poland) .
M. Slemrod (USA) . S. Seelecke (USA) . L. Truskinovsky (France)

IMM is promoted under the auspices of ISIMM (International Society for the Interaction of Mechanics and Mathematics).

## Authors

Marcelo Epstein
Department of Mechanical
& Manufacturing Engineering
The University of Calgary
Calgary, Alberta T2N 1N4
Canada
mepstein@ucalgary.ca

Marek Elżanowski
Department of Mathematics and Statistics
Portland State University
P.O.Box 751
Portland, Oregon 97207
USA
elzanowskim@pdx.edu.

Library of Congress Control Number: 2007930220

ISSN print edition: 1860-6245
ISSN electronic edition: 1860-6253
ISBN    978-3-540-72372-1 Springer Berlin Heidelberg New York

This work is subject to copyright. All rights are reserved, whether the whole or part of the material is concerned, specifically the rights of translation, reprinting, reuse of illustrations, recitation, broadcasting, reproduction on microfilm or in any other way, and storage in data banks. Duplication of this publication or parts thereof is permitted only under the provisions of the German Copyright Law of September 9, 1965, in its current version, and permission for use must always be obtained from Springer. Violations are liable for prosecution under the German Copyright Law.

Springer is a part of Springer Science+Business Media
springer.com
© Springer-Verlag Berlin Heidelberg 2007

The use of general descriptive names, registered names, trademarks, etc. in this publication does not imply, even in the absence of a specific statement, that such names are exempt from the relevant protective laws and regulations and therefore free for general use.

Typesetting: by the authors and Integra using a springer LaTeXmacro package
Cover design: WMX Design, Heidelberg

Printed on acid-free paper    SPIN: 11534907    89/Integra    5 4 3 2 1 0

To the memory of Noé Epstein, z"l,
and Eligiusz Elżanowski

# Preface

With its origins in the theories of continuous distributions of dislocations and of metal plasticity, inhomogeneity theory is a rich and vibrant field of research. The recognition of the important role played by configurational or material forces in phenomena such as growth and remodelling is perhaps its greatest present-day impetus. While some excellent comprehensive works approaching the subject from different angles have been published, the objective of this monograph is to present a point of view that emphasizes the differential-geometric aspects of inhomogeneity theory. In so doing, we follow the general lines of thought that we have propounded in many publications and presentations over the last two decades. Although based on these sources, this book is a stand-alone entity and contains some new results and perspectives. At the same time, it does not intend to present either a historical account of the development of the subject or a comprehensive picture of the various schools of thought that can be encountered by perusing scholarly journals and attending specialized symposia.

The book is divided into three parts, the first of which is entirely devoted to the formulation of the theory in the absence of evolution. In other words, time is conspicuously absent from Part I. It opens with the geometric characterization of material inhomogeneity within the context of simple bodies in Chapter 1, followed by extensions to second-grade and Cosserat media in Chapters 2 and 3. Finally, Chapter 4 deals with a novel generalization of the notions of material uniformity and homogeneity to functionally-graded media (FGM). Throughout these chapters, the required fundamental differential-geometric constructs are introduced and discussed in a rather intuitive fashion, as much as the inherent difficulty of the topic permits. Thus, concepts such as principal frame bundles, G-structures and groupoids are motivated from within the Continuum Mechanics context without paying exaggerated attention to mathematical rigour. The second part of the book deals with the framework underlying the various phenomena of material evolution in time, such as anelasticity, growth and remodelling. The role of material forces, in particular the Eshelby and Mandel stresses, is highlighted. In fact, most of

Chapter 5 deals with various definitions and interpretations of these material forces and their generalizations. Chapter 6 revolves around the formulation of the general principles of the theory of material evolution of simple bodies. Possible applications to phenomena of material growth and remodelling are outlined and discussed. The shorter Chapter 7 formulates the theory of material evolution in the realm of second-grade bodies. An extension to Cosserat media has not been attempted, although it could be carried out along similar lines. An interesting and elegant formulation of material evolution in the context of General Relativity has also been left out for reasons of space. The final part of the book is a rigorous compendium of modern Differential Geometry as necessitated by the mechanical theory presented in Parts I and II. Although many first-rate books on Differential Geometry exist, a comprehensive treatment of all the concepts needed cannot be found in a single source. For this reason, after two foundational chapters (Chapters 8 and 9) dealing with manifolds and connections on principal fibre bundles, we have devoted two chapters to general bundles of frames (Chapter 10) and to connections of higher order (Chapter 11). The book closes with a short appendix on groupoids.

As a result of the particular layout of the book, as described above, many mathematical notions are introduced twice: once within the mechanical context and again in the mathematical compendium. We hope that this feature will be of benefit both to readers whose main background is mechanics and to those who approach the subject form a mathematical viewpoint. To help in the attainment of this objective, cross-referencing with Part III is provided throughout the first and second parts. The analytic index too has been organized with this idea in mind. Finally, a reader wishing to acquire only a working familiarity with the subject, can do so by reading just Chapters 1 and 6, with an occasional incursion into Chapter 5.

Calgary and Portland, *Marcelo Epstein*
March 2007 *Marek Elżanowski*

# Contents

## Part I Inhomogeneity in Continuum Mechanics

**1 An overview of inhomogeneity theory** .................... 3
   1.1 The constitutive equation of a material body ............... 3
      1.1.1 Configurations, deformations and their gradient ....... 3
      1.1.2 Locality, simplicity, elasticity ...................... 4
   1.2 Material uniformity ....................................... 7
      1.2.1 The notion of material isomorphism ................. 7
      1.2.2 Material symmetries and the non-uniqueness of material isomorphisms ............................. 10
      1.2.3 The material archetype ............................ 12
      1.2.4 Local material parallelisms ........................ 16
      1.2.5 Non-uniqueness of the (local) material connection ..... 20
   1.3 The material G-structure and the material groupoid ......... 20
      1.3.1 The material G-structure .......................... 20
      1.3.2 The material groupoid and its associated G-structures . 26
   1.4 Homogeneity ............................................. 30
      1.4.1 Uniformity and homogeneity ........................ 30
      1.4.2 Homogeneity in terms of a material connection ........ 31
      1.4.3 Homogeneity in terms of a material G-structure ....... 33
      1.4.4 Homogeneity in terms of the material groupoid ........ 36
   1.5 Homogeneity criteria ..................................... 36
      1.5.1 Solids ........................................... 36
      1.5.2 Fluids ........................................... 38
      1.5.3 Fluid crystals .................................... 39

**2 Uniformity of second-grade materials** ..................... 41
   2.1 An intuitive picture ...................................... 41
   2.2 The second-grade constitutive law ......................... 43
      2.2.1 Jets of maps ...................................... 43
      2.2.2 Composition of jets ............................... 44

|  |  | 2.2.3 | Second-grade materials | 47 |
|---|---|---|---|---|
|  | 2.3 | Second-grade uniformity | | 47 |
|  |  | 2.3.1 | Material isomorphisms | 47 |
|  |  | 2.3.2 | Second-grade material archetypes | 48 |
|  |  | 2.3.3 | Second-grade symmetries | 49 |
|  |  | 2.3.4 | An example of a nontrivial second-grade symmetry | 50 |
|  | 2.4 | The material second-order G-structures and groupoid | | 52 |
|  |  | 2.4.1 | The second-order frame bundle | 52 |
|  |  | 2.4.2 | The material G-structures | 52 |
|  |  | 2.4.3 | The material groupoid | 54 |
|  | 2.5 | The subgroups of $G^2(n)$ | | 54 |
|  |  | 2.5.1 | The generic subgroup | 54 |
|  |  | 2.5.2 | Toupin subgroups | 57 |
|  |  | 2.5.3 | The subgroups $\{\mathbf{I}, \Sigma_\mathbf{I}\}$ and their conjugates | 58 |
|  | 2.6 | Second-grade homogeneity | | 59 |
|  |  | 2.6.1 | The second-order frames induced by a coordinate system | 59 |
|  |  | 2.6.2 | Homogeneity | 59 |
|  |  | 2.6.3 | Coordinate expressions | 60 |
|  |  | 2.6.4 | Homogeneity in terms of a material $G$-structure | 63 |
|  | 2.7 | Homogeneity in terms of the material groupoid | | 65 |
| 3 | **Uniformity of Cosserat media** | | | 67 |
|  | 3.1 | Kinematics of a Cosserat body | | 67 |
|  | 3.2 | The constitutive law of a simple elastic Cosserat body | | 71 |
|  | 3.3 | Material isomorphisms and uniformity | | 74 |
|  |  | 3.3.1 | Material isomorphisms in a Cosserat body | 74 |
|  |  | 3.3.2 | Uniformity and the Cosserat archetype | 75 |
|  |  | 3.3.3 | Cosserat symmetries | 76 |
|  |  | 3.3.4 | Changing coordinates | 77 |
|  |  | 3.3.5 | Changing the archetype | 78 |
|  | 3.4 | Homogeneity conditions | | 78 |
|  |  | 3.4.1 | Homogeneity of a Cosserat body | 78 |
|  |  | 3.4.2 | The three kinds of material connections of a uniform Cosserat body | 79 |
|  |  | 3.4.3 | Homogeneity conditions | 80 |
|  | 3.5 | The Cosserat material $G$-structures and groupoid | | 81 |
|  |  | 3.5.1 | Frames, and frames of frames | 81 |
|  |  | 3.5.2 | Non-holonomic, semi-holonomic and holonomic frames | 85 |
|  |  | 3.5.3 | The Cosserat material $G$-structures | 90 |
|  |  | 3.5.4 | The Cosserat material groupoid | 91 |
|  | 3.6 | Homogeneity, flatness and integrable prolongations | | 91 |
|  |  | 3.6.1 | Sections of $\bar{F}^2\mathcal{B}$ | 92 |
|  |  | 3.6.2 | Invariant sections and linear connections | 93 |
|  |  | 3.6.3 | Prolongations | 94 |

## 4 Functionally graded bodies ... 97
- 4.1 The extended notion of material isomorphism ... 97
- 4.2 Non-uniqueness of symmetry isomorphisms ... 98
- 4.3 The material $N$-structure ... 99
- 4.4 Homosymmetry ... 100
- 4.5 Unisymmetric homogeneity of elastic solids ... 101
- 4.6 The reduced $N$-structure ... 103
  - 4.6.1 Algebraic preliminaries ... 103
  - 4.6.2 The $\tilde{N}$-structure of a solid functionally-graded unisymmetric body ... 104
- 4.7 Examples ... 106
  - 4.7.1 The isotropic solid ... 106
  - 4.7.2 The transversely isotropic solid ... 106
  - 4.7.3 The $n$-agonal solids ... 106
  - 4.7.4 Orthotropic materials ... 107
- 4.8 Summary ... 107

## Part II Material Evolution

## 5 On energy, Cauchy stress and Eshelby stress ... 111
- 5.1 Preliminary considerations ... 111
- 5.2 The Cauchy stress revisited ... 112
- 5.3 Eshelby's tensor as Cauchy's dual ... 114
- 5.4 Complete expressions of hyperelastic uniformity ... 115
- 5.5 The Eshelby and Mandel Stresses in the Context of Material Uniformity ... 116
- 5.6 Eshelby-stress identities ... 118
  - 5.6.1 Consequences of balance of angular momentum ... 118
  - 5.6.2 Consequences of a continuous symmetry group ... 118
  - 5.6.3 Consequences of the balance of linear momentum ... 119
  - 5.6.4 Inhomogeneity with compact support and the J-integral 120
- 5.7 The Eshelby stress in thermoelasticity ... 122
  - 5.7.1 Thermoelastic uniformity ... 122
  - 5.7.2 The Eshelby stress identity ... 123
  - 5.7.3 Thermal stresses ... 124
  - 5.7.4 The material heat conduction tensor ... 127
- 5.8 On stress, hyperstress and Eshelby stress in second-grade bodies ... 128
- 5.9 On stress, microstress and Eshelby stress in Cosserat bodies ... 130
  - 5.9.1 Equilibrium equations ... 130
  - 5.9.2 Eshelby stresses ... 131
  - 5.9.3 Eshelby stress identities ... 133

## XII   Contents

**6  An overview of the theory of material evolution** .......... 135
  6.1  What is material evolution? .............................. 135
  6.2  A geometric picture ...................................... 137
  6.3  Evolution equations ...................................... 138
    6.3.1  General form ....................................... 138
    6.3.2  Reduction to the archetype ......................... 139
    6.3.3  The principle of actual evolution .................. 141
    6.3.4  Material symmetry consistency ...................... 143
  6.4  The field equations of remodelling and bulk growth ....... 145
    6.4.1  Balance of mass .................................... 146
    6.4.2  Balance of linear momentum ......................... 147
    6.4.3  Balance of angular momentum ........................ 148
    6.4.4  Balance of energy .................................. 148
    6.4.5  The Clausius-Duhem inequality and its consequences .. 149
  6.5  An alternative approach .................................. 156
  6.6  Example: Visco-elasto-plastic theories ................... 162
    6.6.1  A simple non-trivial model ......................... 162
    6.6.2  Some computational considerations .................. 163
    6.6.3  Creep of a bar under uniaxial loading .............. 165
    6.6.4  Evolution, rheological models and the Eshelby stress ... 167
  6.7  Example: Bulk growth ..................................... 168
    6.7.1  Exercise stimulates growth ......................... 169
    6.7.2  A challenge to Wolff's law? ........................ 170
  6.8  Example: Self-driven evolution ........................... 174
    6.8.1  Introduction ....................................... 174
    6.8.2  A solid crystal body ............................... 175
    6.8.3  An isotropic solid ................................. 178

**7  Second-grade evolution** ..................................... 183
  7.1  Introduction ............................................. 183
  7.2  Reduction to the archetype ............................... 184
  7.3  Actual evolution ......................................... 186
  7.4  Material symmetry consistency ............................ 186
  7.5  An example ............................................... 187
  7.6  Concluding remarks ....................................... 188

## Part III Mathematical Foundations

**8  Basic geometric concepts** ................................... 193
  8.1  Differentiable manifolds ................................. 193
  8.2  Lie groups ............................................... 202
  8.3  Fibre bundles ............................................ 204
    8.3.1  Principal fibre bundles ............................ 206
    8.3.2  Associated fibre bundles ........................... 208

            8.3.3   Sections of fibre bundles............................211

9    **Theory of connections** .....................................213
     9.1   Connections on principal $G$-bundles......................213
           9.1.1   Parallelism in a principal $G$-bundle ................216
           9.1.2   Reduction of a connection .........................217
           9.1.3   Structure equation, curvature and holonomy ..........218
           9.1.4   Flat connections ..................................222
     9.2   Linear connections......................................222
     9.3   Connections in an associated bundle ......................226
     9.4   G-structures ...........................................230
           9.4.1   Examples of $G$-structures..........................232

10   **Bundles of linear frames** ..................................235
     10.1 Jet prolongations of fibre bundles........................235
     10.2 Local coordinates on prolongations .......................237
     10.3 Lie groups of jets of diffeomorphisms.....................239
     10.4 Higher-order frame bundles ...............................240

11   **Connections of higher order** ...............................243
     11.1 Fundamental form ........................................243
     11.2 $\mathcal{E}$-connection ................................245
     11.3 Second-order (holonomic) connection......................248
     11.4 Simple connections ......................................252

A    **Groupoids** .................................................257
     A.1   Introduction ...........................................257
     A.2   Groupoids ..............................................257
     A.3   Transitive Lie groupoids and principal bundles .............260

**References**......................................................263

**Index** .........................................................269

# Part I

# Inhomogeneity in Continuum Mechanics

# 1

# An overview of inhomogeneity theory

Broadly speaking, the constitutive equation of a material body $\mathcal{B}$ is a mathematical expression that encapsulates its thermomechanical behaviour. In theory, this mathematical expression may consist of a number of response functionals that specify such physical quantities as the stress, the heat flux, the internal energy and the entropy in terms of the past histories of the configuration and the temperature field of the body as well as that of any number of internal state variables. Although the theory of material inhomogeneity can be in principle formulated within this rather general framework, its essential meaning and the mathematical tools necessary to describe it can be better understood within the context of the simplest non-trivial constitutive law. The loss of generality entailed in the passage from a general non-local history-dependent constitutive law to a "hic et nunc" formulation can be remedied later by a gradual crescendo of the physical and mathematical language. At no point, however, will the most attentive listener detect a change in style or orchestration. Simply put: if properly introduced from the very beginning, the conceptual and instrumental framework remains exactly the same.

## 1.1 The constitutive equation of a material body

We start by reviewing some of the basic concepts of Continuum Mechanics, particularly those that are needed to establish the fundamental notions of inhomogeneity theory.

### 1.1.1 Configurations, deformations and their gradient

We recall that a *configuration* of a material body $\mathcal{B}$ is an embedding[1]:

---

[1] The body itself dwells in the Platonic world of differentiable manifolds, of which we can only see the manifestations in the world of phenomena, amenable to perception, in this case, in the guise of configurations. For a definition of differen-

# 1 An overview of inhomogeneity theory

$$\kappa : \mathcal{B} \longrightarrow \mathbb{E}^3, \tag{1.1}$$

where $\mathbb{E}^3$ is the 3-D Euclidean space of classical mechanics, namely, an affine space supported by a 3-D inner product space. Choosing a *frame* in $\mathbb{E}^3$ (that is, an origin and an orthonormal basis), we can identify $\mathbb{E}^3$ with $\mathbb{R}^3$ (See Box 1.1). When (as is the case in most elementary treatments) the material body is assumed to be a trivial manifold without boundary, a single coordinate chart can be used to cover the whole body. A coordinate chart:

$$\kappa_0 : \mathcal{B} \longrightarrow \mathbb{R}^3. \tag{1.2}$$

is usually identified as a *reference configuration* (see Box 1.1). Once a reference configuration has been fixed, one can associate with any given configuration $\kappa$ the *deformation* $\chi$ (relative to the chosen reference configuration $\kappa_0$) defined as the composition:

$$\chi : \kappa \circ \kappa_0^{-1}. \tag{1.3}$$

Denoting the coordinates in the reference configuration and in physical space by $X^I$ ($I = 1, 2, 3$) and $x^i$ ($i = 1, 2, 3$), respectively, the deformation $\chi$ is represented by three smooth functions:

$$x^i = \chi^i(X^1, X^2, X^3), \tag{1.4}$$

with smooth inverses. The non-singular Jacobian matrix with entries:

$$F^i_I = \frac{\partial x^i}{\partial X^I}, \tag{1.5}$$

evaluated at any given body point $\mathbf{X}$ is the coordinate representation of a (two-point) tensor $\mathbf{F}$, rightly called the *deformation gradient* at $\mathbf{X}$. Indeed, this tensor is nothing but the tangent map $\chi_*$ of the deformation $\chi$ evaluated at the point $\mathbf{X} \in \mathcal{B}$.

## 1.1.2 Locality, simplicity, elasticity

The thermomechanical behaviour of a material body is expressed mathematically by means of one or more *constitutive equations* representing the response of the body to the history of its deformation and temperature. As we have announced, it is our intention in this chapter to present the fundamental notions of inhomogeneity theory within the simplest possible, yet non-trivial, context. To this end, we will start by discarding all thermal effects, confining ourselves to what can be termed a purely mechanical response. Secondly, we will assume that the response of the body is *local*. By this we mean two things: the first one is that the quantities provided by the constitutive functionals have a

---

tiable manifolds, embeddings, tangent maps, tangent space and other differential geometric notions, see Section 8.1

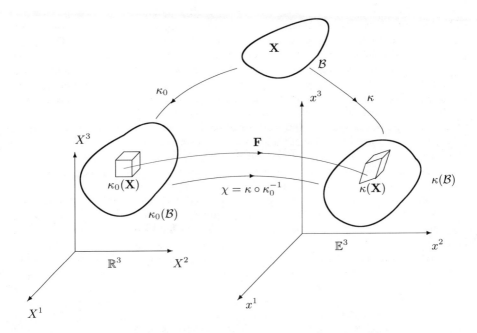

**Fig. 1.1.** The deformation and its gradient

point-wise meaning. In other words, that we can talk of the stress at a point, the internal energy density at a point, and so on. The second implication of locality is that the value of each of these quantities (say, the stress) at a point $\mathbf{X}$ is determined completely by the history of the deformation in an arbitrarily small neighbourhood of $\mathbf{X}$. A particular case of interest (because of its prevalence in most practical applications) is that in which the stress at a point is determined by the history of the deformation gradient at that point. We note that the body would still be local if the history of higher order gradients of the deformation were included in the list of independent variables of the constitutive equation. It is for this reason that the terminology of *simple material* is reserved for a local material whereby only the history of the deformation gradient intervenes in the constitutive law. Finally, a simple body is said to be *elastic* if only the present value of the deformation gradient enters the constitutive equation. In other words, the body exhibits no memory effects. A particular case of elasticity, to which we will confine our attention in this introduction, is that of *hyperelasticity*, whereby the constitutive response is completely determined by a constitutive equation for a single scalar quantity $\psi$. In short, for a simple hyperelastic body, the constitutive equation (expressed in terms of a chosen reference configuration) is of the form:

$$\psi = \psi(\mathbf{F}, \mathbf{X}). \tag{1.6}$$

> **Box 1.1 Body coordinates and reference configurations**
> The main two actors in Continuum Mechanics are the material body $\mathcal{B}$ and the physical space $\mathcal{S}$, both of which are differentiable manifolds of dimension 3. Whereas the physical space $\mathcal{S}$ is endowed with extra structure, the body $\mathcal{B}$ is not. The extra structure possessed by $\mathcal{S}$, according to the principles of Classical Newtonian Physics, can be summarized by saying that $\mathcal{S}$ is an *affine space* supported by an inner-product space. One can then choose an origin and an orthonormal basis, thereby uniquely assigning to each point $\mathbf{x}$ in $\mathcal{S}$ an element of $\mathbb{R}^3$. In this way, we obtain a global coordinate chart on $\mathcal{S}$ called a *Cartesian coordinate system*. Non-Cartesian charts, such as curvilinear coordinates, are also available. When using one of these, however, the theorem of Pythagoras is expressed by means of the components of the metric tensor $\mathbf{g}$ in the chart. These components are obtained by taking the dot products of the base vectors of the natural basis of the coordinate system used. (See Example 9.36).
> The body $\mathcal{B}$ is not canonically endowed with an affine structure. It is just a 3-dimensional smooth trivial manifold. It can, therefore, be covered with a single chart. Now, a chart is a map $\phi: \mathcal{B} \longrightarrow \mathbb{R}^3$, with a smooth inverse defined on its range. A configuration, on the other hand, is an embedding $\kappa: \mathcal{B} \longrightarrow \mathcal{S}$. This means that a chart in $\mathcal{B}$ looks a lot like a configuration. To identify a chart with a configuration, one can make the body inherit the metric structure of one of its charts by declaring, for example, that particular chart to be Cartesian. Once this is done, other charts are judged according to the choice already made. If this policy is adopted, one refers to any of the charts thus related as a *reference configuration*. So, we can say that a reference configuration implies a metric (Riemannian, curvature-free) structure on $\mathcal{B}$. Because of the triviality of $\mathcal{B}$ there exist global charts in which the metric tensor reduces to the identity matrix. These charts are called Cartesian.
> Continuum Mechanics does not need reference configurations to be properly formulated. It can be formulated entirely in terms of embeddings of $\mathcal{B}$ into the affine space $\mathcal{S}$, with its intrinsic metric properties. To write the component version of any equation, all that is needed is to adopt a coordinate chart in the body and a coordinate chart in space. Only for the latter one may occasionally need to invoke its natural metric structure. Nevertheless, it has become customary to work in terms of reference configurations, namely, body charts with an inherited metric structure. We will not depart from this tradition, but will occasionally recall the difference between charts and reference configurations if it is of some relevance to the treatment.

The physical meaning of this scalar quantity $\psi$ is that of an elastic energy density, either per unit spatial volume or per unit mass (assuming that the mass density is known), in which case we denote it by $\psi_\rho$. Another possibility is to consider the elastic energy density $\psi_R$ per unit volume in the reference configuration. By this one means the volume induced on the body by the coordinate chart adopted (see Box 1.2). Be that as it may, it is clear that in the case of a local body one can consistently talk about *the material at each point of the body*, a notion that doesn't make sense otherwise. For a simple body each material point has to be understood as consisting of a body-point $\mathbf{X} \in \mathcal{B}$ accompanied by its "first-order neighbourhood", that is, its *tangent space* $T_\mathbf{X}\mathcal{B}$. Under these conditions, a formula such as (1.6) provides at each material point a constitutive response in terms of the elastic energy density at that point as a function of arbitrary non-singular linear transformations **F** of its tangent space. A convenient pictorial way to visualize these ideas is to imagine the material point as a small die surrounding the point of interest. Then, the elastic energy density is completely determined by the way this die is deformed into an arbitrary non-degenerate parallelepiped (see Figure 1.1).

---

**Box 1.2 Volume forms**

The physical quantity of interest is the total elastic energy $\Psi$ stored in the body as a function of its (spatial) configuration. For a local material response, it is assumed that this quantity is given by an integral over the body $\mathcal{B}$. The object to be integrated, therefore, is a *differentiable 3-form* defined over $\mathcal{B}$, called the *elastic energy 3-form*. If the body is simple and elastic, the value of this 3-form is dictated at each point by the instantaneous value of the deformation gradient at that point. In a given coordinate chart of the body, a 3-form is defined by a single component, but this number is not an invariant scalar. On the other hand, the ratio between the components of two 3-forms is. A 3-form which is always available in $\mathcal{B}$ is the *pull-back* by $\kappa$ of the Cartesian volume form in the physical space. The ratio between the component of the elastic energy 3-form and the component of the pull-back of the spatial volume form is what we have denoted by $\psi$, which is a legitimate scalar. Its physical meaning is the elastic energy stored per unit spatial volume.

On the other hand, it is customary to specify as part of the constitutive law a *mass 3-form* defined over the body $\mathcal{B}$. If we consider, instead of $\psi$, the ratio between the elastic energy 3-form and the mass 3-form, the result is a legitimate scalar $\psi_\rho$. Its physical meaning is the elastic energy stored per unit mass.

Finally, we may consider the component itself of the elastic energy 3-form in a coordinate chart (read "reference configuration"). This component, which we have denoted $\psi_R$, is not an invariant scalar, but rather a so-called scalar density. Its meaning is the elastic energy stored per unit volume induced by the chart (that is, per unit volume in the reference configuration).

---

## 1.2 Material uniformity

### 1.2.1 The notion of material isomorphism

Having established the notion of constitutive response of a material point, the following question makes perfect sense: given two points of the body $\mathcal{B}$, are they made of the same material? Physically, we may reformulate this question in the following intuitive terms. Assume that a body is amenable to experimental observation in some configuration. We look at it under the microscope at two different points and we ask ourselves: do we need to see exactly the same picture in order to assert that the material is the same at both points? The answer is obviously negative. We may very well be observing two points made of, say, table salt, with the chlorine and sodium atoms arranged in the typical alternating way of that material. But it may so happen that the atomic arrangement surrounding one point is rotated with respect to the other, or rotated and distorted. We will, nevertheless, not lose heart by such a discrepancy, since it can be "fixed" by a simple deformation, without detriment to the chemical identity of the substance under consideration at each point. When translating this idea back to the previous picture of the small die surrounding each point, we arrive at the conclusion that the two material points will be made of the same material if the constitutive equation at one of them differs from its counterpart only by a pre-application of a linear transformation. If such a linear transformation exists it is called a *material isomorphism* from the first point to the second. Two points thus related are, accordingly, called *materially isomorphic*.

---

**Box 1.3 A surgical analogy**

The concept of material isomorphism can be compared to a *transplant operation*, a terminology that will occasionally be used. Indeed, if a small piece of face skin has been damaged beyond repair, possibly by burning, a plastic surgeon may resort to a skin graft. To this effect, a small piece of skin is cut from a less visible part of the patient's body and surgically transplanted to the damaged zone to replace the missing part. In doing this, however, it may be necessary to rotate and/or stretch (or perhaps release the stretch of) the skin to be grafted so as to make it match perfectly its new surroundings, which may be differently strained than the old ones. This transplant operation consisting of cutting, rotating and/or distorting and, finally, implanting in the new neighbourhood so as to achieve a perfect graft, is the surgical analog of a material isomorphism. If the graft is perfect, it will be impossible to distinguish it mechanically from the missing original skin.

---

These intuitive ideas can be readily translated into mathematical language. In what follows, we will adopt the practice of identifying the body $\mathcal{B}$ with its image $\kappa_0(\mathcal{B})$ in a given reference configuration $\kappa_0$. This practice, which results in some economy of notation, will be abandoned whenever necessary (for example, when considering more than just one reference configuration). Let $\mathbf{X}_1$ and $\mathbf{X}_2$ be two points of a body $\mathcal{B}$ with constitutive equation (1.6). We say that these points are materially isomorphic if there exists a non-singular linear map, $\mathbf{P}_{12}$, between their tangent spaces:

$$\mathbf{P}_{12} : T_{X_1}\mathcal{B} \longrightarrow T_{X_2}\mathcal{B}, \tag{1.7}$$

such that:

$$\psi(\mathbf{F}\mathbf{P}_{12}, \mathbf{X}_1) = \psi(\mathbf{F}, \mathbf{X}_2), \tag{1.8}$$

identically for all deformation gradients $\mathbf{F}$. In other words, the pre-application of the map $\mathbf{P}_{12}$ renders the material responses of the two points identical to each other[2] (Figure 1.2). The map $\mathbf{P}_{12}$ is called a material isomorphism from

---

[2] As a quantitative measure of material response, we are considering here the energy density, as done in [6]. More generally, though, the Cauchy stress tensor could be

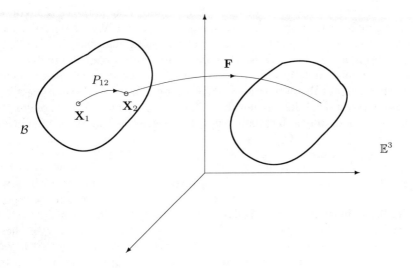

**Fig. 1.2.** A material isomorphism

$\mathbf{X}_1$ to $\mathbf{X}_2$. The points $\mathbf{X}_1$ and $\mathbf{X}_2$ are, respectively, called the *source* and the *target* of the material isomorphism $\mathbf{P}_{12}$.

If $\mathbf{P}_{12}$ is a material isomorphism form $\mathbf{X}_1$ to $\mathbf{X}_2$, then its inverse $\mathbf{P}_{21} = \mathbf{P}_{12}^{-1}$ is a material isomorphism from $\mathbf{X}_2$ to $\mathbf{X}_1$. The truth of this assertion emerges from the following chain of identities:

$$\psi(\mathbf{F}, \mathbf{X}_1) = \psi(\mathbf{F}\mathbf{P}_{12}^{-1}\mathbf{P}_{12}, \mathbf{X}_1) = \psi(\mathbf{F}\mathbf{P}_{12}^{-1}, \mathbf{X}_2), \tag{1.9}$$

where Equation (1.8) has been used. Thus, the mathematical relation of "being materially isomorphic" is a *symmetric* relation. Clearly, this relation is also *reflexive*, since every point is trivially materially isomorphic to itself, via the identity mapping of its tangent space. Finally, it is not difficult to verify that this relation is also *transitive*, namely: if $\mathbf{X}_1$ is materially isomorphic with $\mathbf{X}_2$, via a material isomorphism $\mathbf{P}_{12}$, and in its turn $\mathbf{X}_2$ is materially isomorphic to a point $\mathbf{X}_3$, via a material isomorphism $\mathbf{P}_{23}$, then $\mathbf{X}_1$ is materially isomorphic to $\mathbf{X}_3$ via the material isomorphism:

$$\mathbf{P}_{13} = \mathbf{P}_{23}\mathbf{P}_{12}. \tag{1.10}$$

The proof of this assertion is left to the reader. It follows from these three properties that material isomorphism is an *equivalence relation*.

---

considered. So as not to interrupt the main flow of ideas, we have postponed the discussion of this issue to Chapter 5.

A body with constitutive equation (1.6) is *materially uniform*[3] if all of its points are materially isomorphic to one another. Physically, this means that in a uniform body all points are made of the same material. A materially uniform body is said to be *smoothly uniform* if for each point there exists an open neighbourhood within which the material isomorphisms can be chosen so as to depend smoothly on both source and target. Notice that smoothness is a local property: indeed, in a smoothly uniform body it may not be possible to choose the whole body as a neighbourhood in which the smoothness condition is satisfied.

*Remark 1.1.* A nice illustration of this situation is due to Wang [89, 91]. Consider a long cylindrical body initially straight and made of a material with rhombic symmetry. That is, a point of this material has a configuration possessing three mutually perpendicular preferred axes ($\mathbf{e_1}, \mathbf{e_2}, \mathbf{e_3}$) such that the material properties are preserved after a 180°-rotation about any of these three axes. No other material symmetries are present. We assume that in the initially straight configuration all the points are in this preferred configuration and that the preferred axes are respectively parallel point by point. One of these axes, say $\mathbf{e_1}$, is aligned with the longitudinal axis of the cylinder. In this initial state, the (unique) smooth material isomorphisms are Euclidean translations. We now perform the following operations: first, we twist one end of the cylinder by 180° around the longitudinal axis. Then we bend the cylinder by a full revolution so as to bring its two ends into coincidence. Finally, we "weld" these ends and obtain the desired body $\mathcal{B}$, which Wang suggestively calls a *Moebius crystal*. Clearly, over any simply-connected proper sub-body of $\mathcal{B}$ we have a (unique, in this case) smooth field of material isomorphisms, but not so over the whole body (since, at some point, we must effect a "quantum leap" of 180°). On the other hand, assume that the body is at least transversely isotropic (namely, any rotation about $\mathbf{e_1}$ is a material symmetry in the initially straight configuration). Upon twisting, we can now still choose as material isomorphisms the same Euclidean translations as in the untwisted state (disregarding an unessential longitudinal stretch). As a consequence of this extra (continuous) degree of freedom in the choice of material isomorphisms, when we close the cylinder into a torus, as we did before, we obtain a globally smooth field of material isomorphisms.

### 1.2.2 Material symmetries and the non-uniqueness of material isomorphisms

A *material symmetry* at a point $\mathbf{X}_0 \in \mathcal{B}$ is a *material automorphism*, that is, a material isomorphism between $\mathbf{X}_0$ and itself. We have just pointed out

---

[3] Although the notions of material isomorphism and uniformity had been anticipated by Kondo [59], the formulation and nomenclature of this section follow essentially the work of Noll [75].

> **Box 1.4 The mass-consistency condition**
> So far, the definition of material isomorphism makes no reference to the mass density of the body, and it was not even assumed that such a quantity has been specified. Once the mass 3-form is specified as a (possibly time-dependent) part of the constitutive law, it is desirable to restrict the notion of material isomorphism so as to bring it into consistency with the mass density. In other words, it will be required that the mass 3-form be preserved under the action of the material isomorphisms. This means that the pull-back to $T_{\mathbf{X}_1}\mathcal{B}$ by $\mathbf{P}_{12}$ of the mass 3-form at $\mathbf{X}_2$ is precisely equal to the mass 3-form at $\mathbf{X}_1$. In a given body-chart (i.e., reference configuration), we have the following restriction on the matrix $[\mathbf{P}_{12}]$ representing a material isomorphism:
>
> $$\det[\mathbf{P}_{12}] = \frac{\rho_R(\mathbf{X}_1)}{\rho_R(\mathbf{X}_2)}, \tag{1.11}$$
>
> where $\rho_R(\mathbf{X})$ is the component of the mass 3-form at the point $\mathbf{X}$ in the given chart.

that a material point is always trivially isomorphic to itself, so that the set of material symmetries of a point is never empty. A material symmetry $\mathbf{G}$ at a point $\mathbf{X}_0$ can be seen as a transformation of its neighbourhood leaving the material response unchanged, namely[4]:

$$\psi(\mathbf{FG}, \mathbf{X}_0) = \psi(\mathbf{F}, \mathbf{X}_0), \tag{1.12}$$

for all deformation gradients $\mathbf{F}$. Notice that if the mass-consistency condition is required (see Box 1.4), a symmetry must necessarily have a unit determinant.

The collection $\mathcal{G}_0$ of all material symmetries at $\mathbf{X}_0$ constitutes a group called the *material symmetry group* of the constitutive law at point $\mathbf{X}_0$. The group operation is the composition of tensors (or matrix multiplication), so that in any given reference configuration the symmetry group is always a subgroup of the general linear group in three dimensions (and a subgroup of the unimodular group if the mass-consistency condition is required). The verification that $\mathcal{G}_0$ is indeed a group is rather straightforward. We have already pointed out that the identity map $\mathbf{I}$ is always in $\mathcal{G}_0$. Moreover, if $\mathbf{G}$ is in $\mathcal{G}_0$, its inverse $\mathbf{G}^{-1}$ is also in $\mathcal{G}_0$, as follows from a reasoning similar to that employed in Equation (1.9), but identifying the source and the target points. Finally, from a reasoning similar to that leading to the transitivity of material isomorphisms, it follows that if $\mathbf{G}$ and $\mathbf{H}$ are in $\mathcal{G}_0$, so are the compositions $\mathbf{GH}$ and $\mathbf{HG}$.

Consider now the symmetry groups $\mathcal{G}_1$ and $\mathcal{G}_2$ of two materially isomorphic points, $\mathbf{X}_1$ and $\mathbf{X}_2$, and let $\mathbf{P}_{12}$ be a material isomorphism between them. Choosing an arbitrary symmetry $\mathbf{G}_1 \in \mathcal{G}_1$ we obtain that the map:

$$\mathbf{P}'_{12} = \mathbf{P}_{12}\mathbf{G}_1 \tag{1.13}$$

---

[4] As already pointed out with respect to Equation (1.8), there is a subtle difference between material isomorphisms, and hence symmetries, based on energy and those based on stress. See Chapter 5.

is also a material isomorphism from $\mathbf{X}_1$ to $\mathbf{X}_2$. Indeed:

$$\psi(\mathbf{FP}'_{12}, \mathbf{X}_1) = \psi(\mathbf{FP}_{12}\mathbf{G}_1, \mathbf{X}_1) = \psi(\mathbf{FP}_{12}, \mathbf{X}_1) = \psi(\mathbf{F}, \mathbf{X}_2), \quad (1.14)$$

where Equations (1.8) and (1.12) have been invoked. This simply means that, unless the symmetry group of $\mathbf{X}_1$ is trivial (namely, consists just of the identity $\mathbf{I}$), the material isomorphism between $\mathbf{X}_1$ and any other point of the uniform body to which it belongs is not unique. In fact, the non-uniqueness of the material isomorphisms between two points is governed precisely by the degree of freedom of their respective symmetry groups. More precisely, let $\mathbf{P}_{12}$ and $\mathbf{Q}_{12}$ be two material isomorphisms from the source $\mathbf{X}_1$ to the target $\mathbf{X}_2$. Then the compositions $\mathbf{Q}_{12}^{-1}\mathbf{P}_{12}$ and $\mathbf{Q}_{12}\mathbf{P}_{12}^{-1}$ are, necessarily, members of the symmetry groups $\mathcal{G}_1$ and $\mathcal{G}_2$, respectively. Indeed:

$$\psi(\mathbf{FQ}_{12}^{-1}\mathbf{P}_{12}, \mathbf{X}_1) = \psi(\mathbf{FQ}_{12}^{-1}, \mathbf{X}_2) = \psi(\mathbf{F}, \mathbf{X}_1). \quad (1.15)$$

Denoting by $\mathcal{P}_{12}$ the collection of all material isomorphisms from the source $\mathbf{X}_1$ to the target $\mathbf{X}_2$, it follows from the preceding discussion that this collection can be generated from any particular material isomorphism $\mathbf{P}_{12}$ by any of the following equivalent expressions:

$$\mathcal{P}_{12} = \mathbf{P}_{12}\,\mathcal{G}_1 = \mathcal{G}_2\,\mathbf{P}_{12} = \mathcal{G}_2\,\mathbf{P}_{12}\,\mathcal{G}_1. \quad (1.16)$$

Conversely, the symmetry groups $\mathcal{G}_1$ and $\mathcal{G}_2$ of two materially isomorphic points $\mathbf{X}_1$ and $\mathbf{X}_2$ are necessarily *conjugate*, namely:

$$\mathcal{G}_2 = \mathbf{P}_{12}\,\mathcal{G}_1\,\mathbf{P}_{12}^{-1}. \quad (1.17)$$

where $\mathbf{P}_{12}$ is any member of $\mathcal{P}_{12}$. These ideas are schematically illustrated in Figure 1.3.

### 1.2.3 The material archetype

Since material isomorphism is an equivalence relation, material uniformity can also be regarded as follows: a body $\mathcal{B}$ is materially uniform if, and only if, there exists a material isomorphism $\mathbf{P}(\mathbf{X})$ from a fixed point $\mathbf{X}_0 \in \mathcal{B}$ to each point $\mathbf{X} \in \mathcal{B}$. This fixed point can be arbitrarily chosen as any point in the body, or as any material point made of the same material as the body points. We shall call it an *archetypal material point*[5] and, for the purpose of visualization, we will imagine it placed outside of the body (see Figure 1.4). The material isomorphisms $\mathbf{P}(\mathbf{X})$ from the archetypal material point to the body points will be referred to as *implants*. A collection of such implants (one for each point of the body) is a *uniformity field*. A body is smoothly uniform

---

[5] In earlier works [27, 43, 44, 46, 29] the alternative designation of "crystal of reference" was used. This terminology, however, may be somewhat misleading for non-solid bodies. The term "archetype" has been used already in [42, 30, 31].

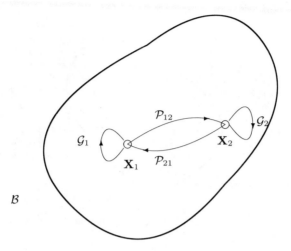

**Fig. 1.3.** Non-uniqueness of the material isomorphisms

if, and only if, for any given archetypal material point a (locally) smooth uniformity field can be chosen. Notice that a change of archetypal material point entails simply a composition of the uniformity field with a fixed linear map to the right, thus not affecting the notion of smoothness.

Just as, for the purpose of actual computations, we place the body in a global reference configuration $\kappa_0$, so too, in order to get a specific form of the archetypal constitutive law, we must place the tangent space of the archetypal material point $\mathbf{X}_0 \in \mathcal{B}$ in $\mathbb{R}^3$ by means of a non-singular linear map[6]:

$$\mu : T_{X_0}\mathcal{B} \longrightarrow \mathbb{R}^3. \tag{1.18}$$

From this point of view, which we shall adopt henceforth, a *material archetype* (or, an *archetype*, for short) will be defined as a (linear) *frame at* $\mathbf{X}_0$. Indeed, the natural basis of $\mathbb{R}^3$ is mapped by $\mu^{-1}$ onto a basis of the tangent space at $\mathbf{X}_0$ and, conversely, this basis at $\mathbf{X}_0$ completely defines the linear map $\mu$.

Let the constitutive equation of the archetype be given by:

$$\psi = \bar{\psi}(\mathbf{F}). \tag{1.19}$$

Then, by virtue of the Equation (1.8), we have:

$$\psi(\mathbf{F}, \mathbf{X}) = \bar{\psi}(\mathbf{FP}(\mathbf{X})), \tag{1.20}$$

which means that in a uniform body the dependence of the constitutive law on the body point $\mathbf{X}$ is severely restricted by a multiplication of $\mathbf{F}$ to the right by an $\mathbf{X}$-dependent tensor.

---

[6] It is the result of this operation what delivers the archetypal "small cube" of material (subtended by the natural basis of $\mathbb{R}^3$) in the intuitive picture.

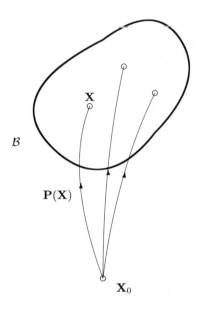

**Fig. 1.4.** The archetype and its implants

*Remark 1.2.* An interesting question is the following [18]: given an arbitrary constitutive law $\psi = \psi(\mathbf{F}, \mathbf{X})$, is there a mathematical criterion to establish whether or not this equation can be brought into the form (1.20)? Although the answer to this question is, in principle, a matter of differentiation and algebraic elimination, an explicit formulation of the obstructions that characterize a uniform constitutive law does not exist.

If the symmetry group $\bar{\mathcal{G}}$ of the archetype is not the trivial identity group, the material implants are not unique. In fact, according to Equation (1.16), the collection $\mathcal{P}_\mathbf{X}$ of material implants (from a given archetype) available at point $\mathbf{X}$ is obtained from any given $\mathbf{P}(\mathbf{X})$ by the equation:

$$\mathcal{P}_\mathbf{X} = \mathbf{P}(\mathbf{X})\bar{\mathcal{G}}, \tag{1.21}$$

or, equivalently, by any of the expressions:

$$\mathcal{P}_\mathbf{X} = \mathcal{G}_\mathbf{X}\mathbf{P}(\mathbf{X})\bar{\mathcal{G}} = \mathcal{G}_\mathbf{X}\mathbf{P}(\mathbf{X}) = \mathbf{P}(\mathbf{X})\bar{\mathcal{G}}, \tag{1.22}$$

where $\mathcal{G}_\mathbf{X}$ is the symmetry group at $\mathbf{X}$. Naturally, as in our earlier discussion, the symmetry group of the archetype is conjugate to the symmetry group of each body point $\mathbf{X}$, the conjugation being achieved by any element of the set $\mathcal{P}_\mathbf{X}$.

## 1.2 Material uniformity

Since the archetype is a frame at a body point $\mathbf{X}_0$, any given uniformity field $\mathbf{P}(\mathbf{X})$ induces, by composition, a basis at (the tangent space of) each point $\mathbf{X}$ of $\mathcal{B}$. A uniformity field can, therefore, be regarded as a *moving frame* (or "repère mobile"), namely, a field of local bases (Figure 1.5). Accordingly, with some abuse of notation, the set $\mathcal{P}(\mathbf{X})$ can be regarded as the collection of all possible bases that a given material archetype induces at $\mathbf{X}$ via all the possible material implants.

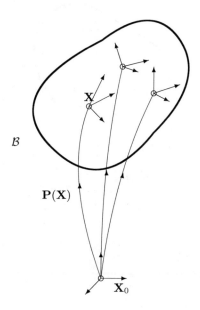

**Fig. 1.5.** A uniformity basis field

A *change of archetype* can be carried out either by keeping the archetypal point $\mathbf{X}_0$ fixed and changing the map $\mu$, as shown in Figure 1.6, or by choosing a new archetypal point $\mathbf{Y}_0 \in \mathcal{B}$ (and, of course, also a new map from its tangent space to $\mathbb{R}^3$). In both cases, the effect upon the uniformity field is of the same type. More specifically, let:

$$\mu' : T_{\mathbf{X}_0}\mathcal{B} \longrightarrow \mathbb{R}^3 \tag{1.23}$$

be a new map (while keeping the archetypal point $\mathbf{X}_0$ fixed). The archetypal constitutive equation (1.19) is replaced by:

$$\psi = \bar{\psi}'(\mathbf{F}) = \bar{\psi}(\mathbf{F} \circ (\mu' \circ \mu^{-1})). \tag{1.24}$$

As a result of this change of archetypal constitutive equation, the old implants $\mathbf{P}(\mathbf{X})$ are replaced by the new implants:

$$\mathbf{P}' = \mathbf{P} \circ (\mu \circ \mu'^{-1}). \tag{1.25}$$

In other words, the effect of this change of archetype on the uniformity field consists of a *multiplication to the right* by a fixed element of the general linear group $GL(3;\mathbb{R})$. With the intermediation of a material isomorphism between $\mathbf{X}_0$ and $\mathbf{Y}_0$, the same result is obtained upon a change of archetypal point.

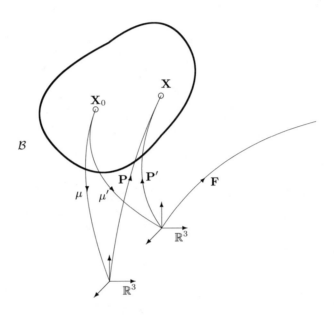

**Fig. 1.6.** Change of archetypal frame

### 1.2.4 Local material parallelisms

We have already remarked that the material isomorphisms and, therefore, the uniformity basis fields in a smoothly uniform body may not be obtainable as smooth fields over the whole body, but only on open sub-bodies. Let us then focus our attention on an open set $\mathcal{U} \subset \mathcal{B}$ within which a smooth uniformity field of bases has been defined. This is represented schematically in Figure 1.7, where we have resorted to draw everything in two (rather than three) dimensions for the sake of clarity.

Having thus defined a frame at each point of $\mathcal{U}$, we are in a position of comparing vectors at different points of $\mathcal{U}$ by introducing the concept of a *local*

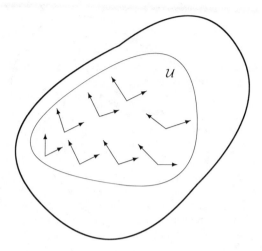

**Fig. 1.7.** A smooth uniformity field on an open neighbourhood

*material parallelism* [7]. We will say that two vectors at two different points $\mathbf{X}_1$ and $\mathbf{X}_2$ of $\mathcal{U}$ are *materially parallel with respect to the given uniformity field*, if they have the same components in the respective local bases of the uniformity field. This situation is illustrated in Figure 1.8, where the vectors $\mathbf{u}$ and $\mathbf{v}$ are materially parallel, since they have the same components $\{1, 2\}$ in their respective bases. In particular, the respective base vectors of a uniformity basis field are materially parallel, since they have all components equal to zero except for the respective unit component. Let now a smooth vector field $\mathbf{w}(\mathbf{X})$ be given in the open set $\mathcal{U}$. We will say that it is *materially constant with respect to the given uniformity field in* $\mathcal{U}$ if $\mathbf{w}(\mathbf{X}_1)$ is materially parallel to $\mathbf{w}(\mathbf{X}_2)$ for every $\mathbf{X}_1$ and $\mathbf{X}_2$ in $\mathcal{U}$. Now let us assume that the vector field $\mathbf{w}(\mathbf{X})$ has been specified in some reference configuration in terms of its components $w^I(\mathbf{X})$ in the (Cartesian) coordinate system $X^I$ ($I = 1, 2, 3$). What are the mathematical conditions for this field to be materially parallel with respect to a given uniformity field in $\mathcal{U}$?

To answer this important question, we will resort to the use of coordinate expressions (but see Box 1.5). Assume that an archetype has been chosen, that is, a body point $\mathbf{X}_0$ and a linear map between its tangent space and $\mathbb{R}^3$. Let $\mathbf{E}_\alpha$ ($\alpha = 1, 2, 3$) be the natural basis of $\mathbb{R}^3$. By means of the uniformity maps $\mathbf{P}(\mathbf{X})$ this basis induces a smooth field of bases in $\mathcal{U}$ (as shown in Figure 1.5), which we will denote by $\mathbf{p}_\alpha(\mathbf{X})$ ($\alpha = 1, 2, 3$). We now adopt a coordinate system in $\mathcal{U}$, which we call $X^I$ ($I = 1, 2, 3$), with natural basis

---

[7] Note that the adjective "local" qualifies the parallelism, not the material. The parallelism is local because it is restricted to a neighbourhood of the body rather than to the whole body.

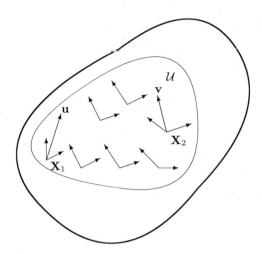

**Fig. 1.8.** The vectors **u** and **v** are materially parallel

$\mathbf{e}_I$ ($I = 1, 2, 3$). Each base vector $\mathbf{p}_\alpha(\mathbf{X})$ will, therefore, be expressible in terms of its components in the coordinate basis, as follows:

$$\mathbf{p}_\alpha = P^I_\alpha \mathbf{e}_I, \qquad (1.26)$$

where the point-dependent matrix with entries $P^I_\alpha$ is precisely the coordinate representation of the material implants. The vector field $\mathbf{w}$ can be expressed in terms of components in either basis, namely:

$$\mathbf{w} = w^\alpha \mathbf{p}_\alpha = w^I \mathbf{e}_I. \qquad (1.27)$$

Combining Equations (1.26) and (1.27), we obtain that these components are related by:

$$w^I = P^I_\alpha w^\alpha, \qquad (1.28)$$

or, equivalently:

$$w^\alpha = (P^{-1})^\alpha_I \, w^I = P^{-\alpha}_I \, w^I. \qquad (1.29)$$

If the given field $\mathbf{w}$ is materially constant, the components $w^\alpha$ are, by definition, constant (independent of $X^I$). Therefore, taking the partial derivative of (1.29) with respect to $X^J$, we obtain:

$$0 = \frac{\partial P^{-\alpha}_I}{\partial X^J} w^I + P^{-\alpha}_I \frac{\partial w^I}{\partial X^J}. \qquad (1.30)$$

Defining the *Christoffel symbols of the local material parallelism* as:

$$\Gamma^K_{IJ} = P^K_\alpha \frac{\partial P^{-\alpha}_I}{\partial X^J}, \qquad (1.31)$$

or, equivalently:

$$\Gamma_{IJ}^K = -P_I^{-\alpha} \frac{\partial P_\alpha^K}{\partial X^J}, \qquad (1.32)$$

we can write Equation (1.30) as:

$$0 = \frac{\partial w^K}{\partial X^J} + \Gamma_{IJ}^K w^I. \qquad (1.33)$$

The expression on the right-hand side of Equation (1.33) is the *material covariant derivative* of the field **w** with respect to the given material parallelism. We have obtained the result that a field is materially constant with respect to a given material parallelism if, and only if, its material covariant derivative vanishes.

---

**Box 1.5 Coordinate expressions**
One of the aims of the use of modern differential-geometric methods in Continuum Mechanics is to provide intrinsic, rather than coordinate dependent, mathematical expressions for the physical phenomena to be described. Thus, for instance, a dislocation in a material medium is an observable physical event whose identity is quite independent of the particular coordinate system in which an observer chooses to describe it. Another clear advantage of the intrinsic geometric approach that we advocate is that it allows for the precise formulation of global concepts. Indeed, a coordinate chart in general covers only a portion of the manifold under consideration, just as a map of the Earth surface can continuously cover only a part of the world. When the manifold is covered with more than one chart, therefore, one has to ensure that appropriate transition conditions are enforced (see Section 8.1). On the other hand, a first encounter with a difficult subject, such as material uniformity, can benefit from a coordinate description, which can be later elevated to the status of an intrinsic formulation.

---

So far, we have discussed the state of affairs in a neighbourhood $\mathcal{U}$ whereby the uniformity field is smooth. Clearly, unless the open set $\mathcal{U}$ can be identified with the whole body $\mathcal{B}$, we will not be able in general to obtain a (smooth) material parallelism over the whole body. Even in this general case, however, it can be shown (by covering the body with open sets in each of which a smooth uniformity field is defined and then using a partition of unity[8]), that one can define globally smooth Christoffel symbols and a corresponding material covariant derivative associated with a *material connection*[9]. With some abuse of terminology, we may say, therefore, that each local material parallelism determines a *local material connection*. The difference between the local and the global constructions, however, is that in the global case the notion of material parallelism may no longer exist. Instead, we have the notion of *parallel transport along curves*. The first case (when we do have a global, or distant,

---

[8] See Section 8.3.3.
[9] For a treatment of connections, parallel transport and curvature see Chapter 9.

parallelism) is characterized by the fact that the material connection has a vanishing *curvature*.[10]

### 1.2.5 Non-uniqueness of the (local) material connection

Let a smooth uniformity field on an open set $\mathcal{U}$ be given. If the symmetry group $\bar{\mathcal{G}}$ of the material happens to be trivial (i.e., it consists of the identity transformation only) it is clear that, having chosen a basis in the archetype, this uniformity field is unique. This conclusion stems directly from Equation (1.21). Consider now the case in which the symmetry group $\bar{\mathcal{G}}$ is not trivial, but is discrete (namely, consists of a finite number of elements). If we change the uniformity basis at one point of $\mathcal{U}$ by premultiplication by an element of $\mathbf{G}$ of $\bar{\mathcal{G}}$, the only way we can change the bases at all other points of $\mathcal{U}$ while retaining the smoothness of the basis field is by premultiplying by exactly the same element $\mathbf{G}$. Indeed, if the group is discrete, there are no other "near" elements of $\bar{\mathcal{G}}$ available! Now, if two basis fields differ by premultiplication by a constant element $\mathbf{G} \in \bar{\mathcal{G}}$, the parallelisms they induce are identical, since the components of vectors in the respective bases are affected by the same matrix at both points. In other words, in the case of a discrete symmetry group, the local material parallelisms are unique. By a similar reasoning, it can be shown that, globally speaking, the material connection is also unique.

In the case of a continuous symmetry group (such as that of an isotropic solid), on the other hand, the previous reasoning does not apply. Indeed, we can now choose a smoothly varying change of basis by exploiting the available smooth degree of freedom of the symmetry group. Correspondingly, for the case of a continuous symmetry group, the material connection is not unique, a fact that will have important implications in the assessment of the existence of distributed inhomogeneities within the body. The non-uniqueness of material connections in the general case reveals that a material connection does not convey the total picture of a materially uniform body. As we will see next, this information is contained in the notion of a *material G-structure*[11].

## 1.3 The material G-structure and the material groupoid

### 1.3.1 The material G-structure

Let a uniformity basis field (Figure 1.5) be specified in a given uniform body. As we have already discussed, this field of bases is neither necessarily unique nor globally smooth. Nevertheless, if the body is smoothly uniform, it can

---

[10] The notion of material connections has been masterfully treated by Wang in [89]. See also Chapter 5 of [91], where various examples are presented of material connections with non-vanishing curvature.

[11] See also Section 9.4.

## 1.3 The material G-structure and the material groupoid

be covered with open sets in each one of which the field can be chosen to be smooth. Consider now the non-empty intersection of two such open sets, $\mathcal{U}$ and $\mathcal{U}'$. At each point of this overlap we have two, in general different, material implants, say $\mathbf{P}(\mathbf{X})$ and $\mathbf{P}'(\mathbf{X})$. By virtue of Equation (1.15), the composition $\mathbf{P}'(\mathbf{X})\,\mathbf{P}^{-1}(\mathbf{X})$ must be in the symmetry group $\mathcal{G}_\mathbf{X}$. Equivalently, the composition $\mathbf{G}(\mathbf{X}) = \mathbf{P}'^{-1}(\mathbf{X})\,\mathbf{P}(\mathbf{X})$ is a member of the symmetry group $\bar{\mathcal{G}}$ of the archetype. We will call $\mathbf{G}(\mathbf{X})$ the *transition map* at $\mathbf{X}$ (from the field $\mathbf{P}(\mathbf{X})$ to the field $\mathbf{P}'(\mathbf{X})$). The transition maps depend smoothly on the points of the overlap, since so do both $\mathbf{P}(\mathbf{X})$ and $\mathbf{P}'(\mathbf{X})$. The transition maps constitute, therefore, a smooth assignment of a member of the group $\bar{\mathcal{G}}$ to each point of the intersection:

$$\mathcal{U} \cap \mathcal{U}' \longrightarrow \bar{\mathcal{G}}. \tag{1.34}$$

**The notion of fibre bundle**

Emboldened by this discussion, let us try to venture even further. Let us attach to each point $\mathbf{X}$ of the smoothly uniform body $\mathcal{B}$ the set $\mathcal{P}(\mathbf{X})$ of the frames induced by all the possible material implants from the given archetype to that point[12]. The result is a set $\mathcal{P}_\mathcal{B}$ roughly consisting of the union of all the $\mathcal{P}(\mathbf{X})$'s. This is schematically indicated in Figure 1.9, where we have represented the body as a (roughly horizontal) curve (the *base manifold*) and each set $\mathcal{P}(\mathbf{X})$ (the *fibre over* $\mathbf{X}$) as a (roughly vertical) line. The set $\mathcal{P}_\mathcal{B}$ thus constructed is called a *fibre bundle*, a happy choice of terminology given the pictorial representation just introduced[13]. We will call it the *material fibre bundle* of the body $\mathcal{B}$. By construction, it is endowed with a *projection map*:

$$\pi : \mathcal{P}_\mathcal{B} \longrightarrow \mathcal{B}, \tag{1.35}$$

which simply indicates to which point of $\mathcal{B}$ any given frame $p$ is attached (see Figure 1.9). Note that, by Equation 1.21, all the fibres are isomorphic to each other and to $\bar{\mathcal{G}}$, which is thus called the *typical (or standard) fibre*.

**Cross sections**

We now ask the obvious question: In this geometrical picture, how is a particular uniformity field to be represented? The answer is obtained by noting that a particular uniformity field is, by definition, a choice at each $\mathbf{X}$ of some element of the corresponding $\mathcal{P}(\mathbf{X})$, namely, a map:

$$\sigma : \mathcal{B} \longrightarrow \mathcal{P}_\mathcal{B}, \tag{1.36}$$

---

[12] Recall that, by abuse of notation, we use $\mathcal{P}(\mathbf{X})$ to denote both the set of material implants at $\mathbf{X}$ and the set of frames generated at $\mathbf{X}$ by a given basis at the archetype.
[13] See Section 8.3.

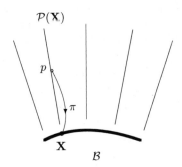

**Fig. 1.9.** A material fibre bundle

with the property:

$$\pi \circ \sigma = id_\mathcal{B}, \qquad (1.37)$$

where $id_\mathcal{B}$ is the identity map of $\mathcal{B}$. This last condition simply ensures that we are assigning to each point a frame at that point, rather than at some other point. A convenient way to express this fact is by means of the following commutative diagram:

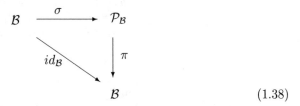 (1.38)

A map satisfying this condition is called a (global) *cross section* of the fibre bundle. The reason for this terminology is, happily again, related to the representation that such a map acquires in our picture, as shown in Figure 1.10. Indeed, the graph of this map $\sigma$ cuts through the fibres.

A word of caution is, however, necessary regarding (again!) the issue of smoothness. It may not be possible to guarantee that a smooth cross section exists. On the other hand, we have already established that (for a smoothly uniform body) this can be done by pieces. This fact is indicated in Figure 1.11, where now each *local* section is assumed to be smooth. The transition maps between these local sections are precisely the ones described above (Equation (1.34)) and, as already pointed out, they belong necessarily to the symmetry group of the archetype. This group $\bar{\mathcal{G}}$ is, accordingly, called *the structure group of the fibre bundle*. Notice that each (smooth) local section represents a local parallelism.

1.3 The material G-structure and the material groupoid

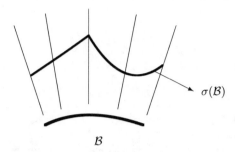

**Fig. 1.10.** A (not necessarily smooth) global section

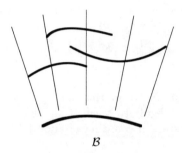

**Fig. 1.11.** Smooth local sections

## The principal bundle of frames

Assume now that we were to look at the body $\mathcal{B}$ as a manifold devoid of any constitutive connotation. In that case, we could consider the fibre bundle $F(\mathcal{B})$ obtained by attaching to each point the collection of *all* frames at that point. This is the so-called *bundle of linear frames*, or *frame bundle*, of the base manifold $\mathcal{B}$. It is also denoted by $F\mathcal{B}$ or $L(\mathcal{B})$. Its structure group is the general linear group in three dimensions, $GL(3;\mathbb{R})$.

More precisely, at each point $\mathbf{X} \in \mathcal{B}$ we form the set $F_\mathbf{X}\mathcal{B}$ of all ordered triples $\{\mathbf{f}\}_\mathbf{X} = (\mathbf{f}_1, \mathbf{f}_2, \mathbf{f}_3)$ of linearly independent vectors $\mathbf{f}_I$ in $T_\mathbf{X}\mathcal{B}$, namely, the set of all bases of $T_\mathbf{X}\mathcal{B}$. Our total space will consist of all ordered pairs of the form $(\mathbf{X}, \{\mathbf{f}\}_\mathbf{X})$ with the obvious projection onto $\mathcal{B}$. The pair $(\mathbf{X}, \{\mathbf{f}\}_\mathbf{X})$ is called *a linear frame at* $\mathbf{X}$. Each basis $\{\mathbf{f}\}_\mathbf{X}$ is expressible uniquely as:

$$\mathbf{f}_I = f_I^J \mathbf{e}_J \tag{1.39}$$

24   1 An overview of inhomogeneity theory

in a coordinate system $X^I$, with natural basis $\mathbf{e}_I$, where $\{f_I^J\}$ is a non-singular matrix. We conclude that the typical fibre in this case is $GL(3;\mathbb{R})$. But so is the structure group. Indeed, in another coordinate system, $Y^I$, with natural basis $\hat{\mathbf{e}}_I$, using the chain rule we have:

$$\mathbf{f}_I = q_I^J \hat{\mathbf{e}}_J, \tag{1.40}$$

where

$$q_I^J = \frac{\partial Y^J}{\partial X^M} f_I^M = b_M^J f_I^M, \tag{1.41}$$

where $b_M^J$ is again some non-singular matrix, namely, an element of the general linear group $GL(3;\mathbb{R})$.

It follows that this is an instance of a *principal fibre bundle*, namely, a fibre bundle whose typical fibre and structure group coincide. The action of the group on the typical fibre is the natural left action of the group on itself. One of the interesting features of a principal bundle is that the structure group has also a natural *right* action on the bundle itself, and this property can be used to provide an alternative definition of principal bundles, as shown in Section 8.3.1. In the case of $F(\mathcal{B})$, for example, the right action $R_a$ is defined, in a given coordinate system $X^I$, by

$$R_a\{\mathbf{f}_J\} = f_I^K a^I{}_J \mathbf{e}_K, \quad J = 1, 2, 3, \tag{1.42}$$

where $a^I{}_J$ are the entries of the non-singular matrix $a \in GL(3;\mathbb{R})$. This right action sends a basis at $\mathbf{X}$ to another basis at $\mathbf{X}$, i.e., the action is fibre-preserving. It is not difficult to verify that this definition of the action is independent of the system of coordinates adopted.[14]

*Example 1.3.* **Bodies with linear microstructure** The idea of endowing bodies with a microstructure represented by affine deformation of "grains" goes back to the pioneering work of the Cosserat brothers [10]. Because the terminology "Cosserat media" is often used in the literature to designate a particular case, we use here the longer and more descriptive title. If each of the "stones" of our previous example along these lines is permitted to undergo just affine deformations (namely, deformations with a constant deformation gradient), it is clear that the extra kinematics can be described in terms of a linear mapping of any basis attached to the stone. Knowing how one basis deforms is enough to determine how all other bases at the same point deform. The choice of basis remaining arbitrary, we are naturally led to the conclusion that the appropriate geometric counterpart of a body $\mathcal{B}$ with linear

---

[14] The principal bundle of linear frames of a manifold is associated to all the tensor bundles, including the tangent and the cotangent bundles of the same manifold. By a direct application of the fundamental existence theorem of fibre bundles [83], we know that the associated principal bundle is defined uniquely up to an equivalence. Many properties of bundles can be better understood by working first on the principal bundle. For further details see Chapter 9.

microstructure (a "general Cosserat body") is the bundle of linear frames $F(\mathcal{B})$.

**The material G-structure**

Clearly, the material frame bundle $\mathcal{P}_\mathcal{B}$ is a subset of the principal frame bundle, since it consists at each point of a subset of all the possible frames. Technically, the material frame bundle is a *reduction* of the principal frame bundle. Because this reduction is governed by a subgroup (in our case: $\bar{\mathcal{G}}$) of the general linear group, the material fibre bundle is called a *G-structure*, which in our case will be referred to as a *material G-structure* [21].[15] A G-structure is itself a principal bundle.

*Remark 1.4.* We have seen that when the symmetry group of the material is the trivial group (consisting of the unit transformation only), the archetype induces a unique uniformity field of bases. By smoothness, we conclude that this unique field must extend smoothly over the whole body. Equivalently, since we have identified a smooth uniformity field on $\mathcal{B}$ with a smooth cross section of the material G-structure, we can affirm that this G-structure has one, and only one, cross section, which is globally smooth. Indeed, the fibre over each point consists of just one element. (Incidentally, this means that for a body with the trivial symmetry group, all smooth local notions become automatically global). When the symmetry group is a non-trivial discrete group, each fibre of the G-structure consists of a discrete collection of points, each one representing a frame. If we consider a smooth local section over an open set $\mathcal{U} \subset \mathcal{B}$, it is clear that all other smooth local sections over this open set must differ from the given one by premultiplication by a *constant* element of the symmetry group (otherwise, the smoothness would be lost, since the choice of group elements is discrete). In other words, one must jump from one smooth local section to other such sections by discrete (constant) steps. Because two local sections that differ by a constant induce, obviously, the same local parallelism, we conclude that the local parallelisms are again unique, just as in the case of the trivial group (except that now globality may have been lost). Finally, when the symmetry group is continuous, the G-structure picture clearly shows that we can alter a local section without losing smoothness by exploiting the fact that each fibre consists of a smooth collection of frames.

We emphasize the fact that the G-structure construct conveys geometrically all the information contained in the constitutive equation, at least as far as the material uniformity of the body is concerned. On the other hand, the material G-structure is not unique, in the sense that it depends on the

---

[15] In some instances, when we wish to emphasize the structure group $\mathcal{G}$ of a given G-structure, we will use the notation $\mathcal{G}$-structure.

choice of an archetype (namely, a frame at a body point). Each material G-structure, however, is *conjugate* to all others (see Box 1.6), as follows from our considerations at the end of Section 1.2.3. It is remarkable that there exists another, more general, differential geometric object that encompasses all these conjugate G-structures under a single umbrella. This object is the *material groupoid*.

---

**Box 1.6 Conjugate $G$-structures**

Two $G$-structures, $F_{G_1}(\mathcal{B})$ and $F_{G_2}(\mathcal{B})$, with structure groups $\mathcal{G}_1$ and $\mathcal{G}_2$, respectively, over the same $n$-dimensional base manifold $\mathcal{B}$, are said to be *conjugate* if there exists a fixed element $g \in GL(n; \mathbb{R})$ such that:

$$F_{G_2}(\mathcal{B}) = R_g(F_{G_1}(\mathcal{B})), \tag{1.43}$$

where $R_g$ is the right action of $g$ on the frame bundle $F(\mathcal{B})$.

The reason for this terminology is the following: Let two elements, $p_1$ and $p_2$, lying on the same fibre of $F_{G_1}(\mathcal{B})$ be related by:

$$p_2 = p_1\, a, \tag{1.44}$$

where $a \in G_1$ and where we have indicated the right action by apposition (i.e., $p_1\, a = R_a(p_1)$). The images of these two points, lying now on the same fibre of $F_{G_1}(\mathcal{B})$, are, respectively,

$$p'_1 = p_1\, g, \tag{1.45}$$

and

$$p'_2 = p_2\, g. \tag{1.46}$$

It follows then that:

$$p'_2 = p'_1\, g^{-1}\, a\, g. \tag{1.47}$$

Refining this argument slightly, it is not difficult to prove that in fact the two structure groups *must be conjugate of each other*. More precisely:

$$G_2 = g^{-1}\, G_1\, g. \tag{1.48}$$

The converse, however, is not true: two $G$-structures (over the same manifold) having conjugate structure groups are not necessarily conjugate. For example, two different global sections of the frame bundle of a trivial manifold can be regarded as two different $G$-structures with the same (trivial) structure group. Nevertheless, unless these two sections happen to differ by the right action of a *constant* element of the general linear group, the two $G$-structures are clearly not conjugate.

---

### 1.3.2 The material groupoid and its associated G-structures

Suppose that a material body $\mathcal{B}$ is given, having a constitutive law of the form (1.6). This body is not necessarily uniform, so that perhaps none, or some, or

## 1.3 The material G-structure and the material groupoid

all of its points are materially isomorphic. Those that are, will necessarily have conjugate material symmetry groups, and their constitutive equations will be related by formulas such as (1.8). Given any pair of body points, $\mathbf{X}_1$ and $\mathbf{X}_2$, we consider the uniquely defined set $\mathcal{P}_{12}$ of all the material isomorphisms with source $\mathbf{X}_1$ and target $\mathbf{X}_2$. As just pointed out, if the body happens not to be uniform, some, or possibly all, of these sets (with $\mathbf{X}_1 \neq \mathbf{X}_2$) are clearly empty. Consider now the Cartesian product $\mathcal{B} \times \mathcal{B}$, that is, the set consisting of all ordered pairs of body points. To each element of this set (namely, to each ordered pair $\{\mathbf{X}_1, \mathbf{X}_2\}$) we attach the corresponding (possibly empty) set $\mathcal{P}_{12}$. In this way, we obtain a new set of what appears to be a rather complicated nature. One way to look at this set is to consider it as the union $\mathcal{Z}$ of all the sets $\mathcal{P}_{ij}$ of material isomorphisms from point $\mathbf{X}_i$ to point $\mathbf{X}_j$ for all $\mathbf{X}_i, \mathbf{X}_j \in \mathcal{B}$, and to endow this new set $\mathcal{Z}$ with two *projection maps* on the body-manifold $\mathcal{B}$. The first (or *source*) projection is the map:

$$\alpha : \mathcal{Z} \longrightarrow \mathcal{B}$$
$$\mathbf{P}_{ij} \mapsto \mathbf{X}_i. \tag{1.49}$$

The second (or *target*) projection is given by:

$$\beta : \mathcal{Z} \longrightarrow \mathcal{B}$$
$$\mathbf{P}_{ij} \mapsto \mathbf{X}_j. \tag{1.50}$$

We note, moreover, that in certain cases two elements of $\mathcal{Z}$ can be composed to yield another element of $\mathcal{Z}$. This is the case whenever the target of the first element happens to be equal to the source of the second. More precisely, given two elements $\mathbf{P}_{ij}$ and $\mathbf{P}_{km}$ of $\mathcal{Z}$ the composition $\mathbf{P}_{ij}\,\mathbf{P}_{km}$ makes sense if, and only if, $i = m$, according to Equation (1.10). We begin to see that, in some sense, the set $\mathcal{Z}$ has properties similar to those of a group. There exists a precisely defined mathematical structure that corresponds exactly to the situation at hand, namely the notion of a *groupoid*. We refer the reader to Appendix A for a review of the basic groupoid terminology. We conclude that to every material body we can associate a groupoid, which we will call the *material groupoid*.

We recall that a groupoid is said to be transitive if for each ordered pair of points of the base manifold there exists at least one groupoid element with one of the points as the source and the other as the target. In the case of the material groupoid, transitivity means, therefore, that each body point is materially isomorphic to every other body point. In other words, the uniformity of a body corresponds exactly to the notion of a *transitive groupoid*, namely, a groupoid whereby none of the sets $\mathcal{P}_{ij}$ is empty. Finally, we recall that a groupoid is said to be a Lie groupoid if the projections and the operations of composition and inverse are smooth.

Putting together the various mechanical and geometrical notions introduced thus far, we conclude that: *A (smoothly) materially uniform body is endowed with the natural structure of a transitive Lie groupoid.* We have called this structure the *material Lie groupoid*. At this point it is worthwhile noticing that if the assumption of smoothness were to be dropped from the definition of material uniformity, then the material groupoid would still exist and would still be transitive, but it would no longer be a Lie groupoid. Such a situation may certainly arise in practice.

*Example 1.5.* **Material discontinuities:** A not very imaginative example of a non-smooth material groupoid is provided by gluing together two identical pieces of wood, but without respecting the continuity of the grain. A somewhat more sophisticated example could be the presence of an isolated edge dislocation in a uniform body.

Consider now a non-uniform body. The material groupoid is still properly definable, except that it loses its transitivity. It may still preserve its smoothness (namely, it may still be a Lie groupoid).

*Example 1.6.* **Functionally graded materials:** A good example of this last situation is provided by the so-called *functionally graded materials*, which are bodies with smoothly varying material properties tailored to specific applications. Under certain circumstances, however, the transitivity of the material Lie groupoid of functionally graded materials can be restored, as we shall show later (in Chapter 4), by modifying the definition of material isomorphism.

To summarize the results of this section we may say that the specification of the constitutive law of any material body gives rise to an associated material groupoid. Within this geometrical context, a body is (smoothly) materially uniform if, and only if, its material groupoid is a transitive Lie groupoid.

To construct the material groupoid, we started by considering the Cartesian product $\mathcal{B} \times \mathcal{B}$ and attaching to each element $\{\mathbf{X}_i, \mathbf{X}_j\}$ of this product the set $\mathcal{P}_{ij}$ of all the possible material isomorphisms from $\mathbf{X}_i$ to $\mathbf{X}_j$. Let us compare this construction with that of the material G-structure, which we now briefly recap. We start by fixing an arbitrary archetype and attaching to each point $\mathbf{X}$ the set $\mathcal{P}_\mathbf{X}$ of all the material isomorphisms from the archetype to the point in question. Since the archetype is a frame at a material point, this procedure is equivalent to attaching to each point of the body the set of frames induced by acting on the archetype with each and every element of $\mathcal{P}_\mathbf{X}$. In this way, we obtain at each point a subset of all the possible bases of the tangent space at that point. The passage from one element of this subset to any other element is clearly governed by the material symmetry group, whereas the passage from an arbitrary basis to any other arbitrary basis (not connected in any way to the constitutive behaviour of the body) is governed by the general linear group in three dimensions. What we obtain, therefore, is a *reduction* of the principal frame bundle of the body manifold to a subbundle, also principal, via a subgroup of the general linear group. We see that this material G-structure can be obtained from the material groupoid of a uniform body by just pinning down one particular basis at one particular point and letting the material isomorphisms do their job of propagating

## 1.3 The material G-structure and the material groupoid

this base throughout the whole body. It follows then that, whereas the material groupoid is uniquely defined, the associated material G-structure is not, since it depends on the choice of archetype. Nevertheless, all the G-structures associated with the material groupoid are conjugate.

To prepare the ground for non-simple materials (whereby principal bundles more general than the bundle of frames are needed as a background to define the material isomorphisms), we may say that, rather than comparing every pair of points, an equivalent way to describe the fact that a body is materially uniform consists of comparing the response of each point with that of an archetype. Without offering any concrete advantages, this point of view nicely corresponds to the idea that all points of a uniform body behave as some ideal piece of material, the only difference being the way in which this piece has been implanted point by point into the body to constitute the whole entity. This point of view is reflected in a precise way in the passage from the (material) groupoid to a principal bundle, as we show in Appendix A. Corresponding to the fact that there is a degree of freedom in the choice of the archetype, we found that a transitive Lie groupoid gives rise not just to one but to a whole family of principal bundles, all of them equivalent to each other.

In view of the result just stated, we define a *material principal bundle* of a materially uniform body $\mathcal{B}$ as any one of the equivalent principal bundles that can be obtained from the material groupoid. We observe, however, that whereas the material groupoid always exists (whether or not the body is uniform), the material principal bundle can only be defined when the body is smoothly uniform. Then, and only then, we have a transitive Lie groupoid to work with. In conclusion, although both geometrical objects are suitable for the description of the material structure of a body, the groupoid representation is the more faithful one, since it is unique and universal.

The structure group of a material principal bundle is, according to the previous construction, nothing but the material symmetry group of the archetype. As expected, it controls the degree of freedom available in terms of implanting this archetype at the diverse points of the body.

The material principal bundle may, or may not, admit (global) cross sections. If it does, the body is said to be *globally uniform*. This term is slightly misleading, since uniformity already implies that *all* the points of the body are materially isomorphic. Nevertheless the term conveys the sense that the material isomorphisms can be prescribed smoothly in a single global chart of the body (which, by definition, always exists). Put in other terms, the existence of a global section implies (in a principal bundle, as we know) that the principal bundle is trivializable. A cross section of a principal bundle establishes, through the right action of the structure group, a global isomorphism between the fibres, also called a *distant (or complete) parallelism*. In our context, this property is called a *material parallelism*. If the structure group is discrete, the material parallelism is unique. Moreover, if the material symmetry group consists of just the identity[16], a uniform body must be globally uniform.

## 1.4 Homogeneity

### 1.4.1 Uniformity and homogeneity

We have interpreted the notion of material uniformity as the formalization of the statement: "all points of the body are made of the same material". Isn't that homogeneity? It is certainly a precondition of homogeneity, but when engineers tell you that a body is homogeneous they certainly mean more than just that. They mean, tacitly perhaps, that the body can be put in a configuration such that a mere translation of the neighbourhood of any point to that of any other point will do as a material isomorphism. In other words, in that particular configuration, all the points are indistinguishable from each other as far as the constitutive equation is concerned. If such a configuration exists and if it were to be used as a reference configuration, the constitutive law (1.6) in a Cartesian coordinate system would become independent of position. We call such a configuration a (globally) *homogeneous configuration*.

The formalization of this notion of homogeneity is not very difficult. We will motivate the derivation by working in coordinates. Notice that, using the notation of Equation (1.26), the uniformity condition (1.20) in terms of tensor components reads:

$$\psi(F^i_{\ I}, X^J) = \bar{\psi}(F^i_{\ I} P^I_\alpha(X^J)). \tag{1.51}$$

We recall that the entries $P^I_\alpha$ represent the components of the base vectors $\mathbf{p}_\alpha$ of a uniformity field in the coordinate system $X^J$. Suppose now that there exists a uniformity field and a coordinate chart such that the natural basis of the coordinate system happens to coincide, at each point, with the uniformity basis at that point. We say that the uniformity basis in this case is *integrable*[17]. Then, clearly, the components $P^I_\alpha$ in this coordinate system become:

$$P^I_\alpha = \delta^I_\alpha, \tag{1.52}$$

whence it follows that the constitutive equation (1.51) in this coordinate system is independent of $\mathbf{X}$. Since we can regard a coordinate system as a reference configuration, it follows that the engineering notion of homogeneity is satisfied in this case. The reverse is clearly true. Recall, however, that both a coordinate system and a smooth uniformity field are, in general, available in open sets that may not be as large as the whole body. If for every point of the body there exists a neighbourhood within which an integrable uniformity basis field can be found, we say that the uniform body is *locally homogeneous*.

---

[16] See Example 9.33.
[17] The terminology *holonomic* can also be used, but we reserve this term for a different concept to be applied to materials of higher grade.

A uniform body is *globally homogeneous*, or simply *homogeneous*, if the previous condition can be satisfied on a neighbourhood consisting of the whole body.

*Remark 1.7.* To illustrate the difference between local and global homogeneity, consider a long cylindrical uniform solid body in a natural (stress-free) configuration. Assume that the material is fully isotropic. This body is clearly homogeneous and the given cylindrical configuration is a homogeneous configuration. If this body is bent into a torus (by welding the two ends together) we obtain a new body $\mathcal{B}$ which, clearly, is not stress-free any longer. Any simply connected chunk of this body can be "straightened" and thus brought back locally into a homogeneous configuration, but not so the whole body. In other words, $\mathcal{B}$ is only locally homogeneous.

A uniform body which is not locally homogeneous is said to be *inhomogeneous* or to contain a distribution of inhomogeneities. Historically, the theory of continuous distributions of inhomogeneities in material bodies stemmed from various attempts at generalizing the geometric ideas of the theory of isolated dislocations arising in crystalline solids. As it should be clear from the previous presentation, however, the theory of uniform and inhomogeneous bodies has a life of its own, whose mathematical and physical apparatus is completely contained in the constitutive law of a continuous medium.

*Remark 1.8.* A simple example of an inhomogeneous elastic body is provided by a uniform thick spherical cap made of a transversely isotropic solid material whose axis of transverse isotropy $\mathbf{e_1}(\mathbf{X})$ is aligned at each point $\mathbf{X}$ with the local radial direction of the sphere. If we assume that in this configuration the material is stress-free, the material isomorphisms between any two points consist of rotations that bring the corresponding axes of transverse isotropy into coincidence. In a (locally) homogeneous configuration we should have that the field $\mathbf{e_1}(\mathbf{X})$ becomes a Euclidean-parallel field over some open subbody. This can be done, but only at the expense of stretching the points within each spherical layer by variable amounts (recall that no portion of a spherical surface can be mapped isometrically into a plane). Thus, any configuration in which the axes of transverse isotropy become parallel in some open neighbourhood, will necessarily contain points in that neighbourhood that are differently stretched. In other words, no (even local) homogeneous configuration can exist.

### 1.4.2 Homogeneity in terms of a material connection

Given a uniformity field on an open set $\mathcal{U}$, how can we ascertain whether or not it is integrable? In other words, given a uniformity field expressed in terms of some coordinate system $X^I$, how can we know whether or not there exists a local coordinate system such that the given uniformity bases are adapted to it? To obtain a necessary condition for this question of *integrability*, let us assume that such a coordinate system, say $Y^M$, does exist and let us write the change of coordinates to our system $X^I$ by means of three smooth and smoothly invertible functions:

$$X^I = X^I(Y^M) \quad (I, M = 1, 2, 3). \tag{1.53}$$

Any given vector field **w** can be expressed in terms of local components in either coordinate basis, say $w^I$ and $\hat{w}^M$, respectively, for the coordinate systems $X^I$ and $Y^M$. These components are related by the formula:

$$\hat{w}^M = w^I \frac{\partial Y^M}{\partial X^I}. \tag{1.54}$$

When it comes to each of the three vectors $\mathbf{p}_\alpha$ ($\alpha = 1, 2, 3$) of the given uniformity field, with components $P^I_\alpha$ in the $X^I$-coordinate system, their components in the $Y^M$-coordinate system are, according to our starting assumption, given by $\delta^M_\alpha$ (since they coincide with the natural basis of this system). By Equation (1.54), therefore, we must have:

$$\delta^M_\alpha = P^I_\alpha \frac{\partial Y^M}{\partial X^I}. \tag{1.55}$$

Taking the partial derivative of this equation with respect to $X^J$, we obtain:

$$0 = P^I_\alpha \frac{\partial^2 Y^M}{\partial X^J \partial X^I} + \frac{\partial P^I_\alpha}{\partial X^J} \frac{\partial Y^M}{\partial X^I}, \tag{1.56}$$

or, multiplying through by $P^{-\alpha}_K$:

$$0 = \frac{\partial^2 Y^M}{\partial X^J \partial X^K} + P^{-\alpha}_K \frac{\partial P^I_\alpha}{\partial X^J} \frac{\partial Y^M}{\partial X^I}. \tag{1.57}$$

Using the definition (1.32) of the Christoffel symbols of the given material parallelism, we obtain:

$$\frac{\partial^2 Y^M}{\partial X^J \partial X^K} = \Gamma^I_{KJ} \frac{\partial Y^M}{\partial X^I}. \tag{1.58}$$

But, by the equality of mixed partial derivatives, the left-hand side is symmetric, implying that so must the Christoffel symbols be, namely:

$$\Gamma^I_{KJ} = \Gamma^I_{JK}, \tag{1.59}$$

throughout $\mathcal{U}$. We have only shown that this is a necessary condition for integrability, but it can be proven that under certain restrictions (such as simple-connectedness of the domain $\mathcal{U}$) the symmetry of the Christoffel symbols is also a sufficient condition for the existence of an adapted coordinate system. We conclude that the body is locally homogeneous if, and only if, there exists for each point a neighbourhood and a (local) material parallelism whose Christoffel symbols are symmetric.

A more geometric way to view the local homogeneity condition just discovered is to define the *torsion* $\tau$ of a connection as the tensor with coordinate components given by:

$$\tau^I_{JK} = \Gamma^I_{JK} - \Gamma^I_{KJ}. \tag{1.60}$$

It can be shown, indeed, that although the Christoffel symbols are not the components of a tensor, their skew-symmetrized components (1.60) are. For a body to be locally homogeneous, therefore, it will be required that for each point there exists a neighbourhood in which a (local) material parallelism can be defined with vanishing torsion.

We have already seen that, if the symmetry group of the uniform body is discrete, the material connection is unique. This means that for such materials the question of homogeneity is settled once and for all by checking the torsion of the unique material connection. Equivalently, we may say that these geometric quantities are a true measure of the presence of inhomogeneities. For materials with a continuous symmetry group, on the other hand, the fact that a given material connection has a non-vanishing torsion is not necessarily an indication of inhomogeneity. Indeed, the non-vanishing of the torsion may be due to an unhappy choice of material connection. For materials with continuous symmetry groups, therefore, the criteria for homogeneity are mathematically more sophisticated, as we shall see.

*Remark 1.9.* An important remark needs to be made regarding the choice of an archetype, that is, an archetypal material point and a basis therein. Could it happen that perhaps the frame fields induced by a certain archetype satisfy the homogeneity criteria, whereas the frame fields generated by a different archetype and frame do not? The answer in the case of simple materials is negative. Indeed, if a coordinate system exists such that its natural basis is a uniformity field of bases for some archetype, then any linear change of coordinates immediately provides a new coordinate system whose natural basis corresponds to a linear transformation of the archetype. The situation is different in the case of higher-grade materials to be studied in the next chapter.

### 1.4.3 Homogeneity in terms of a material G-structure

Since we have repeatedly said that a genuine representation of the inhomogeneity of a material body is not given by the (possibly non-unique) local material connections but rather by any one of its conjugate material G-structures, it is incumbent upon us now to explicitly show what feature of a material G-structure corresponds exactly to the notion of local homogeneity. To this end, and recalling that a G-structure is a principal fibre bundle, we start by noticing that a particularly simple example of a principal bundle is given by $F(\mathbb{R}^n)$, the bundle of frames of $\mathbb{R}^n$. Since $\mathbb{R}^n$ is endowed at each point with a standard natural basis, there exists a natural global parallelism on $\mathbb{R}^n$. In other words, $F(\mathbb{R}^n)$ admits smooth global constant sections. We will, accordingly, call $F(\mathbb{R}^n)$ the *standard frame bundle* (in $n$ dimensions).

Given the frame bundle of an $n$-dimensional manifold and a subgroup $\mathcal{G}$ of the general linear group $GL(n;\mathbb{R})$, it is not necessarily true that the frame bundle is reducible to this subgroup. The standard frame bundle $F(\mathbb{R}^n)$, however, is reducible to *any* subgroup $\mathcal{G}$ of $GL(n;\mathbb{R})$. To see why this is the case, we need only consider at each point of $\mathbb{R}^n$ the standard natural basis and construct the set of all bases obtained from it by the right action of all elements of $\mathcal{G}$. The collection of bases thus obtained can be regarded as a reduction of $F(\mathbb{R}^n)$ to the subgroup $\mathcal{G}$, namely, as a $\mathcal{G}$-structure, since (by construction) it is stable under the action of this subgroup. Note that the availability of a distinguished global section is crucial for this procedure to work. We will call the $\mathcal{G}$-structure obtained in this way the *standard flat $\mathcal{G}$-structure* (in $n$ dimensions).

Having thus established the existence of these standard $\mathcal{G}$-structures, we are in a position of inquiring whether or not a given non-standard $\mathcal{G}$-structure "looks" (at least locally) like its standard counterpart. This question is a particular case of the more general question of *local equivalence* between $\mathcal{G}$-structures, which we now tackle. We commence by remarking that, given two $n$-dimensional manifolds, $\mathcal{B}$ and $\mathcal{B}'$, a diffeomorphism:

$$\phi : \mathcal{B} \longrightarrow \mathcal{B}', \tag{1.61}$$

maps, through its tangent map $\phi_*$, every frame $\{e_i\}$ ($i = 1, ..., n$) at each point $b \in \mathcal{B}$ to a frame $\{e_i'\} = \{\phi_*(e_i)\}$ ($i = 1, ..., n$) at the image point $b' = \phi(b)$. Let now $\mathcal{P}$ be a $\mathcal{G}$-structure over $\mathcal{B}$. Clearly, the procedure just described generates at each point $b' \in \mathcal{B}'$ a collection of frames. The union $\mathcal{P}'$ of all these image bases can be considered as a $\mathcal{G}$-structure over $\mathcal{B}'$ having *the same structure group $\mathcal{G}$* as the original $\mathcal{G}$-structure. The $\mathcal{G}$-structure thus constructed is said to be *induced* or *dragged* by the diffeomorphism $\phi$ between the base manifolds. The induced $\mathcal{G}$-structure is denoted by $\mathcal{P}' = \phi_*(\mathcal{P})$. Clearly, this is a particular example of a principal-bundle morphism.

*Remark 1.10.* The reader may verify that the set of bases $\phi_*(\mathcal{P})$ is indeed a $\mathcal{G}$-structure and that its structure group is indeed the original $\mathcal{G}$. In other words, that the set is stable under the right action of this group.

Given two $\mathcal{G}$-structures, $\mathcal{P}$ and $\mathcal{P}'$, defined, respectively, over the base manifolds $\mathcal{B}$ and $\mathcal{B}'$, we say that they are (globally) *equivalent* if there exists a diffeomorphism $\phi : \mathcal{B} \longrightarrow \mathcal{B}'$ such that $\mathcal{P}' = \phi_*(\mathcal{P})$. Two $\mathcal{G}$-structures are said to be *locally equivalent* at the points $b \in \mathcal{B}$ and $b' \in \mathcal{B}'$ if there exists an open neighbourhood $\mathcal{U}$ of $b$ and a diffeomorphism $\phi : \mathcal{B} \longrightarrow \mathcal{B}'$ with $\phi(b) = b'$ such that:

$$\mathcal{P}'|_{\phi(\mathcal{U})} = \phi_*(\mathcal{P}|_\mathcal{U}), \tag{1.62}$$

where $\mathcal{P}|_\mathcal{U}$ denotes the sub-bundle of $\mathcal{P}$ obtained by restricting the base manifold to $\mathcal{U}$.

A $\mathcal{G}$-structure is called *locally flat* if it is locally equivalent to the standard flat $\mathcal{G}$-structure. From the material point of view, the diffeomorphisms $\phi : \mathcal{B} \longrightarrow \mathbb{R}^3$ can be seen as changes of reference configuration of the body $\mathcal{B}$. We conclude, therefore, that local flatness of a material $\mathcal{G}$-structure means that a mere change of reference configuration renders the material isomorphisms (within an open neighbourhood) Euclidean parallelisms, which is precisely what local homogeneity is intended to convey! So, local flatness of a material $\mathcal{G}$-structure is the precise mathematical expression of local homogeneity.

*Remark 1.11.* Although the material $\mathcal{G}$-structure is not unique (depending as it does on the choice of archetype), local flatness is independent of the particular $\mathcal{G}$-structure chosen. To prove this assertion, we recall that all the possible material $\mathcal{G}$-structures are conjugate of each other. Suppose that one of the material $\mathcal{G}$-structures is locally flat. Any other material $\mathcal{G}$-structure can, therefore (see Box 1.6), be obtained from the given one through the right action $R_g$ of a fixed element $g$ of $GL(3;\mathbb{R})$. This element can be represented as a $3 \times 3$ matrix, which we denote by $[g]$. We now construct the following affine transformation of $\mathbb{R}^3$:

$$\psi : \mathbb{R}^3 \longrightarrow \mathbb{R}^3$$
$$\{x\} \mapsto \{a\} + [g]\{x\}, \qquad (1.63)$$

where $\{x\} = <x^1, x^2, x^3>^T$ is the generic position vector in $\mathbb{R}^3$, and $\{a\} = <a^1, a^2, a^3>$ is a fixed vector. Let $\phi$ denote a local diffeomorphism (such as in Equation (1.62)) used in establishing the local flatness of the original $\mathcal{G}$-structure. Then, the map $\psi^{-1} \circ \phi$ can be used in establishing the local flatness of the new $\mathcal{G}$-structure, as it can be verified by a direct calculation. A quick symbolic way to carry out this calculation is the following: Let $\{\mathbf{f}\}$ be a frame belonging to the original $\mathcal{G}$-structure. Its counterpart in the new (conjugate) $\mathcal{G}$-structure is then $R_g\{\mathbf{f}\}$. This frame is dragged by $\psi^{-1} \circ \phi$ according to:

$$(\psi^{-1} \circ \phi)_*(R_g\{\mathbf{f}\}) = (\psi^{-1})_* \circ \phi_* \circ R_g(\{\mathbf{f}\}) = L_{g^{-1}} \circ R_g(\phi(\{\mathbf{f}\})). \qquad (1.64)$$

This equation shows that the previous standard $\mathcal{G}$-structure moves (always within the frame bundle of $\mathbb{R}^3$) to the standard $\mathcal{G}$-structure corresponding to the new (conjugate) group, as desired[18].

To connect this characterization of local homogeneity with the coordinate-based one presented in the previous section, we need only note that a diffeomorphism $\phi$ of an open neighbourhood in $\mathcal{B}$ with an open neighbourhood in $\mathbb{R}^3$ can be regarded as a coordinate chart $x^i = \phi^i(b)$. Moreover, the inverse image by $\phi$ of the standard basis in $\mathbb{R}^3$ provides, at each point $b \in \mathcal{B}$, the natural basis $\{\frac{\partial}{\partial x^i}\}$ associated with this coordinate chart. In other words, a

---

[18] Notice that the corresponding standard $\mathcal{G}$-structures are not conjugate of each other, since, by definition, they all share the standard (unit) section.

$\mathcal{G}$-structure is locally flat if, and only if, there exists an atlas $\{\mathcal{U}_\alpha, \phi_\alpha\}$ such that its natural frames belong to the $\mathcal{G}$-structure[19].

### 1.4.4 Homogeneity in terms of the material groupoid

As we know, all the conjugate material $\mathcal{G}$-structures of a material body $\mathcal{B}$ are implicitly contained in the material groupoid $\mathcal{Z}$. Now, this groupoid is necessarily a sub-groupoid[20] of the tangent groupoid $\mathcal{T}(\mathcal{B})$. We recall that the tangent groupoid of a differentiable manifold $\mathcal{B}$ is obtained by considering, for each pair of points $a, b \in \mathcal{B}$, the collection of all the non-singular linear maps between the tangent spaces $T_a\mathcal{B}$ and $T_b\mathcal{B}$. The material groupoid $\mathcal{Z}$ is obtained by retaining only those linear maps that represent material isomorphisms.

Let $\mathcal{Z}$ be a transitive Lie groupoid with base manifold $\mathcal{B}$. Given an atlas of $\mathcal{B}$, every chart (via the natural basis of the coordinate system) induces a local parallelism. In other words, for every pair of points in the domain of the chart there exists a distinguished (chart-related) element of the tangent groupoid $\mathcal{T}(\mathcal{B})$. We say that the groupoid $\mathcal{Z}$ is *locally flat* if there exists an atlas such that, for every pair of points in the domain of every chart of the atlas, this distinguished chart-induced element of $\mathcal{T}(\mathcal{B})$ belongs to $\mathcal{Z}$. We conclude, therefore, that a material body $\mathcal{B}$ is locally homogeneous if, and only if, its material groupoid is locally flat.

## 1.5 Homogeneity criteria

### 1.5.1 Solids

We recall that, by definition, an elastic material point is a *solid point* if its material symmetry group is a conjugate of a subgroup of the orthogonal group $\mathcal{O}$. Solid points, moreover, usually possess *natural states*, that is, configurations in which they are free of stress. Given a natural state of a solid point, all the other natural states are obtained by applying to the given natural state an arbitrary orthogonal transformation. The symmetry group of a solid point in any natural state is a subgroup of the orthogonal group. An elastic body is said to be a *solid body* if all its points are solid points. In a uniform solid body it is always possible and often convenient to adopt an archetype that is in a natural state.

Consider now a uniform solid body $\mathcal{B}$ in which a natural-state archetype has been chosen with symmetry group $\bar{\mathcal{G}} \subset \mathcal{O}$. Let $\mathbf{P}(\mathbf{X})$ be a uniformity field

---

[19] For this to be the case, according to Frobenius' theorem, we need that the Lie brackets of pairs of base vectors vanish identically, which is precisely the same as the vanishing of the torsion of the local material connections induced by these bases in their respective neighbourhoods

[20] A subset of a groupoid is called a sub-groupoid if it is a groupoid with respect to the law of composition of the original groupoid.

## 1.5 Homogeneity criteria

(that is: a field of "implants" of the archetype over the body). We claim that the standard (Cartesian) inner product at the archetype induces a unique inner product in each of the tangent spaces $T_\mathbf{X}\mathcal{B}$. In other words, we claim that a uniform solid body is automatically endowed with a specific materially induced *intrinsic Riemannian metric structure* **g**. Let $\mathbf{u}, \mathbf{v} \in \mathbf{T_X}\mathcal{B}$ be two vectors at a point $\mathbf{X} \in \mathcal{B}$. We define their inner product $*$ by the following formula:

$$\mathbf{u} * \mathbf{v} \equiv (\mathbf{P}^{-1}\mathbf{u}) \bullet (\mathbf{P}^{-1}\mathbf{v}), \qquad (1.65)$$

where $\bullet$ denotes the ordinary Cartesian inner product at the archetype. We need to show that this new inner product is independent both of the particular uniformity field chosen and of the particular natural state of the archetype. As shown in Section 1.2.3, in each case the change in the uniformity field consists on a multiplication of **P** to the right by a member of the archetypal symmetry group (for the degree of freedom permitted by the material symmetry) or a member of the general linear group (for the degree of freedom permitted by an arbitrary change of archetype). Since we are only considering natural-state archetypes, the latter degree of freedom boils down to a multiplication by an arbitrary member of the orthogonal group. Moreover, since the body is an elastic solid, the former degree of freedom is a subgroup of the orthogonal group. But the right-hand side of Equation (1.65) is invariant under multiplication of **P** to the right by any member of the orthogonal group, which proves our claim.

The material Riemannian metric **g** will, in general, have a non-vanishing Riemann-Christoffel curvature tensor $R_g$ (see Equation (9.53)). The special case in which $R_g \equiv 0$ deserves further consideration. In this case, every body point has an open neighbourhood for which an appropriate change of configuration renders the metric **g** equal to the Cartesian identity. In other words, in this configuration every material frame (as induced by all possible transplants of the archetypal frame) is an orthonormal frame in the Cartesian sense. From the physical point of view, we have that the neighbourhood in question can be brought to a configuration in which every point in the neighbourhood is in a natural state. This special case of inhomogeneity is called (local) *contorted aelotropy*.[21] Since, clearly, if the body enjoys local contorted aelotropy the curvature tensor of the intrinsic material metric vanishes, we have that:

$$R_g = 0 \Leftrightarrow \text{Contorted aelotropy}. \qquad (1.66)$$

Assume now that the body is isotropic, namely: $\bar{\mathcal{G}} = \mathcal{O}$. Then, in any of the special configurations of contorted aelotropy, we can make use of the material symmetry to give a further rotation to the material frames so as to render

---

[21] This terminology is due to Noll [75]. A somewhat more general case, in which the archetype is not necessarily in a natural state, has been referred to as a body endowed with states of *constant strain* (see [27]). The common feature is that the body can be brought locally into configurations in which all the material isomorphisms are orthogonal transformations.

them all parallel to each other (in the Euclidean sense). In other words, we have just proved that for an isotropic solid:

$$R_g = 0 \Leftrightarrow \text{Local homogeneity.} \tag{1.67}$$

*Remark 1.12.* The case of the intrinsic metric just presented is a particular case of a *characteristic object*. For a detailed treatment of this concept and its relevance to the problem of finding integrability conditions of $G$-structures, see the following works: [21, 89]. See also Chapter 5 of [91]. For the related concept of a $G$-structured generated by a tensor, see [52].

At this point it is worth recalling that for solid bodies with a discrete symmetry group $\bar{\mathcal{G}}$, since the local material connections are uniquely defined, the criterion of local homogeneity simply boils down to the vanishing of the torsion thereof. Besides the case of full isotropy already discussed, the only other possible continuous symmetry group of a solid point consists of *transverse isotropy*, whereby the symmetry group is (a conjugate of) the group of rotations about a fixed unit vector **e**. Choosing a natural-state archetype, it is clear that the condition $R_g = 0$ is necessary, but no longer sufficient, to ensure local homogeneity. The unit vector **e** at the archetype gives rise, via the implants $\mathbf{P}(\mathbf{X})$, to a smooth vector field in the neighbourhood of each point $\mathbf{X}$. If the covariant derivative of this vector field (with respect to the symmetric metric connection generated by **g**) vanishes identically over this neighbourhood, at a configuration of contorted aelotropy we will have that these vectors become actually parallel (in the Euclidean sense). Using now the material symmetry at hand so as to give a further rotation about this axis at each point, we can obtain a field of parallel material frames, so that the body is locally homogeneous. The converse is clearly true. We have shown that for a transversely isotropic body:

$$R_g = 0, \nabla_g \mathbf{e} = \mathbf{0} \Leftrightarrow \text{Local homogeneity.} \tag{1.68}$$

### 1.5.2 Fluids

An *elastic fluid point* is an elastic material point whose symmetry group is the unimodular group, namely, the group of all matrices with unit determinant. This group is the same in all configurations, since a conjugation of the unimodular group by any non-singular matrix delivers again the unimodular group. For physical reasons (e.g., cavitation), elastic fluid points do not have natural (i.e., stress-free) states. An elastic body is said to be an *elastic fluid* if all its points are elastic fluid points. It is not difficult to prove that the constitutive law of a fluid point boils down to a dependence on the determinant of the deformation gradient. Let $\kappa_0$ be a reference configuration of a uniform elastic fluid body $\mathcal{B}$ and let $\mathbf{P}(\mathbf{X})$ denote a field of implants from some archetype. The constitutive law of the body relative to this reference configuration must be of the form:

$$\psi_0(\mathbf{F}, \mathbf{X}) = \bar{\psi}(\det(\mathbf{FP}(\mathbf{X}))), \quad (1.69)$$

where, as usual, we have denoted by $\bar{\psi}$ the constitutive law of the archetype. Let $\kappa_1$ denote another reference configuration and let:

$$\lambda = \kappa_1 \circ \kappa_0^{-1} : \kappa_0(\mathcal{B}) \longrightarrow \kappa_1(\mathcal{B}), \quad (1.70)$$

denote the corresponding change of reference configuration. This map $\lambda$ is necessarily a diffeomorphism and the corresponding Jacobian determinant is a smooth function of position. The constitutive law of the body relative to the reference configuration $\kappa_1$ is given by:

$$\psi_1(\mathbf{F}, \lambda(\mathbf{X})) = \psi_0(\mathbf{F}\lambda_*, \mathbf{X}) = \bar{\psi}(\det(\mathbf{F}\lambda_*\mathbf{P}(\mathbf{X}))), \quad (1.71)$$

where $\lambda_*$ denotes the derivative map of $\lambda$ at $\mathbf{X}$. Obviously, the body is homogeneous if, and only if, a change of reference configuration, $\lambda$, can be found such that:

$$\det(\lambda_*\mathbf{P}(\mathbf{X})) = \text{constant}. \quad (1.72)$$

Moser's Lemma [73] guarantees that this can always be done, at least locally. We conclude, therefore, that for an elastic fluid:

$$\text{Uniformity} \Rightarrow \text{Local homogeneity}. \quad (1.73)$$

### 1.5.3 Fluid crystals

An elastic material point that is not a solid is, in Wang's terminology [89], a *fluid crystal point*. A body all of whose points are fluid crystal points is called a *fluid crystal* body. As a consequence of a theorem in group theory [74, 3], a fluid crystal cannot be isotropic, unless it is a fluid point. In other words, the symmetry group of a proper fluid crystal point (one that is neither a solid nor a fluid) cannot contain the orthogonal group (or a conjugate thereof).

As an example of a fluid crystal point, consider a material with a preferred direction. Its symmetry group consists of all unimodular transformations that leave that direction invariant (i.e., all unimodular matrices with that direction as an eigenvector, with an arbitrary eigenvalue). Given a uniform fluid crystal body of this type in some reference configuration, we can always find a change of reference configuration that renders the field of preferred directions parallel in a neighbourhood of a given point. After this has been achieved, a further unimodular transformation keeping this field of directions intact can be used to bring all points in the neighbourhood to the same state. We conclude, therefore, that for this type of fluid crystal the implication (1.73) holds. In other words, every uniform fluid crystal of this type is automatically locally homogeneous.

A different type of fluid crystals of interest consists of material points with a preferred plane. The symmetry group in this case consists of all unimodular transformations having that plane as an eigenspace. Attaching to each point the preferred plane, we obtain a two-dimensional distribution over the body manifold. It can be shown [21] that a uniform body of this type is locally homogeneous if, and only if, this distribution is involutive. A detailed study of the local homogeneity conditions for all possible types of fluid crystals can be found in [67].

# 2

# Uniformity of second-grade materials

A *second-grade material* is characterized by a mechanical response that is sensitive to both the first and the second gradient of the deformation. Theories of uniformity and homogeneity of second- and higher-grade bodies [71, 62, 63, 64] are considerably more difficult than in the case of simple materials. The extra difficulty is due not only to the obvious enlargement of the number of independent variables entering the constitutive law, but also to the fact that the law of composition of second derivatives brings into the picture a material symmetry group with a new kind of composition law. In order to keep the geometric ideas as close as possible to the case of simple materials, it is convenient to introduce first a formulation in terms of jets of maps, and then to express the results in terms of matrices of derivatives.

## 2.1 An intuitive picture

Before embarking on the study of second-grade uniformity, it may prove useful to trigger an intuitive picture with no mathematical content or rigour whatsoever. Let us first revisit the first-grade (simple-material) situation. A first-grade archetype, as we know, can be intuitively conceived as a small die whose elastic energy is a function of its linear transformations into arbitrary parallelepipeds. The uniform body itself can then be seen as an ensemble of implanted deformed archetypes which, in general, do not fit together, much as the pieces of a poorly solved jigsaw puzzle (Figure 2.1). If, on the other hand, the implants can be chosen (within the degree of freedom permitted by the symmetry group of the archetype) in such a way that all the pieces fit well (within any specified neighbourhood), then the material is (locally) homogeneous (Figure 2.2 ). For clarity, we draw two-dimensional pictures, whereby the archetype and its implants look like parallelograms, rather than parallelepipeds.

42    2 Uniformity of second-grade materials

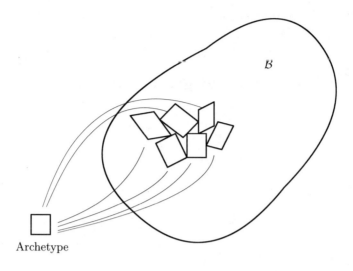

**Fig. 2.1.** First-grade uniformity

For second-grade materials, it would appear that we should work with curvilinear implanted archetypes to represent somehow the second derivatives of the deformation. An alternative representation (devoid, as has been stated, of any mathematical rigour) can be obtained by resorting to a possible physical interpretation of second-grade behaviour in terms of molecular interactions. In a simple (first-grade) solid, only the first neighbours (up to one lattice unit away, say) affect the mechanical response of an average point. If this idea is conveyed by means of a parallelepiped, then the corresponding idea that second neighbours (up to two lattice units away) affect the response at a point of a second-grade material can be conveyed by a collection of 27 parallelepipeds stacked in a Rubik's cube fashion. In the two-dimensional picture, we would have 9 quadrilaterals (Figure 2.3). A first-grade structure, as we will learn, is always buried inside a second-grade one. This induced first-grade structure is now represented by the central parallelepiped. We may have an inhomogeneous second-grade body which still has a homogeneous first-grade behaviour, in which case the central units of the adjoining implants fit well, while the surrounding ones do not (Figure 2.4). Finally, second-grade homogeneity corresponds to a perfect fit, as shown in Figure 2.5. For clarity, each of these figures depicts just two implanted second-grade archetypes.

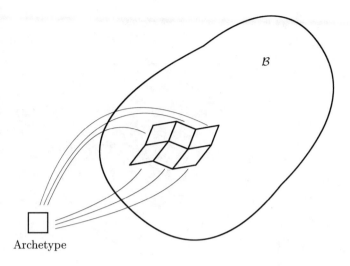

**Fig. 2.2.** First-grade homogeneity

## 2.2 The second-grade constitutive law

### 2.2.1 Jets of maps

Given two smooth manifolds, $\mathcal{M}$ and $\mathcal{N}$ (of dimensions $m$ and $n$, respectively), we denote by $C^\infty(\mathcal{M},\mathcal{N})$ the collection of all $C^\infty$-maps (i.e., smooth maps) $f : \mathcal{M} \longrightarrow \mathcal{N}$. We say that two maps $f, g \in C^\infty(\mathcal{M},\mathcal{N})$ have the same $k$-jet at a point $\mathbf{X} \in \mathcal{M}$ if: (i) f($\mathbf{X}$)=g($\mathbf{X}$); (ii) in a coordinate chart in $\mathcal{M}$ containing $\mathbf{X}$ and a coordinate chart in $\mathcal{N}$ containing the image $f(\mathbf{X})$, all the partial derivatives of $f$ and $g$ up to and including the order $k$ are respectively equal.

Although the above definition is formulated in terms of charts, it is not difficult to show by direct computation that the property of having the same derivatives up to and including order $k$ is in fact independent of the coordinate systems used in either manifold. Notice that, in order for this to work, it is imperative to equate *all* the lower-order derivatives. If, for example, we were to equate just the second derivatives, without regard to the first, the equality of the second derivatives would not be preserved under arbitrary coordinate transformations.

*Remark 2.1.* Two functions have the same $k$-jet at $\mathbf{X}$ if, and only if, they map every smooth curve through $\mathbf{X}$ to curves having a $k$-th order contact at the (common) image point. This property can be used as an alternative (coordinate independent) definition.

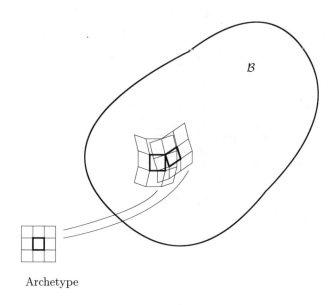

Archetype

**Fig. 2.3.** Second-grade uniformity

The property of having the same $k$-jet at a point is, clearly, an equivalence relation. The corresponding equivalence classes are called *$k$-jets at* **X**. Any function in a given $k$-jet is then called a *representative* of the $k$-jet. The $k$-jet at **X** of which a given function $f : \mathcal{M} \longrightarrow \mathcal{N}$ is a representative is denoted by $j_{\mathbf{X}}^k f$. The collection of all $k$-jets at $\mathbf{X} \in \mathcal{M}$ is denoted by $J_{\mathbf{X}}^k(\mathcal{M}, \mathcal{N})$. The point **X** is called the *source* of $j_{\mathbf{X}}^k f$ and the image point $f(\mathbf{X})$ is called its *target*.

From an intuitive point of view, we can view a $k$-jet at a point as the result of placing at that point a myopic observer, who does not distinguish between functions as long as they are sufficiently close to each other in a small neighbourhood of the point. The worst-case scenario is $k = 0$, which corresponds to identifying all functions that just have the same value at the point. Next (for $k = 1$) comes the identification of all functions with the same tangent map at the point. With better prescription glasses, the range of vision (and the corresponding $k$) can be increased.

### 2.2.2 Composition of jets

Jets of the same order can be composed whenever the target of one jet coincides with the source of the next. Let $\mathcal{M}, \mathcal{N}$ and $\mathcal{P}$ be smooth manifolds of dimension $m$, $n$ and $p$, respectively, and let $f \in C^\infty(\mathcal{M}, \mathcal{N})$ and $g \in C^\infty(\mathcal{N}, \mathcal{P})$. For any point $\mathbf{X} \in \mathcal{M}$, the composition of the jets $j_X^k f$ and $j_{f(X)}^k g$ is defined by:

## 2.2 The second-grade constitutive law

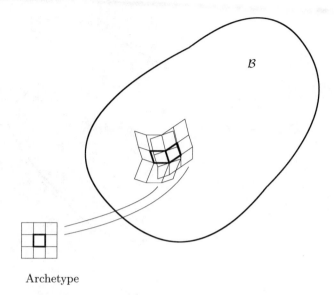

Archetype

**Fig. 2.4.** First-grade homogeneity within second-grade uniformity

$$j^k_{f(X)}g \circ j^k_X f = j^k_X(g \circ f). \tag{2.1}$$

This definition is independent of the representatives $f$ and $g$ of each jet. Thus, to compose two jets, we choose any representative of the first jet and compose it with any representative of the second, thus obtaining a map from the first manifold ($\mathcal{M}$) to the third ($\mathcal{P}$). We then calculate the jet of this composite function and obtain, by definition, the composition of the given jets.

To illustrate this operation, consider the case $k = 2$. Let coordinate systems $X^I$ ($I = 1, ..., m$), $x^i$ ($i = 1, ..., n$) and $z^\alpha$ ($\alpha = 1, ..., p$) be chosen, respectively, around the points $\mathbf{X} \in \mathcal{M}$, $f(\mathbf{X}) \in \mathcal{N}$ and $g(f(\mathbf{X})) \in \mathcal{P}$. The functions $f : \mathcal{M} \longrightarrow \mathcal{N}$ and $g : \mathcal{N} \longrightarrow \mathcal{P}$ are then given locally in coordinates by:

$$x^i = x^i(X^1, ..., X^m), \quad z^\alpha = z^\alpha(x^1, ..., x^n), \tag{2.2}$$

where the indices take values in the appropriate ranges. The jet $j^2_X f$ is then given by the following coordinate expressions:

$$x^i(X^1, ..., X^m), \tag{2.3}$$

$$\left[\frac{\partial x^i}{\partial X^I}\right]_X, \tag{2.4}$$

$$\left[\frac{\partial^2 x^i}{\partial X^J \partial X^I}\right]_X, \tag{2.5}$$

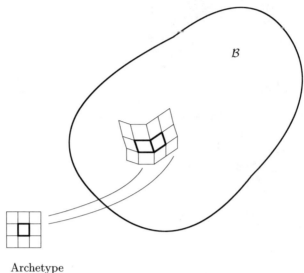

Archetype

**Fig. 2.5.** Second-grade homogeneity

a total of $n + mn + m^2n$ numbers. Similarly, the jet $j^2_{f(X)}g$ is given in coordinates by the following $p + np + n^2p$ numbers:

$$z^\alpha(x^1, ..., x^n), \tag{2.6}$$

$$\left[\frac{\partial z^\alpha}{\partial x^i}\right]_{f(X)}, \tag{2.7}$$

$$\left[\frac{\partial^2 z^\alpha}{\partial x^j \partial x^i}\right]_{f(X)}. \tag{2.8}$$

To obtain the components of the composite jet $j^2_{f(X)}g \circ j^2_X f$, we consider the composite functions $z^\alpha(x^i(x^I))$ and take its first and second derivatives. The result consists of the following $p + mp + m^2p$ numbers:

$$z^\alpha(x^i(X^I)), \tag{2.9}$$

$$\left[\frac{\partial z^\alpha}{\partial x^i}\right]_{f(X)} \left[\frac{\partial x^i}{\partial X^I}\right]_X, \tag{2.10}$$

$$\left[\frac{\partial^2 z^\alpha}{\partial x^j \partial x^i}\right]_{f(X)} \left[\frac{\partial x^j}{\partial X^J}\right]_X \left[\frac{\partial x^i}{\partial X^I}\right]_X + \left[\frac{\partial z^\alpha}{\partial x^i}\right]_{f(X)} \left[\frac{\partial^2 x^i}{\partial X^J \partial X^I}\right]_X. \tag{2.11}$$

It is this peculiar composition formula for second (and indeed higher) derivatives that confirms our previous observation as to the necessity of including all lower order derivatives in the definition of jets.

### 2.2.3 Second-grade materials

A local hyperelastic *second-grade body* is characterized by a constitutive law of the form:
$$\psi = \psi(j_X^2 \kappa), \tag{2.12}$$
where we have economized the indication of the explicit dependence on **X** since it is already included in the definition of the jet. In coordinates, this equation reads as follows:
$$\psi = \psi(F^i_{\,I}, F^i_{\,IJ}, X^I), \tag{2.13}$$
where we have used the notation:
$$F^i_{\,IJ} \equiv \frac{\partial F^i_{\,I}}{\partial X^J} = \frac{\partial^2 x^i}{\partial X^J \partial X^I} = F^i_{\,JI}. \tag{2.14}$$

Notice that in the coordinate expression (2.13) we are not including the explicit spatial argument $x^i$ (which is certainly included in the definition of jet) because of the requirement of translation invariance of the energy. Under a coordinate transformation in the body (change of reference configuration) of the form:
$$Y^M = Y^M(X^I), \quad (I, M = 1, 2, 3), \tag{2.15}$$
Equation (2.13) transforms according to:
$$\psi = \hat{\psi}(F^i_{\,M}, F^i_{\,MN}, Y^M)$$
$$= \psi(F^i_{\,M} \frac{\partial Y^M}{\partial X^I}, \ F^i_{\,MN} \frac{\partial Y^N}{\partial X^J} \frac{\partial Y^M}{\partial X^I} + F^i_{\,M} \frac{\partial^2 Y^M}{\partial X^J \partial X^I}, \ Y^M(X^I)). \tag{2.16}$$

## 2.3 Second-grade uniformity

### 2.3.1 Material isomorphisms

Let us go back for a moment to the case of simple bodies. Within the language of jets, we can say that a simple body is the same as a first-grade body. In the hyperelastic case, it is characterized by a constitutive law of the form:
$$\psi = \psi(j_X^1), \tag{2.17}$$
which is clearly equivalent to Equation (1.6). When defining a material isomorphism between two points of such a body, $\mathbf{X}_1$ and $\mathbf{X}_2$ say, we motivated the use of a linear map $\mathbf{P}_{12}$ between their respective tangent spaces to compare their responses. The map $\mathbf{P}_{12}$, on the other hand, can itself be regarded as a 1-jet. Indeed, consider a map $p_{12}$ of the body into itself such that a neighbourhood of $\mathbf{X}_1$ is mapped diffeomorphically onto a neighbourhood of

$\mathbf{X}_2$, with $p_{12}(\mathbf{X}_1) = \mathbf{X}_2$. What we mean by "diffeomorphically" is simply that the map is smooth and has a smooth inverse as far as the neighbourhoods are concerned. We call the map $p_{12} : \mathcal{B} \longrightarrow \mathcal{B}$ a *local diffeomorphism* from $\mathbf{X}_1$ to $\mathbf{X}_2$. The 1-jet $j^1_{\mathbf{X}_1} p_{12}$ is then precisely equivalent to a linear map $\mathbf{P}_{12}$ from the tangent space at $\mathbf{X}_1$ to the tangent space at $\mathbf{X}_2$. Conversely, for every linear map $\mathbf{P}_{12}$ a jet $j^1_{\mathbf{X}_1} p_{12}$ can be constructed that is equivalent to it. In terms of jets, we can then rephrase the definition of material isomorphism and its attendant condition (1.8) by stating that two material points of a simple hyperelastic body are materially isomorphic if there exists a local diffeomorphism $p_{12}$ from the first to the second such that the condition:

$$\psi(j^1_{\mathbf{X}_2}\kappa \circ j^1_{\mathbf{X}_1} p_{12}) = \psi(j^1_{\mathbf{X}_2}\kappa), \tag{2.18}$$

is satisfied identically for all $j^1_{\mathbf{X}_2}\kappa$.

Seen in the way just described, the definition of material isomorphism can be immediately lifted to second-grade materials (and, in fact, to materials of any finite grade). We say that two material points $\mathbf{X}_1$ and $\mathbf{X}_2$ of a second-grade hyperelastic body are *materially isomorphic*, if there exists a local diffeomorphism $p_{12}$ such that:

$$\psi(j^2_{\mathbf{X}_2}\kappa \circ j^2_{\mathbf{X}_1} p_{12}) = \psi(j^2_{\mathbf{X}_2}\kappa), \tag{2.19}$$

is satisfied identically for all $j^2_{\mathbf{X}_2}\kappa$. All we have done is change the order of the jets from 1 to 2. From the physical point of view, this definition achieves precisely what we want, namely, a material comparison of second-grade material points that takes into consideration the sensitivity of the material to the second gradients of the transplants.

According to the general coordinate representation of 2-jets, we can now obtain a coordinate representation of Equation (2.19). Namely, two points $\mathbf{X}_1$ and $\mathbf{X}_2$ of a second-grade hyperelastic body are materially isomorphic if there exist quantities $P^I_M$ and $Q^I_{MN} = Q^I_{NM}$ such that:

$$\psi(F^i_I P^I_M, F^i_{IJ} P^I_M P^J_N + F^i_I Q^I_{MN}, \mathbf{X}_1) = \psi(F^i_I, F^i_{IJ}, \mathbf{X}_2) \tag{2.20}$$

identically for all non-singular $\{F^i_I\}$ and all symmetric $\{F^i_{IJ}\}$.

A second-grade body is *materially uniform* if all its points are mutually materially isomorphic.

### 2.3.2 Second-grade material archetypes

In a uniform second-grade body we can choose any body point $\mathbf{X}_0$ as a second-grade archetypal point, just as we did for the first-grade case. The 2-jets (implants) from the archetypal point to a point $\mathbf{X}$ in the body will be denoted by $j^2_0 p_X$. The uniformity condition (2.19) in terms of the archetypal constitutive function $\bar\psi$ becomes:

$$\psi(j^2_X \kappa) = \bar\psi(j^2_X \kappa \circ j^2_0 p_X). \tag{2.21}$$

## 2.3 Second-grade uniformity

In the first-grade case, we were able to obtain a component expression of this uniformity condition by fixing, apart from a reference configuration of the body, a placement of the archetypal point in $\mathbb{R}^3$. This process, as we have seen, is equivalent to choosing a frame at the archetypal point, and we agreed to call an archetype precisely the choice of a frame at a point of the body. In order to emulate this idea for the case of second-grade bodies, we need to introduce the concept of a *second-order frame* at a point $\mathbf{X}_0$ of a manifold $\mathcal{B}$.

Consider a local diffeomorphism $\phi$ from the origin $O$ of $\mathbb{R}^3$ to $\mathbf{X}_0$. The 1-jet $j_O^1\phi$ can be seen as a linear map from $\mathbb{R}^3$ to the tangent space $T_{\mathbf{X}_0}\mathcal{B}$. This map sends the canonical basis of $\mathbb{R}^3$ to three linearly independent vectors, namely, a basis of $T_{\mathbf{X}_0}\mathcal{B}$, also called a *linear frame* at $\mathbf{X}_0$. It is natural, therefore, to consider the 2-jet $j_O^2\phi$, and to identify it, by definition, with a second-order frame[1] at $\mathbf{X}_0$. A material isomorphism between $\mathbf{X}_0$ and a point $\mathbf{X}$ can now be pulled back (by composition to the right with the 2-jet of $\phi^{-1}$) to a second-order frame at $\mathbf{X}$. Given a system of coordinates in $\mathcal{B}$ and the natural coordinate system in $\mathbb{R}^3$, an implant from the archetype to $\mathbf{X}$ consists of matrices $P_\alpha^I(\mathbf{X})$ and $Q_{\alpha\beta}^I(\mathbf{X})$. The component expression of Equation (2.21) for the chosen second-order frame at the archetype becomes:

$$\psi(F_I^i, F_{IJ}^i, \mathbf{X}) = \bar{\psi}(F_I^i P_\alpha^I, F_{IJ}^i P_\alpha^I P_\beta^J + F_I^i Q_{\alpha\beta}^I). \tag{2.22}$$

Just as in the first-grade case, we see that a second-order frame at a point $\mathbf{X}_0$ induces, via a field of material implants, a second-order frame field (or second-order moving frame) on the body. A second-grade body is *smoothly uniform* if for each point there exists an open neighbourhood on which the frame field can be chosen smoothly.

### 2.3.3 Second-grade symmetries

A second-grade symmetry at $\mathbf{X}$ is a second-grade material automorphism, namely, a second-grade material isomorphism whose source and target are $\mathbf{X}$. In a given coordinate system, therefore, a symmetry $H$ will consist of a pair $(G_J^I, S_{LM}^K)$ with $S_{LM}^K = S_{ML}^K$. Given two symmetries, $H$ and $H'$, the operation:

$$HH' = (G_J^I, S_{LM}^K)(G_B'^A, S_{DE}'^C) \equiv (G_J^I G_B'^J, S_{LM}^K G_D'^L G_E'^M + G_J^K S_{DE}'^J), \tag{2.23}$$

stemming from the law for composition of 2-jets, endows the collection $\mathcal{H}_X$ of all second-order symmetries at $\mathbf{X}$ with a group structure. In particular, the symmetry group of the archetype is denoted by $\bar{\mathcal{H}}$.

We denote by $\mathcal{R}_X$ the set of second-order frames at $\mathbf{X}$ induced by all possible material isomorphisms from a given archetype to $\mathbf{X}$. In components, each element of $\mathcal{R}_X$ consists of a pair $(P_\alpha^I(\mathbf{X}), Q_{\beta\gamma}^J(\mathbf{X}))$. All the considerations about non-uniqueness of material isomorphisms that were made for the case of simple materials translate verbatim to the second-grade case.

---

[1] Also called a *holonomic frame*.

The classification and study of second-grade symmetry groups will be carried out later. At this point, we merely point out that the projection of the group $\bar{\mathcal{H}}$ on its first component (that is, the totality of matrices $G^I_J$ appearing as first components of elements of $\bar{\mathcal{H}}$) constitutes a subgroup of the 3-D general linear group. We will assume that the projected group satisfies the mass consistency condition, so that it becomes a subgroup of the unimodular group.

The ideas of material groupoid and material $G$-structure can be generalized to second-grade bodies. This formulation will be developed later in this chapter.

### 2.3.4 An example of a nontrivial second-grade symmetry

As we have just remarked, the projection of a second-grade symmetry group on its first component constitutes a first-grade symmetry group. We can, therefore, consider a material for which this induced first-grade group is trivial. In other words, we want to construct an example in which we focus our attention entirely on genuine second-grade symmetries. To arrive at the simplest possible (but applications-wise meaningful) example, we will consider a material whose second-grade behaviour is characterized by a dependence of the energy density $\psi$ on the spatial gradient of the density $\rho$ (per unit spatial volume)[2]. In terms of the density $\rho_R$ of the reference configuration, we can write:

$$\rho = J_F^{-1} \rho_R. \tag{2.24}$$

It follows that to capture the desired density gradient we need only include a dependence on the referential gradient of the determinant of $\mathbf{F}$. Indeed:

$$\rho_{,i} = \left((J_F^{-1})_{,I}\, \rho_R + J_F^{-1} (\rho_R)_{,I}\right) F^{-I}_{\phantom{-I},i}, \tag{2.25}$$

where we can clearly see that the dependence of the energy on $\mathbf{F}$ and the knowledge of the referential density take care of all other terms. More explicitly, we have:

$$\rho_{,i} = J_F^{-1} \left(-F^{-M}_{\phantom{-M}m} F^m_{M,I}\, \rho_R + (\rho_R)_{,I}\right) F^{-I}_{\phantom{-I}i}, \tag{2.26}$$

where we have made use of the formula for the derivative of a determinant, viz.:

$$(J_F)_{,I} = J_F F^{-M}_{\phantom{-M}m} F^m_{M,I}. \tag{2.27}$$

We are dealing, then, with an energy density of the form:

$$\psi = \psi(j^2_X \kappa) = \psi(F^i_I, (J_F)_{,I}; X^I). \tag{2.28}$$

By definition, a symmetry $h$ at $\mathbf{X}$ is an $\mathbf{X}$-preserving automorphism $h$ of a neighbourhood of $\mathbf{X}$ such that:

---

[2] This is indeed the assumed driving force in phenomena of mass diffusion.

## 2.3 Second-grade uniformity

$$\psi(j_X^2 \kappa \circ j_X^2 h) = \psi(j_X^2 \kappa), \tag{2.29}$$

identically for all deformations $\kappa$. Since we have assumed that the induced first-grade symmetry group is trivial, we need only consider those automorphisms $h$ that preserve at $\mathbf{X}$ the value of the first jet $j_X^1 h$. In terms of components, therefore, we may write Equation (2.29) as:

$$\psi(F_I^i, (J_F J_h)_{,I}; X^I) = \psi(F_I^i, (J_F)_{,I}; X^I), \tag{2.30}$$

where we have denoted by $J_h$ the determinant of the derivative of $h$ evaluated at $\mathbf{X}$. Clearly, the symmetry group of this material point will contain at least all the 2-jets of automorphisms $h$ such that $j_X^1 h$ is the identity and such that:

$$(J_F J_h)_{,I} = (J_F)_{,I}. \tag{2.31}$$

But, since the determinant $J_h$ at $\mathbf{X}$ is equal to 1, we obtain the condition:

$$(J_h)_{,I} = 0. \tag{2.32}$$

By the formula for the derivative of a determinant and, always recalling that we have assumed that $j_X^1 h$ is the identity, we obtain the condition:

$$h_{J,I}^J = 0, \tag{2.33}$$

where, for simplicity, we have denoted by $h_J^I$ the components of the gradient of $h$. In other words, we conclude that the symmetry group of this material point contains all pairs $H = \{\mathbf{G}, \mathbf{S}\}$ of the form:

$$H = \{\mathbf{I}, \mathbf{S}\} = \{\delta_J^I, S_{JK}^I\} \quad \text{with } S_{IK}^I = 0. \tag{2.34}$$

It is not difficult to verify that all such "traceless" elements form a group under the composition law (2.23). This symmetry group is the smallest possible symmetry group of a material point whose first-grade symmetry group is trivial and whose second-grade behaviour is characterized by a dependence on the spatial density gradient. Moreover, it is not difficult to verify that this group transforms into itself under conjugation by any pair of the form $\{A_J^I, B_{JK}^I\}$, where $A_J^I$ are the entries of any non-singular matrix and where $B_{JK}^I = B_{KJ}^I$. Physically, this means that this group is the same in all reference configurations. We note, finally, that (because of the assumed identity value of the first component) this turns out to be an additive (commutative) group, as can be verified directly from Equation (2.23).

*Remark 2.2.* Notice that the following converse is true: If the symmetry group of a hyperelastic second-grade point is of the form (2.34), then the dependence of the energy density on the second gradient of the deformation must be limited to a dependence on the referential gradient of the determinant of $\mathbf{F}$. The proof of this assertion is left to the reader.

## 2.4 The material second-order G-structures and groupoid

### 2.4.1 The second-order frame bundle

We have defined a second-order (holonomic)[3] frame at a point $\mathbf{X} \in \mathcal{B}$ as the 2-jet of a local diffeomorphism $\phi$ from the origin $O$ of $\mathbb{R}^3$ to $\mathbf{X}$. The intuitive justification of this definition stems from the fact that a usual (first-order) frame can be clearly identified with the 1-jet of $\phi$. Indeed, such 1-jet sends the standard basis of $\mathbb{R}^3$ to a basis of the tangent space $T_X \mathcal{B}$. The second-order frame is, therefore, a natural generalization of this idea.

We now consider the collection $F^2(\mathcal{B})$ of all second-order frames at all points of $\mathcal{B}$ and call it the *second-order frame bundle of* $\mathcal{B}$.[4] This nomenclature already suggests that this collection has the structure of a differentiable fibre bundle. In fact, it can be shown[5] that $F^2(\mathcal{B})$ is the total space of two different principal bundles whose base manifolds are, respectively, $\mathcal{B}$ and $F(\mathcal{B})$ (see Box 2.7 and Section 10.4). We are interested mainly in the first one, namely the one with base manifold $\mathcal{B}$, whose bundle projection we denote by $\pi^2$. It being a principal bundle, its structure group can be found by looking into the nature of any fibre. More precisely, we consider the *standard second-order frame bundle* $F^2(\mathbb{R}^3)$ and look into the fibre at the origin: $(\pi^2)^{-1}(O)$, since a similar procedure can be used to reveal the structure group of a first-order frame bundle. Clearly, since the passage from one second-order frame to another is achieved by composition of 2-jets, the fibre $(\pi^2)^{-1}(O)$ is a group, which we denote by $G^2(3)$, whose elements are pairs of the form $A = \{A_J^I, A_{JK}^I\}$ (with $A_{JK}^I = A_{KJ}^I$) and whose composition law and inverse are given, respectively, by the following formulas:

$$AB = \{A_L^I B_J^L,\ A_L^I B_{JK}^L + A_{LM}^I B_J^L B_K^M\} \tag{2.35}$$

and

$$A^{-1} = \{A_J^{-I},\ -A_{MN}^L A_L^{-I} A_J^{-M} A_K^{-N}\}. \tag{2.36}$$

### 2.4.2 The material G-structures

Let $\mathcal{G}$ be a Lie subgroup of $G^2(3)$. A reduction of $F^2(\mathcal{B})$ to the subgroup $\mathcal{G}$ is called a *second-order $\mathcal{G}$-structure* over the manifold $\mathcal{B}$. As in the first-order case, the standard second-order frame bundle $F^2(\mathbb{R}^3)$ is reducible to any subgroup of $G^2(3)$, but for an arbitrary manifold $\mathcal{B}$ this may not be the case.

---

[3] In the next chapter, we will generalize the notion of second-order frame to include the so-called non-holonomic frames.
[4] See Chapter 10.
[5] For a detailed treatment of this topic see [9].

## 2.4 The material second-order G-structures and groupoid

> **Box 2.7 The two bundle structures of $F^2(\mathcal{B})$**
>
> The total space $F^2(\mathcal{B})$ consists of the collection of all second-order frames at all points of the manifold $\mathcal{B}$. Therefore, if $p$ is an element of $F^2(\mathcal{B})$, there exists a local diffeomorphism $\phi$ from the origin $O$ of $\mathbb{R}^3$ to the point $\mathbf{X} = \phi(O) \in \mathcal{B}$ such that:
>
> $$p = j_O^2 \phi. \tag{2.37}$$
>
> The first bundle structure of $F^2(\mathcal{B})$ is based on the canonical projection $\pi^2$ defined as follows:
>
> $$\begin{aligned} \pi^2 : F^2(\mathcal{B}) &\longrightarrow \mathcal{B} \\ p &\mapsto \phi(O). \end{aligned} \tag{2.38}$$
>
> In other words, this projection assigns to each second-order frame the point of $\mathcal{B}$ at which it is attached. In local coordinates, $p$ is given by $\{X^I, P_\alpha^I, Q_{\alpha\beta}^I = Q_{\beta\alpha}^I\}$, and the projection $\pi^2$ is given by:
>
> $$\pi^2(\{X^I, P_\alpha^I, Q_{\alpha\beta}^I\}) = \{X^I\}. \tag{2.39}$$
>
> The second bundle structure of $F^2(\mathcal{B})$ is obtained by means of the canonical projection $\pi_1^2$ defined as:
>
> $$\begin{aligned} \pi_1^2 : F^2(\mathcal{B}) &\longrightarrow F(\mathcal{B}) \\ p &\mapsto j_O^1 \phi. \end{aligned} \tag{2.40}$$
>
> In other words, this projection takes advantage of the fact that a second-order frame necessarily contains a first-order frame, which is obtained by taking the 1-jet of any one of the local diffeomorphisms entailed in the definition of the second-order frame. In local coordinates we have:
>
> $$\pi_1^2(\{X^I, P_\alpha^I, Q_{\alpha\beta}^I\}) = \{X^I, P_\alpha^I\}. \tag{2.41}$$

The choice of a second-grade archetype will induce at each point $\mathbf{X}$ of the body $\mathcal{B}$, via all the material isomorphisms ("implants") from the archetype to the point, a collection of second order frames at $\mathbf{X}$. Indeed, a second-order frame at the archetype is the 2-jet of a local diffeomorphism from the origin of $\mathbb{R}^3$ to the archetype (a small piece of material), and a material implant is the 2-jet of a local diffeomorphism from the archetype to the body-point $\mathbf{X}$. These two jets can, therefore, be composed to provide the 2-jet of a local diffeomorphism from the origin of $\mathbb{R}^3$ to $\mathbf{X}$, that is, a second-order frame at $\mathbf{X}$. The situation becomes entirely similar to that of simple materials: the material implants are not unique, but their non-uniqueness is governed by the (second-grade) material symmetry group $\bar{\mathcal{G}}$ of the archetype. As a result, the induced second-order frames at each point are only a subset of the corresponding fibre of $F^2(\mathcal{B})$. The collection of all these subsets is precisely a reduction of $F^2(\mathcal{B})$ to $\bar{\mathcal{G}}$, namely, a second-order $\mathcal{G}$-structure which is called a *second-grade material $\mathcal{G}$-structure*.

The choice of archetype and of a second-order frame therein has, naturally, a bearing upon the material $\mathcal{G}$-structure obtained. Nevertheless, a change of archetype will affect the implants (and, therefore, the induced second-order frames) by composition (to the right) with a *constant* 2-jet. Two $\mathcal{G}$-structures thus related are said to be *conjugate* (see Box 1.6). As in the case of simple materials, one can define a *material groupoid*, which is the repository of all these conjugate material $\mathcal{G}$-structures.

### 2.4.3 The material groupoid

For a finite-dimensional manifold, such as a body $\mathcal{B}$, we consider the collection $\Pi^k(\mathcal{B})$ of the $k$-jets of local diffeomorphisms of all pairs of points. It is not difficult to verify that the set thus obtained has the structure of a transitive Lie groupoid with projections given by: $\alpha(j_X^k\phi) = X$ and $\beta(j_X^k\phi) = \phi(X)$. We call this groupoid the *$k$-jet groupoid* of $\mathcal{B}$. Clearly, for $k = 1$ we recover the tangent groupoid.

If we now restrict the local diffeomorphisms to those whose 2-jets are material isomorphisms between points of the second-grade body $\mathcal{B}$, we obtain a subgroupoid of $\Pi^2(\mathcal{B})$, which we call the *material groupoid* of $\mathcal{B}$. This entity can always be defined, whether the second-grade body is uniform or not. The uniformity of the body is mirrored by the transitivity of its material groupoid. Moreover, if the body is smoothly uniform, we obtain a transitive Lie groupoid. These concepts are identical to their counterparts in the first-grade case.

## 2.5 The subgroups of $G^2(n)$

### 2.5.1 The generic subgroup

For simple materials, whereby the symmetry groups are subgroups of the general linear group (or, more precisely, of its unimodular subgroup), a complete classification of all possible symmetry groups (and their Lie algebras) exists [91]. For second-grade materials (as well as for Cosserat-type media), on the other hand, it appears that only particular cases have been considered in the literature. Without attempting a complete classification, we will present a discussion of the general kinds of subgroups of $G^2(3)$ (or, more generally, of $G^2(n)$) that one may expect and their relevance to applications in Continuum Mechanics[6].

The elements of $G^2(n)$ are pairs of the form $\{A_J^I, A_{JK}^I\}$ with $A_{JK}^I = A_{KJ}^I$ and $I, J, K = 1, ..., n$. For convenience of notation, we will sometimes indicate such a pair as $\{\mathbf{A}, \mathbb{A}\}$, with $\mathbf{A} \in GL(n; \mathbb{R})$ and $\mathbb{A} \in S^2(n)$ (where $S^2(n)$ denotes the real vector space of symmetric $\mathbb{R}^n$-valued bilinear forms on $\mathbb{R}^n$). In short, $G^2(n)$ may be identified with the semi-direct product $GL(n; \mathbb{R}) \times S^2(n)$

---

[6] See also [62, 39].

## 2.5 The subgroups of $G^2(n)$

endowed with the composition law (2.35). We denote by $\mathbf{I}$ and $\mathbf{0}$, respectively, the identity elements of $GL(n;\mathbb{R})$ and of the additive group $S^2(n)$. A preliminary observation consists of determining in which way the general linear group $GL(n;\mathbb{R})$ and its subgroups, on the one hand, and the additive group $S^2(n)$ and its subgroups, on the other hand, "sit" within $G^2(n)$. A direct computation shows that for any subgroup $\mathcal{G}$ of $GL(n;\mathbb{R})$, the product $\mathcal{G} \times \{\mathbf{0}\}$ is a subgroup of $G^2(n)$. More explicitly, the collection of pairs of the from $\{\mathbf{A}, \mathbf{0}\}$ (with $\mathbf{A} \in \mathcal{G}$) as a subset of $G^2(n)$ is closed under the group operations (2.35) and (2.36). Similarly, one can easily verify that for any subgroup $\mathcal{S}$ of $S^2(n)$ the product $\{\mathbf{I}\} \times \mathcal{S}$ is a subgroup of $G^2(n)$.

Let $\mathcal{H}$ be a subgroup of $G^2(n)$. Since $\mathcal{H}$ is ultimately a collection of pairs of the form $\{\mathbf{A}, \mathbb{A}\}$, namely a subset of the Cartesian product $GL(n;\mathbb{R}) \times S^2(n)$, it is legitimate to consider its projections onto the first and second factors. The projection over the first factor, $\mathcal{G}_H = pr_1 \mathcal{H}$, namely the collection of all the $\mathbf{A}$'s that appear in some element of $\mathcal{H}$, turns out to be a subgroup of $GL(n;\mathbb{R})$, as can be seen directly from the first part of the composition law (2.35). In fact, this projection is a group homomorphism. The projection on the second factor, on the other hand, is not. From the point of view of Continuum Mechanics what this means is that in every second-grade material there lies hidden a simple (first-grade) material, whose symmetries are simply the projection of the total symmetries, disregarding, as it were, the second-gradient dependence. But it is not possible to obtain a "pure" second-gradient behaviour without involving also the first gradient. The second-gradient symmetries are intertwined with the first. Thus, in our example of Section 2.3.4, we were able to separate the second-grade behaviour only by eliminating completely the non-trivial first-grade symmetries (that is, by prescribing $\mathcal{G}_H = \{\mathbf{I}\}$).

For a given subgroup $\mathcal{H} \subset G^2(n)$, let us denote:

$$\Sigma_{\mathbf{A}} = \{\mathbb{A} \in S^2(n) \mid \{\mathbf{A}, \mathbb{A}\} \in \mathcal{H}\}. \tag{2.42}$$

Put in words: for a given $\mathbf{A} \in \mathcal{G}_H$, the set $\Sigma_{\mathbf{A}}$ consists of all the elements of $S^2(n)$ that appear together with $\mathbf{A}$ in some element of the subgroup $\mathcal{H}$ under consideration. It is not difficult to verify directly from the composition law that the set $\Sigma_{\mathbf{I}}$ is an additive subgroup of $S^2(n)$ and that, therefore, the subset $\{\mathbf{I}\} \times \Sigma_{\mathbf{I}}$ is a subgroup of $G^2(n)$. This fact was clearly exploited in our example of Section 2.3.4, as already pointed out.

Although for a fixed $\mathbf{A} \neq \mathbf{I}$, in general, neither is $\Sigma_{\mathbf{A}}$ a subgroup of $S^2(n)$ nor is $\{\mathbf{A}\} \times \Sigma_{\mathbf{A}}$ a subgroup of $G^2(n)$, we will now prove that there exists a bijection between $\Sigma_{\mathbf{A}}$ and $\Sigma_{\mathbf{I}}$. Intuitively, the "number" of elements of $S^2(n)$ paired with each element of $GL(n;\mathbb{R})$ in a given subgroup $\mathcal{H}$ of $G^2(n)$ is exactly the same. To prove this assertion, consider an element $\{\mathbf{A}, \mathbb{A}_0\}$. By definition, $\mathbb{A}_0$ must belong to $\Sigma_{\mathbf{A}}$. Let $\mathbb{A}$ represent an arbitrary element of $\Sigma_{\mathbf{A}}$. Then, according to the group laws (2.35) and (2.36), we can calculate the product:

$$\{\mathbf{A}, \mathbb{A}_0\}^{-1} \{\mathbf{A}, \mathbb{A}\} = \{\mathbf{I}, \mathbf{A}^{-1}(\mathbb{A} - \mathbb{A}_0)\}, \tag{2.43}$$

with an obvious notation. In other words, this product belongs to $\{\mathbf{I} \times \Sigma_\mathbf{I}\}$. We have thus explicitly constructed a one-to-on map:

$$\phi : \Sigma_\mathbf{A} \longrightarrow \Sigma_\mathbf{I}, \tag{2.44}$$

defined by:

$$\phi(\mathbb{A}) = \mathbf{A}^{-1}\left(\mathbb{A} - \mathbb{A}_0\right). \tag{2.45}$$

Conversely, let $\mathbb{B}$ be an arbitrary element of $\Sigma_\mathbf{I}$. We have:

$$\{\mathbf{A}, \mathbb{A}_0\}\{\mathbf{I}, \mathbb{B}\} = \{\mathbf{A}, \mathbb{A}_0 + \mathbf{A}\mathbb{B}\}. \tag{2.46}$$

We have again constructed a one-to-one map:

$$\psi : \Sigma_\mathbf{I} \longrightarrow \Sigma_\mathbf{A}, \tag{2.47}$$

defined by:

$$\psi(\mathbb{B}) = \mathbb{A}_0 + \mathbf{A}\mathbb{B}. \tag{2.48}$$

This map is, in fact, the inverse of $\phi$. This completes the proof. Pictorially, we can represent the general appearance of a subgroup $\mathcal{H}$ of $G^2(n)$ as shown in Figure 2.6.

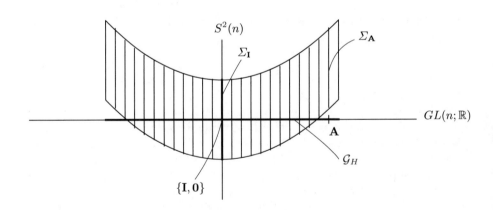

**Fig. 2.6.** Pictorial representation of a general subgroup $\mathcal{H}$ of $G^2(n)$

### 2.5.2 Toupin subgroups

There exists a special type of subgroups of $G^2(n)$ whereby the second-grade symmetries are subordinated to the first. Since, in his classic paper on couple-stress elasticity [86], Toupin considered only this restricted type of second-grade symmetry, we call these subgroups of $G^2(n)$ *Toupin subgroups*.

Let $\mathcal{G}$ be a given subgroup of $GL(n;\mathbb{R})$ and let $\mathbb{A}_0$ be a fixed element of $S^2(n)$. The Toupin subgroup $\mathcal{T}$ associated with the linear subgroup $\mathcal{G}$ and the symmetric element $\mathbb{A}_0$ is defined as the following conjugation:

$$\mathcal{T} := \{\mathbf{I}, \mathbb{A}_0\} \, \{\mathcal{G}, \mathbf{0}\} \, \{\mathbf{I}, \mathbb{A}_0\}^{-1}. \tag{2.49}$$

A direct computation shows that the first-order projection of this group is precisely $\mathcal{G}$, as expected. Moreover, in the pictorial representation of Figure 2.6, the two curves coalesce into one curve passing through the origin. The word "curve" is to be understood with caution, since the horizontal thick line in Figure 2.6 does not necessarily represent a one-parameter subgroup of $GL(n;\mathbb{R})$. On the other hand, it is clear that since in a Toupin subgroup the set $\Sigma_\mathbf{I}$ is a singleton (consisting of just the zero element of $S^2(n)$), it follows that every $\Sigma_\mathbf{A}$ in a Toupin subgroup is also a singleton (thus explaining why we just said that the two "curves" coalesce into one).

The question now arises as to whether the reverse is true, namely: is a subgroup with $\Sigma_\mathbf{I} = \{\mathbf{0}\}$ necessarily a Toupin subgroup? We will now show, by constructing an explicit counterexample, that this is not the case. We will, therefore, call any subgroup whose $\Sigma_\mathbf{I}$ is a singleton a *generalized Toupin subgroup* of $G^2(n)$. To construct our counterexample, let $\mathcal{H}_t$ be a smooth one-parameter ($t$) subgroup of $G^2(n)$. As we know, the one parameter subgroups of a Lie group are in one-to-one correspondence with the elements of the Lie algebra of the group, namely, with the tangent vectors at the group identity (cf. Section 8.2). In the case of our group $G^2(n)$, the elements of the Lie algebra look very much like the elements of the group itself, except that the first matrix entry may now be singular. Consider, therefore, first a case in which the one-parameter group $\mathcal{H}_t$ does not correspond to an element of the Lie algebra with a zero first entry. If $\mathcal{H}_t$ were a Toupin subgroup, then it should be a conjugate by an element of the form $\{\mathbf{I}, \mathbb{A}_0\}$ of a smooth one-parameter subgroup, $\mathcal{G}_t$ say, of $GL(n;\mathbb{R})$. More precisely, according to equation (2.49), we should have

$$\mathcal{H}_t = \{\mathbf{I}, \mathbb{A}_0\} \, \{\mathcal{G}_t, \mathbf{0}\} \, \{\mathbf{I}, \mathbb{A}_0\}^{-1}. \tag{2.50}$$

Differentiating this equation with respect to the parameter $t$ at $t = 0$, we must obtain elements of the respective Lie algebras, namely:

$$\{\mathbf{B}, \mathbb{B}\} = \{\mathbf{I}, \mathbb{A}_0\} \, \{\mathbf{A}, \mathbf{0}\} \, \{\mathbf{I}, \mathbb{A}_0\}^{-1}, \tag{2.51}$$

where (as it was assumed) $\mathbf{B} \neq \mathbf{0}$. It follows now from the composition laws that:

$$\mathbf{A} = \mathbf{B}, \tag{2.52}$$

58    2 Uniformity of second-grade materials

and that
$$\mathbb{B} = \mathbb{A}_0(\mathbf{A}, \mathbf{A}) - \mathbf{A}\mathbb{A}_0. \tag{2.53}$$

Recalling that this construction can be carried out with an arbitrary one-parameter subgroup $\mathcal{H}_t$, we may choose, for example, the one-parameter subgroup associated with a vector $\{\mathbf{I}, \mathbb{B}\}$ with $\mathbb{B} \neq \mathbf{0}$. But in that case, equation (2.53) would necessarily yield that $\mathbb{B} = \mathbf{0}$, which is a contradiction. We conclude that such a one-parameter subgroup cannot be a Toupin subgroup, in spite of the fact that $\Sigma_\mathbf{I} = \mathbf{0}$.

We have just proved that not all generalized Toupin subgroups are Toupin subgroups. In the process of the construction of the counterexample, however, the impression may have been conveyed that all one-parameter subgroups of $G^2(n)$ are generalized Toupin subgroups. That this is not the case can be observed by considering, for some fixed $\mathbb{A}_0 \neq \mathbf{0}$, the one-parameter subgroup:
$$\mathcal{H}_t = \{\mathbf{I}, t\mathbb{A}_0\}. \tag{2.54}$$

In conclusion, the Toupin subgroups of $G^2(n)$ are very particular cases of subgroups, ruled, as it were, by the first-order projection. Thus, in an application to Continuum Mechanics, a change of reference configuration and/or of archetype is the only second-grade control one can exert (via conjugation) on a material with this type of symmetry. The material symmetries are, essentially, of the first-grade only.

### 2.5.3 The subgroups $\{\mathbf{I}, \Sigma_\mathbf{I}\}$ and their conjugates

The counterpart of the Toupin subgroups is given by subgroups of the form $\{\mathbf{I}, \Sigma_\mathbf{I}\}$, in the sense that while the former emphasize the first-grade symmetries, the latter manage to isolate the second-grade ones. It is interesting to notice that the conjugation of such a subgroup by an arbitrary element of $G^2(n)$ picks up only the first component of this element. Indeed:
$$\{\mathbf{A}, \mathbb{A}\}\{\mathbf{I}, \Sigma_\mathbf{I}\}\{\mathbf{A}, \mathbb{A}\}^{-1} = \{\mathbf{I}, \mathbf{A}\Sigma_\mathbf{I}(\mathbf{A}^{-1}, \mathbf{A}^{-1})\}. \tag{2.55}$$

As we have already seen, $\Sigma_\mathbf{I}$ must be an additive subgroup of $S^2(n)$. We start by noticing that, as an additive group, $S^2(n)$ is isomorphic (as a vector space) to $\mathbb{R}^{n(n+1)/2}$. Using a result of Morris [72] in the context of $\mathbb{R}^n$, one can conclude that the closed subgroups of $S^2(n)$ are either discrete subgroups, or vector subspaces, or products thereof. For applications to Continuum Mechanics, one may discard the non-trivial subgroups for which $\Sigma_\mathbf{I}$ is either discrete or mixed. Any such subgroup may be interpreted as a periodicity of the material response in terms of its dependence on second derivatives of the deformation, a periodicity that may be rejected on physical grounds. If this reasoning is correct, we are left with just the continuous additive subgroups, such as the subgroup of "traceless" elements of $S^2(n)$ already used in the example of Section 2.3.4.

## 2.6 Second-grade homogeneity

A local coordinate chart induces smoothly at each point of its domain a second-order frame. We have seen, on the other hand, that a second-order archetype of a smoothly uniform second-grade body induces a second-order frame smoothly, at least on an open set around each point of the body. We say that a smoothly uniform second-grade body is locally homogeneous if there exists a coordinate chart around each point and a choice of implants of the archetype such that the respectively induced frame fields coincide with each other. If this can be accomplished with a single chart over the whole body, the body is said to be (globally) homogeneous.

These definitions coincide in spirit and detail with their counterparts for first-grade materials. Nevertheless, there exists a subtle difference between the two cases when it comes to the availability of (locally) homogeneous configurations, namely, configurations where the constitutive law is independent of the body points in a neighbourhood. In the case of first-grade materials, once a body is homogeneous, it is possible to find configurations that will place all the points in any desired state of strain, a luxury that is no longer affordable in the second-grade case.

### 2.6.1 The second-order frames induced by a coordinate system

The space $\mathbb{R}^3$ is endowed with canonical *translation* maps $t_\mathbf{Z} : \mathbb{R}^3 \longrightarrow \mathbb{R}^3$ for every $\mathbf{Z} \in \mathbb{R}^3$. The map $t_\mathbf{Z}$ sends every ordered triple to another ordered triple in which each entry has been incremented by the corresponding entry of $\mathbf{Z}$. Let $\phi : \mathcal{U} \longrightarrow \mathbb{R}^3$ be a coordinate system $(X^I)$ on an open set $\mathcal{U} \subset \mathcal{B}$. Without loss of generality, assume that $\phi(\mathcal{U})$ contains the origin $O \in \mathbb{R}^3$. The 2-jet $j_O^2 \phi^{-1}$ is, therefore, a second-order frame at a point $\mathbf{X}_0 = \phi^{-1}(O)$ of $\mathcal{U}$ with coordinates $X^1 = X^2 = X^3 = 0$. Consider any other point $\mathbf{X} \in \mathcal{U}$. The map:
$$\tau_\mathbf{X} = \phi^{-1} \circ t_{\phi(\mathbf{X})} \circ \phi : \mathcal{V}_0 \longrightarrow \mathcal{U}, \tag{2.56}$$
is well defined for a sufficiently small neighbourhood $\mathcal{V}_0 \subset \mathcal{U}$ of $\mathbf{X}_0$. We will call it a $\phi$-*translation*. By construction $\tau_\mathbf{X}(\mathbf{X}_0) = \mathbf{X}$. Clearly, this map is a local diffeomorphism, since its coordinate expression coincides with a canonical translation of $\mathbb{R}^3$. The composition $j_{\mathbf{X}_0}^2 \tau_\mathbf{X} \circ j_O^2 \phi^{-1}$ is then a second-order frame at $\mathbf{X}$. In this way, we can uniquely and smoothly assign to each point of $\mathcal{U}$ a second-order frame induced by the given coordinate system. By construction, the coordinate expression of this frame field is everywhere equal to $(\{\delta_J^I\}, \{\mathbf{0}\})$.

### 2.6.2 Homogeneity

Let the point $\mathbf{X}_0$ of the previous construction be adopted as an archetype, with the coordinate-induced frame $j_O^2 \phi^{-1}$. Since the body is smoothly uniform,

60    2 Uniformity of second-grade materials

there exist neighbourhoods of $\mathbf{X}_0$ where this frame induces smooth second-order frame fields via material isomorphisms. We choose one such field and assume, without loss of generality, that this field of material isomorphisms attains the identity value at $\mathbf{X}_0$, so that the frame therein remains unchanged. If the symmetry group is discrete, this field is uniquely defined. Otherwise, there exists a degree of freedom governed by the continuous symmetry group. We say that the second-grade body is locally homogeneous if for each point there exists a neighbourhood and a coordinate system such that the second-order frame induced by this coordinate system is also a second-order frame field constructed in the manner just described via material isomorphisms. If the neighbourhood where this is true happens to be the body itself, we have a case of global homogeneity. A locally homogeneous body with a trivial symmetry group is, of necessity, globally homogeneous.

Another way to state the preceding local homogeneity condition is to refrain from adopting any archetype and to refer directly to the material isomorphisms between pairs of points. We notice that a coordinate system $\phi$ induces, via its $\phi$-translation maps, a (not necessarily material) unique second-grade isomorphism between all pairs of points in its domain. The body is locally homogeneous if for each point there exists a neighbourhood and a coordinate system such that these isomorphisms are actually material. This characterization is equivalent to the previous one. On the other hand, should we adopt a different frame at $\mathbf{X}_0$, it may very well happen that, although the body is locally homogeneous, none of the materially induced frame fields coincides with a coordinate-induced field. In other words, the thought-procedure to check for homogeneity so far is as follows: (i) choose a coordinate system and any point $\mathbf{X}_0$ in its domain; (ii) check whether the coordinate-induced frame field in some neighbourhood of $\mathbf{X}_0$ can also be obtained *from the coordinate-induced frame at $\mathbf{X}_0$* via material isomorphisms.

In practice, the constitutive law of a body is given in some coordinate system. We would like, therefore, to obtain a criterion that detects local homogeneity and is, at the same time, valid in an arbitrary coordinate system.

### 2.6.3 Coordinate expressions

We have already established, as per Equation (2.22), that in a coordinate system and for a choice of frame at an archetypal point, a uniformity field boils down to a pair $(\{P_\alpha^I\}, \{Q_{\beta\gamma}^J\})$. Upon a coordinate transformation $Y^M = Y^M(X^I)$, the components of this pair transform according to:

$$\hat{P}_\alpha^M = \frac{\partial Y^M}{\partial X^I} P_\alpha^I, \tag{2.57}$$

and

$$\hat{Q}_{\beta\gamma}^N = \frac{\partial^2 Y^N}{\partial X^J \partial X^I} P_\beta^I P_\gamma^J + \frac{\partial Y^N}{\partial X^I} Q_{\beta\gamma}^I. \tag{2.58}$$

It follows from this last equation that the quantities defined by:

## 2.6 Second-grade homogeneity

$$\Lambda^I_{JK} \equiv -Q^I_{\alpha\beta} (P^{-1})^\alpha_J (P^{-1})^\beta_K, \tag{2.59}$$

can be regarded as Christoffel symbols of a symmetric linear connection $\Lambda$ on $\mathcal{U}$. On the other hand, as we know from the first-grade case, the expressions:

$$\Gamma^I_{JK} = P^I_\alpha (P^{-1})^\alpha_{J,K}, \tag{2.60}$$

are the Christoffel symbols of a generally non-symmetric (first-grade) material connection $\Gamma$. The difference:

$$\mathbf{D} = \Gamma - \Lambda \tag{2.61}$$

is, therefore, a tensor that will be called *the second-grade inhomogeneity tensor* associated with the given archetype and uniformity field.[7]

We now claim that a second-grade body is locally homogeneous if, and only if, each point has neighbourhood in which a uniformity field can be found with vanishing inhomogeneity tensor $\mathbf{D}$. To give a rough proof of this assertion, we start by remarking that if $\mathbf{D}$ vanishes, since $\Lambda$ is always symmetric, it turns out that $\Gamma$ must also be symmetric. By the homogeneity conditions of a first-grade case, we know that this implies the existence of a coordinate system adapted to the field $P^I_\alpha$. In this coordinate system, the components of $\mathbf{P}$ are precisely $\delta^I_\alpha$. Moreover, since $\mathbf{D}$ is a tensor, it must vanish in all coordinate systems and, in particular, in this adapted one. This means that we have found a coordinate system such that the components of the materially induced second-order frames are $(\{\delta^I_J\}, \{\mathbf{0}\})$, which proves that these frames are those induced by translations of this coordinate system. The proof of the converse is trivial.

The main difference with the first-grade case, as already noted, is that, even if the symmetry group is trivial, the choice of frame at the archetypal point plays a role in the vanishing of the inhomogeneity tensor $\mathbf{D}$, so that there is a degree of freedom even in the trivial case. On the other hand, this degree of freedom can be used to pin down a more stringent definition of homogeneity, one that has a clear physical meaning. Let us assume that we want that the homogeneous configurations (that is, the configurations whereby in a neighbourhood the constitutive law is independent of position) be in a particular state of strain (for example, we may wish that these be stress-free configurations). In that case, we fix an archetype and a frame a priori to satisfy the desired condition. We then check the corresponding inhomogeneity tensor $\mathbf{D}$. If it vanishes, we are assured of the existence of the desired configurations. Otherwise, even if the body is locally homogeneous according to the general definition, such configurations are unattainable. We can, therefore, say that a body is locally homogeneous *with respect to a given archetype* if it possesses frame fields (unique in the case of a trivial group), induced by that particular archetype, with vanishing inhomogeneity tensor $\mathbf{D}$.

---

[7] See also Section 11.4.

## 62    2 Uniformity of second-grade materials

Assume that the condition $\mathbf{D} = 0$ is satisfied for some choice of frame in the archetypal point. How will this condition look if we change the frame at the archetypal point, as shown in Figure 2.7? By composition of jets, we calculate the new tensor $\mathbf{D}'$ corresponding to a change of archetype frame as:

$$D'^I_{JK} = D^I_{JK} + B^\alpha_{\beta\gamma}(A^{-1})^\beta_\rho (A^{-1})^\gamma_\sigma P^I_\alpha (P^{-1})^\rho_J (P^{-1})^\sigma_K. \tag{2.62}$$

If $\mathbf{D} = 0$, therefore, it follows that the quantities

$$D^\alpha_{\beta\gamma} \equiv D'^I_{JK} P^J_\beta P^K_\gamma (P^{-1})^\alpha_I = B^\alpha_{\rho\sigma}(A^{-1})^\rho_\beta (A^{-1})^\sigma_\gamma. \tag{2.63}$$

are constant. We conclude that, for the general case, we can replace the vanishing of $\mathbf{D}$ with the condition that the pull-back of $\mathbf{D}$ to the archetype, as given by Equation (2.63), be symmetric and constant over $\mathcal{U}$.

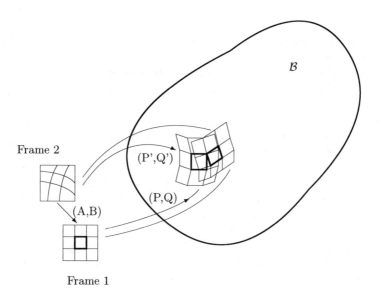

**Fig. 2.7.** Change of archetypal frame: $(P', Q') = (P, Q) \circ (A, B)$

*Remark 2.3.* As an exercise, one can show that for a uniform body with a symmetry group of the type discussed in Section 2.3.4 (that is, one for which the energy density depends on the referential gradient of the determinant of $\mathbf{F}$ and having a trivial first-grade symmetry group) the local homogeneity condition can be expressed as follows: the inhomogeneity tensor $\mathbf{D}$ calculated on the basis of *any* uniformity field and a given archetype is symmetric and traceless. Or, in the case of an arbitrary archetype: the pull-back to the archetype (as per Equation (2.63)) of the uniformity field **D** calculated on the basis of any uniformity field is symmetric and of constant trace. [Hint: show that if **D** is symmetric the vanishing of its trace is preserved under changes of reference configuration.]

### 2.6.4 Homogeneity in terms of a material $\mathcal{G}$-structure

As we have already remarked in Section 1.4.3, the standard second-order frame-bundle $F^2(\mathbb{R}^n)$ is canonically reducible to any subgroup $\mathcal{G}$ of the structure group $G^2(n)$. These *standard flat $\mathcal{G}$-structures* are obtained by letting the group $\mathcal{G}$ act on the canonical constant section which assigns to each point of $\mathbb{R}^n$ its standard second-order frame. The existence of these standard second-order frames is a luxury that is not available in the case of an arbitrary base manifold $\mathcal{B}$. Therefore, just as in the case of first-order $G$-structures, we will exploit the concept of *equivalence* between $G$-structures so as to be able to compare any given second-order $G$-structure with its standard flat counterpart.

Given two $n$-dimensional manifolds, $\mathcal{B}$ and $\mathcal{B}'$, and a diffeomorphism:

$$\phi : \mathcal{B} \longrightarrow \mathcal{B}', \tag{2.64}$$

the 2-jet $j_b^2 \phi$ drags every second-order frame at $b \in \mathcal{B}$ to a second-order frame at the image point $\phi(b) \in \mathcal{B}'$. Indeed, a second-order frame at $b$ is the 2-jet of a local diffeomorphism $\psi : \mathbb{R}^n \longrightarrow \mathcal{B}$ from a neighbourhood of the origin of $\mathbb{R}^n$ to a neighbourhood of $b$. The composition $\phi \circ \psi$ is, therefore, a local diffeomorphism whose 2-jet is a second-order frame at $\phi(b)$, as desired. As a consequence of this canonical dragging of second-order frames, we conclude that (just as in Section 1.4.3) every second-order $\mathcal{G}$-structure $\mathcal{R}$ on $\mathcal{B}$ gives rise, via the diffeomorphism $\phi$, to an induced second-order $\mathcal{G}$-structure on $\mathcal{B}'$, denoted by $j^2\phi(\mathcal{R})$. Two second-order $\mathcal{G}$-structures $\mathcal{R}$ and $\mathcal{R}'$ defined, respectively, over the base manifolds $\mathcal{B}$ and $\mathcal{B}'$ are *locally equivalent* at the points $b \in \mathcal{B}$ and $b' \in \mathcal{B}'$ if there exists an open neighbourhood $\mathcal{U}$ of $b$ and a diffeomorphism $\phi : \mathcal{B} \longrightarrow \mathcal{B}'$ such that $\phi(b) = b'$ and such that $\mathcal{R}'|_{\phi(\mathcal{U})} = j^2\phi(\mathcal{R}|_\mathcal{U})$. A $\mathcal{G}$-structure is called *locally flat* if (at each point) it is locally equivalent to the standard (flat) $\mathcal{G}$-structure. Since a local diffeomorphism from $\mathcal{B}$ to $\mathbb{R}^n$ can be regarded as coordinate chart on $\mathcal{B}$, local flatness of a second-order $\mathcal{G}$-structure can also be regarded as the existence of an atlas of $\mathcal{B}$ such that its induced second-order frames (as defined in Section 2.6.1) belong to the $\mathcal{G}$-structure.

There is an interesting difference between first- and second-order $\mathcal{G}$-structures as far as the properties of their conjugates. We recall (see Box 1.6) that, to obtain a conjugate $\mathcal{G}$-structure to a given one, we exert the right action $R_g$ of a *constant element* $g$ of the structure group of the underlying frame bundle. If this element does not belong to the structure group of the given $\mathcal{G}$-structure, then all the frames are moved to other frames "outside" of the original $\mathcal{G}$-structure. The new frames, however, constitute the elements of a new $\mathcal{G}'$-structure with a structure group $\mathcal{G}'$ which is conjugate to the original one. This situation is pictorially demonstrated in Figure 2.8. Let us effect this conjugation, for instance, upon a standard flat $\mathcal{G}$-structure. What we obtain is a $\mathcal{G}$-structure which could have been generated in a manner similar to the standard one, but starting from a constant cross section of the standard frame bundle of $\mathbb{R}^n$ other than the canonical one. For simplicity, imagine that we are working with the trivial $\mathcal{G}$-structure, namely, that for which the structure group $\mathcal{G}$ is the trivial (identity) group. In the case of the first-order frame bundle of $\mathbb{R}^n$, after applying the right action of a fixed element of $GL(n;\mathbb{R})$, we obtain that at each point of $\mathbb{R}^n$ we have a constant (but not necessarily orthonormal) basis which belongs to the new $\mathcal{G}'$-structure. Clearly, an affine change of coordinates of $\mathbb{R}^n$ can always be found that is adapted to this system of bases. Consider now instead the case of the second-order frame bundle of $\mathbb{R}^n$. By the right action of a constant element of $G^2(n)$ we again obtain at each point of $\mathbb{R}^n$ a constant second-order frame. Now, a second-order frame always contains a first-order one, since equality of 2-jets implies equality of 1-jets. To "fix" the skewness of this first-order frame we clearly need to effect an affine change of coordinates, like before. But once we do this, we lose the ability to "fix" the remaining part of the second-order frame (which will, in general, require curvilinear coordinates). As a result of this simple example we conclude that, in contradistinction to the first-order case (see Remark 1.11), *the local flatness of a second-order $\mathcal{G}$-structure does not imply the local flatness of its conjugates.* Instead, the local flatness of a second-order $\mathcal{G}$-structure implies that its conjugate $\mathcal{G}$-structures are locally equivalent to a second-order $\mathcal{G}$-structure over $\mathbb{R}^n$ generated by the right action of a conjugate of the structure group $\mathcal{G}$ on *some* constant section of $F^2(\mathbb{R}^n)$.

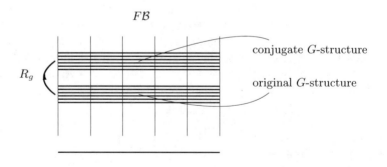

**Fig. 2.8.** Conjugate $G$-structures

When the preceding concepts are translated to the language of material homogeneity, it is clear that we recover precisely the same picture that was presented in Section 2.6.3, according to which a restricted notion of local homogeneity *with respect to a specific archetype* is contrasted with the more general notion of local homogeneity *with respect to some archetype*. Since the choice of archetype corresponds to the pinning down of one of all the possible conjugate $\mathcal{G}$-structures representing the material uniformity of the body, we may say that the local flatness of a material $\mathcal{G}$-structure is the mathematical counterpart of the restricted notion of local homogeneity. The general notion of local homogeneity, on the other hand, corresponds to the local equivalence of the material $\mathcal{G}$-structure with a second-order $\mathcal{G}$-structure over $\mathbb{R}^n$ generated by the action of the structure group $\mathcal{G}$ on some constant section of $F^2(\mathbb{R}^n)$ (not necessarily the standard (unit) section) .

## 2.7 Homogeneity in terms of the material groupoid

As we already know, the material groupoid is not prejudiced to this or that particular $\mathcal{G}$-structure. Indeed, a transitive Lie groupoid is not even concerned about point-wise entities, but rather about entities ("arrows") that exist smoothly between any two points. In the case of the groupoid of 2-jets, $\Pi^2(\mathcal{B})$, and its subgroupoids, for example, the elements of the groupoid are *not* the second-order frames themselves. In terms of the material groupoid, therefore, one cannot expect a property of the groupoid (such as local flatness) to depend in any way on an archetype (which, after all, is a second-order frame at a point).

Via the induced translation maps in its domain, $\mathcal{U}$, a coordinate chart clearly induces a section of $\Pi^2(\mathcal{U})$, that is, a 2-jet for every pair of points of $\mathcal{U}$. By definition, a subgroupoid of $\Pi^2(\mathcal{B})$ is locally flat if there exists an atlas such that the induced sections are local sections of the material groupoid, namely, if each of the induced 2-jets belongs to the material groupoid. It follows, then, that the local flatness of the material groupoid corresponds exactly to the more general notion of local homogeneity. Specific instances of homogeneity criteria for second-grade bodies can be found in [39].

# 3
# Uniformity of Cosserat media

In this chapter we establish the concepts of uniformity and homogeneity for bodies endowed with a certain kind of internal structure. The origin of this concept can be traced back to the first decade of the twentieth century when the brothers E. and F. Cosserat published a treatise on the theory of deformable bodies [10]. Neglected for several decades, this revolutionary book anticipated not only the ideas of modern Continuum Mechanics but also some concepts of modern Differential Geometry. In their original work, the Cosserats intended to represent the microstructure ("internal degrees of freedom") by means of a rigid orthonormal triad attached to each point of the body. In extending the theory in a variety of ways to a deformable triad, different authors have used various terms to designate the object of their study and to emphasize the desired physical overtones[1]. We will, however, use the terminology of "Cosserat medium" to designate a body to which a triad of arbitrarily deformable vectors (or *directors*) is attached at each point. A good intuitive picture can be derived from a material such as concrete, where a standard body (the cement matrix) is augmented by a densely distributed collection of grains. In the limit, each point of the matrix can be considered as carrying a grain, whose deformation can be considered in first approximation (due to its smallness) as homogeneous. Other types of microstructure, not necessarily describable by means of a deformable triad, could be considered as well [4]. The treatment in this chapter follows [37, 39].

## 3.1 Kinematics of a Cosserat body

A *Cosserat body* is, by definition, the frame bundle $F\mathcal{B}$ of an ordinary body $\mathcal{B}$, usually called the *macromedium*, the *matrix* or the *underlying body*. We

---

[1] A historical and technical account of the various theories can be found in the encyclopedic works of Truesdell and Toupin [88], Truesdell and Noll [87], and Eringen and Kafadar [50].

recall that the frame bundle of a differentiable manifold consists roughly of the collection of all frames at each point of the manifold. Thus, in the case of a three-dimensional manifold $\mathcal{B}$, we obtain the collection of all triads of linearly independent vectors forming all possible bases of the tangent spaces of $\mathcal{B}$. Since each triad is attached to a particular point of $\mathcal{B}$, we have a *projection map*:

$$\pi : F\mathcal{B} \longrightarrow \mathcal{B}, \tag{3.1}$$

which assigns to each triad the point at which it is attached. In the terminology of Differential Geometry, the macromedium $\mathcal{B}$ is known as the *base manifold* of the frame bundle. Given a point $\mathbf{X} \in \mathcal{B}$, the inverse image $\pi^{-1}(\mathbf{X})$ is called the *fibre* at $\mathbf{X}$. It consists of all the possible bases of the tangent space $T_\mathbf{X}\mathcal{B}$. In the physical picture, the fibre is the carrier of the information about the events taking place at the "grain" level. Since, as conceived by the Cosserats, any particular basis (rather that the whole collection thereof) should carry that very information, we will take this fact into consideration when defining the concept of configuration and deformation of a Cosserat body.

Assume now that a coordinate chart with coordinates $X^I$ ($I = 1, 2, 3$) is specified on an open set $\mathcal{U}$ of the base manifold $\mathcal{B}$. The natural basis of this chart:

$$\mathbf{E}_I = \frac{\partial}{\partial X^I}, \quad I = 1, 2, 3, \tag{3.2}$$

determines, at each point of $\mathcal{U} \subset \mathcal{B}$, a basis of the tangent space. In other words, the coordinate chart induces a smooth local section of the frame bundle. Any frame $\mathbf{H}_I$ ($I = 1, 2, 3$) within the domain $\mathcal{U}$ can be expressed in terms of components in the coordinate-induced frame by means of a matrix, viz.:

$$\mathbf{H}_I = H_I^J \mathbf{E}_J. \tag{3.3}$$

We can say, therefore, that, as far as the domain of the chart is concerned, every element of $F\mathcal{B}$ can be represented uniquely by the twelve numbers $(X^I, H_J^K)$. In fact one can prove that the frame bundle $F\mathcal{B}$ is itself a differentiable manifold of dimension 12 and that the numbers just described constitute admissible coordinates of this manifold.[2] If we should consider a different coordinate system, $Y^I$ say, on an open set $\mathcal{V} \subset \mathcal{B}$ such that $\mathcal{U} \cap \mathcal{V} \neq \emptyset$, the natural bases of both systems can be related point-wise by an arbitrary non-singular $3 \times 3$-matrix, that is, by an arbitrary member of the general linear group $GL(3; \mathbb{R})$. This means that fibre-wise the coordinate transformations are governed by this group, which is, therefore, called the *structure group* of the bundle $F\mathcal{B}$. On the other hand, for a fixed basis at a point, all the elements in the fibre, according to Equation (3.3), are precisely spanned by the collection of non-singular $3 \times 3$-matrices. As we have seen, this special situation, whereby the nature of the fibres and the nature of the structure group are identical, is described in Differential Geometry by saying that the bundle of frames of a manifold is a principal fibre bundle.

---

[2] Cf. Example 8.12.

*Remark 3.1.* We have demonstrated the manifold character of the frame bundle by means of fibre-wise coordinates which consist of components of the frames in terms of the natural basis of a coordinate system of the base manifold. It should be clear, however, that we could as well have singled out at each point any basis of the tangent space of the base manifold, not necessarily a coordinate basis, and expressed the fibre-wise coordinates in terms of the matrix of components of the frames in that particular basis.

In a principal bundle we have at our disposal a special operation called the *right action* of the structure group on the principal bundle. We will describe this operation for the particular case at hand. Let $\mathbf{M}$ belong to our structure group. $\mathbf{M}$ can, therefore, be regarded as a nonsingular matrix with entries $\{M^I_J\}$. We want to define the right action $R_M$ of $\mathbf{M}$ as it applies to each element of the principal bundle $F\mathcal{B}$ to produce another element of $\mathcal{B}$. We will do this as follows: let $(X^I, H^I_J)$ be the components of an element of $F\mathcal{B}$ in some coordinate system. Then the image of this element by the right action of $\mathbf{M}$ is given, by definition, as the element of $F\mathcal{B}$ with components $(X^I, H^I_J M^J_K)$ in the same coordinate system. It is not difficult to prove that this definition, although expressed in a particular chart, is in fact independent of the chart chosen. Note that a frame at a point is always mapped to another frame at the same point, so that the right action just defined is fibre preserving.

We now seek an appropriate definition of a configuration of a Cosserat body. To this end, we start by noting that the physical space (which we have identified with $\mathbb{R}^3$) is itself a differentiable manifold and, therefore, it has a naturally defined frame bundle $F\mathbb{R}^3$ with projection $\pi_R$. We want to define a configuration of a Cosserat body as a map $K$ between these two principal bundles, namely:

$$K : F\mathcal{B} \longrightarrow F\mathbb{R}^3. \tag{3.4}$$

But it is clear that an arbitrary map will not do, so this concept needs further clarification. When we map a principal bundle into another, there are three elements at play. Firstly, there are the two base manifolds, which in our case are $\mathcal{B}$ and $\mathbb{R}^3$. Secondly, there are the fibres at each point of these manifolds. And finally, there are the two structure groups. We will assume that the configuration $K$ incorporates an ordinary configuration of the base manifold (the macromedium) $\mathcal{B}$, that is, an embedding:

$$\kappa : \mathcal{B} \longrightarrow \mathbb{R}^3. \tag{3.5}$$

This map is, as we know, smooth and has a smooth inverse defined on the image $\kappa(\mathcal{B})$. Secondly, we want that fibres don't get mixed up: a frame at a point of $\mathbf{X} \in \mathcal{B}$ must be mapped to a frame at the image point $\kappa(\mathbf{X})$. In the physical picture, we want each point in the matrix to carry its own "grain" in the process of deformation. Mathematically, this means that the map $K$ must satisfy the equation:

$$\pi_R \circ K = \kappa \circ \pi. \tag{3.6}$$

This restriction is nicely represented in the following commutative diagram:

$$\begin{array}{ccc} F\mathcal{B} & \xrightarrow{K} & F\mathbb{R}^3 \\ \pi \downarrow & & \downarrow \pi_R \\ \mathcal{B} & \xrightarrow{\kappa} & \mathbb{R}^3 \end{array} \qquad (3.7)$$

But we are not done yet, and this is because in a principal bundle we also have to take into consideration the structure groups and provide an appropriate map between them. Since in our particular case the two structure groups are identical, namely $GL(3;\mathbb{R})$, we will agree that the map between them is just the identity map. Finally, we will require that the right action of the structure group must commute with the map between fibres. This can be represented by the following commutative diagram:

$$\begin{array}{ccc} F\mathcal{B} & \xrightarrow{K} & F\mathbb{R}^3 \\ R_M \downarrow & & \downarrow R_M \\ F\mathcal{B} & \xrightarrow{K} & F\mathbb{R}^3 \end{array} \qquad (3.8)$$

Physically, this means that the deformation of a grain is an intrinsic quantity independent of the particular triad that one chooses to represent that grain. This is precisely the consistency condition that reconciles the original Cosserat picture (one frame representing the grain at a point) with the principal-bundle picture (the collection of all frames at a point representing the same grain).

In the terminology of Differential Geometry, with all the above restrictions, the map between $F\mathcal{B}$ and its image $K(F\mathcal{B}) \subset F\mathbb{R}^3$ is called a *principal-bundle isomorphism*. In terms of components in a given coordinate system in the body and in space, a configuration of a Cosserat body is defined by twelve smooth functions:

$$x^i = \kappa^i(X^J), \qquad (3.9)$$

and

$$K^i{}_I = K^i{}_I(X^J). \qquad (3.10)$$

We see that in a Cosserat body there exist two independent mechanisms, as it were, of dragging vectors by means of a deformation (Figure 3.1): The first mechanism is the ordinary dragging of vectors by means of the deformation gradient of the macromedium, represented by the matrix with entries $F^i{}_I = x^i{}_{,I}$. The second mechanism is the one associated with the deformation of the "micromedium" or grain, and is represented by the matrix with entries $K^i{}_I$. In a second-grade body these two mechanisms are identified with each other.

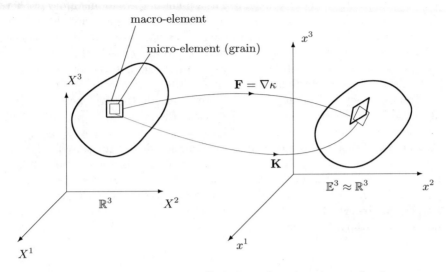

**Fig. 3.1.** The two kinematically independent dragging mechanisms

We have been freely talking about configurations and deformations almost interchangeably. Indeed, since principal-bundle morphisms can be inverted and composed, we can clearly adopt a reference configuration of a Cosserat body and define the notion of a deformation in the same way as we have done for ordinary bodies.

## 3.2 The constitutive law of a simple elastic Cosserat body

A Cosserat body is said to be simple (or first-grade) if its mechanical response is local and depends only on the 1-jet of the configuration. If, moreover, we disregard memory effects, just as in the case of ordinary bodies, we obtain a simple elastic Cosserat body. Although we have just emulated the definition of an ordinary simple elastic body, by using the terminology of jets, it is apparent that the jet of a configuration $K$ of a Cosserat body is a more complicated entity than the jet of a configuration $\kappa$ of the base manifold. In coordinates, the 1-jet of a configuration at a point $\mathbf{X}$ *of the base manifold* consists, in addition to the configuration itself as given by Equations (3.9) and (3.10), of the following quantities:

$$F^i_I = \frac{\partial x^i}{\partial X^I}, \tag{3.11}$$

and

$$K^i{}_{I,J} = \frac{\partial K^i{}_I}{\partial X^J}. \tag{3.12}$$

72    3 Uniformity of Cosserat media

The reason why the 1-jet is well defined at a point of the base manifold (rather than, as one might have expected, at each point of the larger manifold $F\mathcal{B}$), is that the maps between fibres are governed, as we have established, by a single matrix at each point of the base manifold. This was precisely the meaning of the group-consistency condition expressed in the commutative diagram (3.8) and reflected in the coordinate expression (3.10).

From the physical point of view, a simple elastic Cosserat point is characterized by a constitutive law that feels both the macro and micro dragging mechanisms (as depicted in Figure 3.1), as well as the gradient of the latter (Figure 3.2). So, a "simple" Cosserat medium is sensitive not only to the deformation of the grains but also to the gradient of this deformation. When the deformation of the grains is identified with the deformation gradient of the matrix, we recover a second-grade material without microstructure (see Remark 3.3 below). In this way, a simple Cosserat medium can give rise to a non-simple ordinary medium.

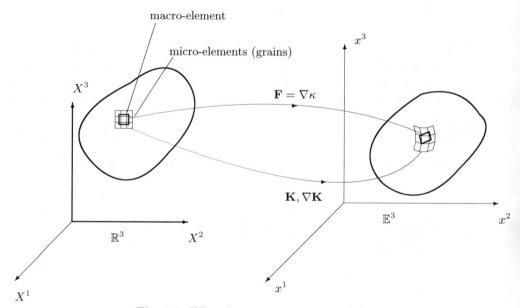

**Fig. 3.2.** What the constitutive equation feels

*Remark 3.2.* In Equation (3.12) we have taken the ordinary partial derivative rather than the covariant derivative. We faced a similar situation in the case of second-grade materials, where we took the ordinary second partial derivative of the deformation. This would seem to preclude the use of curvilinear coordinates in the reference configuration. Nevertheless, the covariant derivative of

$\{K^i{}_I\}$ differs from the ordinary one only in additional terms involving just the variables $\{K^i{}_I\}$, which are already included in the list of variables conforming the 1-jet.

*Remark 3.3.* This remark has to do with the relationship between the theory of Cosserat media, on the one hand, and the theory of second-grade bodies, on the other. We already spoke of the two vector-dragging mechanisms present in a Cosserat body (see Figure 3.1). If we deliberately choose to identify them, namely if we impose the extra condition: $K^i{}_I = F^i{}_I$ identically, then it follows that $K^i{}_{I,J} = F^i{}_{I,J} = \kappa^i{}_{,IJ}$, and we recover the second-grade case. Notice that in the Cosserat body, we have in general $K^i{}_{I,J} \neq K^i{}_{J,I}$.

If we assume the body to be hyperelastic, its constitutive behaviour will be completely encompassed in a single scalar quantity $\psi$ measuring the elastic energy content per unit spatial volume:

$$\psi = \psi(j_X^1 K), \tag{3.13}$$

or in components[3]:

$$\psi = \psi(K^i{}_I, F^i{}_I, K^i{}_{I,J}, X^I). \tag{3.14}$$

It is interesting to record the way in which this constitutive equation changes in form upon a change of reference configuration. In the compact notation of jets, this issue is handled as follows. Let $K_1$ and $K_2$ be two reference configurations, and let $K$ be an arbitrary actual configuration. Then the deformations $\Xi_1$ and $\Xi_2$ of $K$ with respect to the two reference configurations are given, respectively, by the compositions:

$$\Xi_1 = K \circ K_1^{-1}, \tag{3.15}$$

and

$$\Xi_2 = K \circ K_2^{-1}. \tag{3.16}$$

Similarly, the change of reference from $K_1$ to $K_2$ is given by:

$$K_{2|1} = K_2 \circ K_1^{-1}, \tag{3.17}$$

whence:

$$\Xi_1 = \Xi_2 \circ K_{2|1}. \tag{3.18}$$

---

[3] In principle, the 1-jet includes also the local value of the macromedium deformation itself, as given by Equation (3.9). In the component expression, however, we eliminate this dependence, which is rendered unnecessary by the required translational invariance of the constitutive law. Notice also that the ordering of the remaining arguments follows the natural order of the arguments in a 1-jet. Thus, the ordinary deformation gradient $F^i{}_I$, being itself a derivative, comes after the micromedium deformation, which is not. This policy, however, was not followed in [37]. See also [22].

Denoting the constitutive law expressed in terms of the respective reference configurations by $\psi_1$ and $\psi_2$, we obtain:

$$\psi_2(j_Y^1 \Xi_2) = \psi_1(j_Y^1 \Xi_2 \circ j_X^1 K_{2|1}) \qquad (3.19)$$

where $Y = \kappa_{2|1}(X)$. Equation (3.19) is to be satisfied identically for all deformations $\Xi_2$.

To obtain an explicit coordinate representation of this compact expression afforded by the language of jets, we start by adopting coordinate charts $X^I$ and $Y^A$ in the base manifolds of reference configurations 1 and 2, respectively. The component expression of the deformation $K_{2|1}$ between the two reference configurations is notated as $(Y^A(X^K), K_I^A(X^K))$, while the spatial deformation with respect to each reference is notated, respectively, as $(x_1^i(X^K), K^i{}_I(X^K))$ and $(x_2^i(Y^A), K^i{}_B(Y^A))$. This notational scheme, using the same letters for similar entities while reserving the indices $A, B, C$ for the components of the second reference configuration and retaining the indices $I, J, K$ for the first, will be used in the following equations. The two spatial deformations are related by:

$$x_1^i(X^I) = x_2^i(Y^A(X^I)), \qquad (3.20)$$

and

$$K^i{}_I = K^i{}_A\, K^A_I. \qquad (3.21)$$

By the chain rule of differentiation, we finally obtain:

$$\psi = \psi_2(K^i{}_A,\ F^i{}_A,\ K^i{}_{A,B},\ Y^C)$$
$$= \psi_1(K^i{}_A K^A_I,\ F^i{}_A F^A_I,\ K^i{}_{A,B} K^A_I F^B_J + K^i{}_A K^A_{I,J},\ X^K(Y^C)). \quad (3.22)$$

This equation should be compared with its counterpart (2.16) for second-grade bodies, to which it reduces when we systematically identify the fibre deformation with the deformation gradient at the base point.

## 3.3 Material isomorphisms and uniformity

### 3.3.1 Material isomorphisms in a Cosserat body

The notion of material isomorphism between two points, $\mathbf{X}_1$ and $\mathbf{X}_2$, of a body is always conceptually the same, regardless of the type of material body under consideration. It consists always of a local diffeomorphism[4] $p_{12}$ that brings the constitutive responses of both points into coincidence. In each case, however, we need to capture that particular feature of the local diffeomorphism that the material "feels". In the case of a simple material, it was enough to retain the 1-jet of $p_{12}$, since it would be futile to retain higher order jets

---

[4] For a previous explanation of this concept, see Section 2.3 of Chapter 2.

to which the material is, by definition, insensitive. Similarly, for a second-grade body, it is the 2-jet that plays the meaningful role. For the case of a Cosserat medium, following the same line of thought, we need to consider a *local principal-bundle isomorphism* $p_{12}$ between $\mathbf{X}_1$ and $\mathbf{X}_2$. We already know what a principal-bundle isomorphism is. By adding the qualification "local", we are only requiring that there exist neighbourhoods, $\mathcal{U}_1$ and $\mathcal{U}_2$, of the two points such that the map $p_{12}$ is a principal-bundle isomorphism between the restricted frame bundles $F\mathcal{U}_1$ and $F\mathcal{U}_2$. In other words, we don't need $p_{12}$ to do a nice smooth job over the whole of the body, but only on the neighbourhoods in question. In keeping with the policy used in previous cases, we will consider the 1- jet of $p_{12}$ as the operation representing the transplant, since the Cosserat material is sensitive to this quantity only.

We state, therefore, that two points $\mathbf{X}_1$ and $\mathbf{X}_2$ of the base manifold $\mathcal{B}$ of a Cosserat body F $\mathcal{B}$ with constitutive equation (3.13) are materially isomorphic if there exists a local principal-bundle isomorphism $p_{12}$ from $\mathbf{X}_1$ to $\mathbf{X}_2$ such that the equation:

$$\psi(j^1_{X_2}K \circ j^1_{X_1}p_{12}) = \psi(j^1_{X_2}K), \tag{3.23}$$

is satisfied identically for all configurations $K$. In components, the 1-jet of $p_{12}$ will consist of the map between the base neighbourhoods, its gradient $P^I_J$, the map between the fibres $Q^I_J$ and its gradient $R^I_{JK}$, all evaluated at $\mathbf{X}_1$. Correspondingly, the condition given in Equation (3.23) reads:

$$\psi(K^i_I Q^I_M,\ F^i_I P^I_M,\ K^i_{I,J} Q^I_M P^J_N + K^i_I R^I_{MN},\ (X_1)^I)$$
$$= \psi(K^i_M,\ F^i_M,\ K^i_{M,N},\ (X_2)^I). \tag{3.24}$$

### 3.3.2 Uniformity and the Cosserat archetype

A Cosserat body is materially uniform if all the points of the base manifold are materially isomorphic in the sense of Equation (3.23) or (3.24). Choosing a point $\mathbf{X}_0$, and denoting the implant from this point to an arbitrary point $\mathbf{X}$ by $j^1_0 p_X$, the uniformity condition can be written as:

$$\psi(j^1_X K) = \psi(j^1_X K \circ j^1_0 p_X). \tag{3.25}$$

If we should want to define an archetype (and to write Equation (3.25) in components), as we have done in previous chapters, we need to handle its definition with great care. In the case of a simple material, the archetype was characterized by means of a frame, in the ordinary sense of the word, namely, a basis of the tangent space of $\mathbf{X}_0$. In the case of second-grade materials, we extended this concept to that of a second-order frame. This entity was defined as the 2-jet of a local diffeomorphism from the origin of $\mathbb{R}^3$ to $\mathbf{X}_0$. In the case of a Cosserat material, we are led in a natural way to consider 1-jets of local principal-bundle isomorphisms between the frame bundles $F\mathbb{R}^3$ and $F\mathcal{B}$. This idea, which we will pursue later in Section 3.5, gives rise to the so-called *non-holonomic frames of second order* in the manifold $\mathcal{B}$. From the physical point of view, a non-holonomic frame of second order at $\mathbf{X}_0$ is nothing more than (the inverse of) a *localized configuration* at this point. This is obtained by choosing a Cosserat configuration and retaining the values of its 1-jet at $\mathbf{X}_0$ in order to express once and for all the constitutive law at $\mathbf{X}_0$ in component form. At this point, we only need to realize that once such a frame is adopted at an archetype such as $\mathbf{X}_0$, the uniformity condition (3.25) boils down to the existence of three fields, $(Q^I_\alpha(\mathbf{X}), P^I_\alpha(\mathbf{X}), R^I_{\alpha\beta}(\mathbf{X}))$, representing the implants from the archetype to point $\mathbf{X}$, such that:

$$\psi(K^i_I, F^i_I, K^i_{I,J}, X^I)$$
$$= \bar{\psi}(K^i_I Q^I_\alpha, F^i_I P^I_\alpha, K^i_{I,J} Q^I_\alpha P^J_\beta + K^i_I R^I_{\alpha\beta}), \qquad (3.26)$$

where we have denoted by $\bar{\psi}$ the constitutive equation of the archetype. We assume the above implant fields to be smooth.

*Remark 3.4.* We note that given a local diffeomorphism $\kappa$ between two points of a manifold $\mathcal{B}$, we can "lift" $\kappa$ to a local principal-bundle isomorphism, $F\kappa$ say, in the frame bundle $F\mathcal{B}$. This is simply accomplished by considering the tangent (or derivative) map of $\kappa$ and applying it to each of the vectors of a frame to obtain the image frame. The same idea applies for maps between different manifolds. If we restrict the principal-bundle isomorphisms used to define our non-holonomic frames of second order to just this particular type of map, we recover the definition of the second-order frames that we used in the case of second-grade materials. Those frames are, therefore, also called holonomic frames of second order. It is interesting to point out that there exists an intermediate situation, the so-called semi-holonomic frames, treated in detail in Chapter 10.

### 3.3.3 Cosserat symmetries

A symmetry $g$ of a Cosserat material point $\mathbf{X} \in \mathcal{B}$ is a material isomorphism from the point to itself. In other words, in condition (3.23) we identify the source $\mathbf{X}_1$ and the target $\mathbf{X}_2$ with $\mathbf{X}$ to obtain:

$$\psi(j^1_X K \circ j^1_X g) = \psi(j^1_X K), \qquad (3.27)$$

for all configurations $K$. In components, therefore, a symmetry at $\mathbf{X}$ consists of a triple $(H^I_J, G^I_J, L^I_{JK})$ such that

$$\psi(K^i_I H^I_J, F^i_I G^I_K, K^i_{I,J} H^I_M G^J_N + K^i_I L^I_{MN}, X^I)$$
$$= \psi(K^i_J, F^i_K, K^i_{M,N}, X^I). \qquad (3.28)$$

The collection $\mathcal{G}_X$ of symmetries at the point $\mathbf{X}$ constitutes a group whose operation is the composition of jets. If $(H^I_J, G^I_J, L^I_{JK})$ and $(\bar{H}^I_J, \bar{G}^I_J, \bar{L}^I_{JK})$ are the components of two symmetries, we obtain the following expression for their product:

$$(H^I_{\ J},\ G^I_{\ J},\ L^I_{\ JK})(\bar{H}^M_{\ N},\ \bar{G}^M_{\ N},\ \bar{L}^M_{\ NP})$$
$$= (H^I_{\ J}\bar{H}^J_{\ N},\ G^I_{\ J}\bar{G}^J_{\ N},\ L^I_{\ JK}\bar{H}^J_{\ N}\bar{G}^K_{\ P} + H^I_{\ J}\bar{L}^J_{\ NP}). \qquad (3.29)$$

The inverse of an element with components $(H^I_{\ J}, G^I_{\ J}, L^I_{\ JK})$ is given by:

$$(H^I_{\ J},\ G^I_{\ J},\ L^I_{\ JK})^{-1} = (H^{-I}_{\ J},\ G^{-I}_{\ J},\ -L^M_{\ NP}H^{-I}_{\ M}H^{-N}_{\ J}G^{-P}_{\ K}). \qquad (3.30)$$

We point out that, just as in the case of second-grade bodies, the collection of all the second entries $G^I_{\ J}$ constitutes a group, which we will call the *projected group* representing the symmetries of the macromedium.

The considerations leading to Equations (1.13-1.17), as well as the equations themselves, are applicable to Cosserat bodies.

### 3.3.4 Changing coordinates

We have already found that on changing reference configuration the components of a deformation change according to the composition rule embodied in Equation(3.22). If instead of changing reference configuration we just change coordinates in the base manifold according to some functions:

$$Y^I = Y^I(X^J), \qquad (3.31)$$

the components of the deformation will change according to the following formulas:

$$\bar{F}^i_{\ K} = F^i_{\ I}\frac{\partial X^I}{\partial Y^K}, \qquad (3.32)$$

$$\bar{K}^i_{\ K} = K^i_{\ I}\frac{\partial X^I}{\partial Y^K}, \qquad (3.33)$$

and

$$\bar{K}^i_{\ K,L} = K^i_{\ I,J}\frac{\partial X^J}{\partial Y^L}\frac{\partial X^I}{\partial Y^K} + K^i_{\ I}\frac{\partial^2 X^I}{\partial Y^K \partial X^L}, \qquad (3.34)$$

where superimposed bars denote values in the $Y$-coordinate system.

If we now introduce these values in the uniformity condition (3.26), we obtain the following transformation laws for the components of a given uniformity field:

$$\bar{P}^K_{\ \alpha} = P^I_{\ \alpha}\frac{\partial Y^K}{\partial X^I}, \qquad (3.35)$$

$$\bar{Q}^K_{\ \alpha} = Q^I_{\ \alpha}\frac{\partial Y^K}{\partial X^I}, \qquad (3.36)$$

and

$$\bar{R}^K_{\ \alpha\beta} = R^I_{\ \alpha\beta}\frac{\partial Y^K}{\partial X^I} + P^J_{\ \beta}Q^I_{\ \alpha}\frac{\partial^2 Y^K}{\partial X^I \partial X^J}. \qquad (3.37)$$

We observe that the law of transformation of $R^I_{\ \alpha\beta}$ resembles that of a Christoffel symbol, as per Equation (9.36). Specifically, the combinations:

78    3 Uniformity of Cosserat media

$$\Lambda^K_{LM} = -R^K_{\alpha\beta} Q^{-\alpha}_L P^{-\beta}_M, \qquad (3.38)$$

are the Christoffel symbols of a linear connection on the base manifold $\mathcal{B}$.

The transformation equations just derived should not be confused with the more general transformations of a given uniformity field induced by a change of reference configuration. In the notation of Equations (3.20-3.22), these more general transformations are:

$$\bar{P}^A_\alpha = P^I_\alpha F^A_I, \qquad (3.39)$$

$$\bar{Q}^A_\alpha = Q^I_\alpha K^A_I, \qquad (3.40)$$

and

$$\bar{R}^A_{\alpha\beta} = R^I_{\alpha\beta} K^A_I + P^J_\beta Q^I_\alpha K^A_{I,J}. \qquad (3.41)$$

The equations of transformation under a mere coordinate change of the base can be obtained as a particular case of Equations (3.39-3.41) by effecting a holonomic change of reference configuration, that is, a change which is a lift of a change of coordinates.

### 3.3.5 Changing the archetype

A change of archetype, (in particular, a change of (nonholonomic) basis in a given archetypal point), alters the implant fields $(Q^I_\alpha, P^I_\alpha, R^i_{\alpha\beta})$ by composition with a constant field. Indeed, let the change of basis be represented by the quantities $(B^\beta_\alpha, A^\beta_\alpha, C^\gamma_{\alpha\beta})$. Then, the implant fields transform according to:

$$\hat{P}^I_\alpha = P^I_\beta A^\beta_\alpha, \qquad (3.42)$$

$$\hat{Q}^I_\alpha = Q^I_\beta A^\beta_\alpha, \qquad (3.43)$$

$$\hat{R}^I_{\alpha\beta} = R^I_{\gamma\rho} B^\gamma_\alpha A^\rho_\beta + Q^I_\gamma C^\gamma_{\alpha\beta}, \qquad (3.44)$$

where hatted quantities refer to the new implants.

## 3.4 Homogeneity conditions

### 3.4.1 Homogeneity of a Cosserat body

A (reference) configuration of a Cosserat body, by identifying it with $F\mathbb{R}^3$, provides us with a collection of unique local fibre-bundle isomorphisms between every pair of points of the body $\mathcal{B}$, namely, the canonical fibre-bundle isomorphisms arising from Euclidean translations. If these happen to be also material isomorphisms, we say that the body is (globally) homogeneous and that the configuration enjoying this property is a homogeneous configuration. If, on the other hand, for every point of $\mathcal{B}$ there exists a configuration such that the above property holds for a neighbourhood of the point, we say that the Cosserat body is locally homogeneous and the configuration is locally homogeneous. These definitions are completely analogous to their counterparts in first- and second-grade materials. In a (locally) homogeneous configuration (using, say, Cartesian coordinates) the (local) transplant fields appearing in Equation (3.24) become $P^I_J = \delta^I_J$, $Q^I_J = \delta^I_J$ and $R^I_{MN} = 0$. This condition is independent of the archetype chosen, which was not even invoked in the definition of homogeneity. If an archetype is used, then clearly the implant fields $(Q^I_\alpha, P^I_\alpha, R^I_{\alpha\beta})$ in a homogeneous configuration become merely *constant*, not necessarily $(\mathbf{I}, \mathbf{I}, \mathbf{0})$. A change of the archetype, however, can be used to bring the constants to these special values. Just as was the case in second-grade bodies, we may use this accident to suggest a more stringent definition of homogeneity, namely, homogeneity with respect to a given archetype. We may wish, for example, to reserve the homogeneous distinction for bodies that can be brought to a homogeneous configuration whereby all the points are free of stress. We then choose the archetype to be stress-free and check for the existence of a homogeneous configuration such that the implants from this archetype are of the form $(\mathbf{I}, \mathbf{I}, \mathbf{0})$.

### 3.4.2 The three kinds of material connections of a uniform Cosserat body

Restricting our attention to coordinate transformations on the base manifold $\mathcal{B}$, and referring to a given (locally smooth) uniformity field of implants $(Q^I_\alpha, P^I_\alpha, R^I_{\alpha\beta})$ from a given archetype, we know from the case of simple materials that the quantities:

$$\Gamma^I_{JK} = P^I_\alpha \, P^{-\alpha}_{J,K}, \tag{3.45}$$

are the Christoffel symbols of a connection $\Gamma$. This connection is naturally associated with the homogeneity (or lack thereof) of the macromedium $\mathcal{B}$.

It is not difficult to verify that the quantities:

$$\Delta^I_{JK} = Q^I_\alpha \, Q^{-\alpha}_{J,K}, \tag{3.46}$$

transform as Christoffel symbols of a, generally different, connection $\Delta$ on the base manifold. Notice that in the case of second-grade bodies, these two connections coincide.

Finally, we have already found out that there exists a third connection $\Lambda$ on the base manifold whose Christoffel symbols are given by Equation (3.38). The connection $\Delta$ and $\Lambda$, as we shall see, carry important information concerning the homogeneity of the Cosserat superstructure.

A legitimate question consists of inquiring whether or not these connections are independent of the archetype chosen. According to the rules of transformation of the uniformity fields given in Equations (3.42-3.44), it can be verified that the connections $\Gamma$ and $\Delta$ are independent of the choice of

80    3 Uniformity of Cosserat media

archetype (or the second-order nonholonomic frame therein). The third connection (3.38), on the other hand, is affected by a change of archetype according to the formula:

$$\hat{\Lambda}^I{}_{JK} = \Lambda^I{}_{JK} - Q^I{}_\alpha Q^{-\gamma}{}_J P^{-\rho}_K C^\alpha{}_{\beta\lambda} B^{-\beta}{}_\gamma A^{-\lambda}{}_\rho. \tag{3.47}$$

### 3.4.3 Homogeneity conditions

We seek a smooth uniformity (implant) field $(Q^I{}_\alpha, P^I{}_\alpha, R^I{}_{\alpha\beta})$ and a change of reference configuration $(Y^A(X^K), K^A_I(X^K))$ such that:

$$P^I{}_\alpha F^A_I = a^A{}_\alpha, \tag{3.48}$$

$$Q^I{}_\alpha K^A_I = b^A{}_\alpha, \tag{3.49}$$

and

$$R^I{}_{\alpha\beta} K^A_I + P^J{}_\beta Q^I{}_\alpha K^A_{I,J} = c^A{}_{\alpha\beta}, \tag{3.50}$$

where the quantities appearing on the right-hand sides are constant and where we have used Equations (3.39-3.41) for the transformation of implants under a change of reference configuration. For local homogeneity, these equations need to be valid only on a neighbourhood of any given point. In fact, both the (smooth) implant field and the appropriate change of reference configuration may not exist for the body as a whole.

Equation (3.48) is exactly the same as for a first-grade body, whence we conclude that its satisfaction is equivalent to the existence of a material connection of the $\Gamma$-type with vanishing torsion. Physically, this means that the macromedium itself has to be locally homogeneous, regardless of the Cosserat microstructure imposed upon it. Recalling that the $\Gamma$-connection is independent of the archetype, we can state that the first condition of local homogeneity is the existence of a local uniformity field such that:

$$\tau^I_{JK} \equiv \Gamma^I_{JK} - \Gamma^I_{KJ} = 0. \tag{3.51}$$

If the projected symmetry group is discrete, the $\Gamma$-connection is unique, so that this test is definitive for the discrete case, as far as the macromedium is concerned.

We now observe that Equation (3.49) does not impose any restriction, since it simply prescribes that $K^A_I(X^K)$, which except for smoothness can be chosen freely, be equal to a constant matrix times the inverse of the matrix $Q^I{}_\alpha$. Introducing this condition into Equation (3.50), we obtain:

$$(\Delta^K_{IJ} - \Lambda^K_{IJ}) Q^I{}_\alpha P^J{}_\beta Q^{-\gamma}{}_K = c^A{}_{\alpha\beta} b^{-\gamma}{}_A, \tag{3.52}$$

where we have used Equations (3.38) and (3.46). We define the following tensor:

$$D \equiv \Delta - \Lambda, \tag{3.53}$$

as the *Cosserat inhomogeneity tensor* relative to a given uniformity field and a given archetype. In conclusion, the second, and final, condition for local homogeneity of a Cosserat body is the existence of a local uniformity field such that:

$$D_{IJ}^K \, Q^I_\alpha \, P^J_\beta \, Q^{-\gamma}_K = \text{constant}. \tag{3.54}$$

This last condition can be further reduced by effecting a change of archetype $(B^\beta_\alpha, A^\beta_\alpha, C^\gamma_{\alpha\beta})$ such that:

$$C^\gamma_{\alpha\beta} B^\alpha_\lambda A^\beta_\rho = -D_{IJ}^K \, Q^I_\lambda \, P^J_\rho \, Q^{-\gamma}_K, \tag{3.55}$$

where the right-hand side is evaluated at any point within the neighbourhood. This can be achieved in many different ways, such as choosing $A^\alpha_\beta = B^\alpha_\beta = \delta^\alpha_\beta$ and, therefore, $C'^\gamma_{\lambda\rho}$ equal to the right-hand side of (3.55). According to the transformation equations under a change of archetype, we obtain that the new value of the Cosserat inhomogeneity tensor $D$ at the point in question is zero. We conclude, therefore, that the second condition of local homogeneity can be formulated as the existence of a field of implants such that the inhomogeneity tensor with respect to the new archetype vanishes in the neighbourhood.

## 3.5 The Cosserat material *G*-structures and groupoid

The treatment so far in this chapter has been presented in terms of the matrix components of the implant fields. Although such treatment has led us all the way from the concept of material isomorphism to conditions for local homogeneity of Cosserat media in terms of restrictions imposed on certain Christoffel symbols, it is fair to claim that we have not yet given a clear geometric picture. The purpose of the treatment that follows is to recast the theory within the framework of the notions of material *G*-structures and material groupoid, concepts with which we are familiar from the cases of simple and second-grade materials. The fact that the two kinematical dragging mechanisms present in a Cosserat medium are mutually independent, however, leads to a higher level of detail and abstraction and, one would hope, to a deeper understanding of the theory. In particular, the theory of second-grade uniformity and homogeneity will emerge as a clear particular case of the more general theory of Cosserat bodies.

### 3.5.1 Frames, and frames of frames

#### Three ways to define a frame

A frame $f$ at a point $b$ of an $n$-dimensional manifold $\mathcal{B}$ is defined as a basis of the tangent space $T_b\mathcal{B}$. In other words, $f$ is an ordered set of $n$ linearly independent vectors in $T_b\mathcal{B}$:

$$f = \{\mathbf{f}_1, \mathbf{f}_2, ..., \mathbf{f}_n\} \quad \mathbf{f}_\alpha \in T_b\mathcal{B} \quad (\alpha = 1, ..., n). \tag{3.56}$$

In a local coordinate system $X^I$, $(I = 1, ..., n)$, there exist $n^2$ numbers $f_\alpha^J$, uniquely defined, such that:

$$\mathbf{f}_\alpha = f_\alpha^J \frac{\partial}{\partial x^J}. \tag{3.57}$$

As a consequence of the assumed linear independence, the matrix $\{f_\alpha^J\}$ is non-singular.

Another way to look at the notion of a frame is to define it as a linear isomorphism:

$$\tilde{f} : \mathbb{R}^n \longrightarrow T_b\mathcal{B}. \tag{3.58}$$

To relate this definition with the previous one, we need only recognize that, considered as a vector space, $\mathbb{R}^n$ has a canonical basis $e$ consisting of the vectors:

$$\mathbf{e}_1 = \{1, 0, 0, ..., 0\}$$
$$\mathbf{e}_2 = \{0, 1, 0, ..., 0\}$$
$$\cdot$$
$$\cdot$$
$$\mathbf{e}_n = \{0, 0, 0, ..., 1\}. \tag{3.59}$$

We now define:
$$\mathbf{f}_\alpha \equiv \tilde{f}(\mathbf{e}_\alpha) \quad (\alpha = 1, ..., n), \tag{3.60}$$

so that each linear isomorphism $\tilde{f}$ determines a frame $f$ in the sense of Equation (3.56). Conversely, given a basis $f = \{\mathbf{f}_1, \mathbf{f}_2, ..., \mathbf{f}_n\}$ of $T_b\mathcal{B}$, we construct the following linear isomorphism:

$$\tilde{f}(\mathbf{r}) = r^\alpha \mathbf{f}_\alpha \quad \forall \, \mathbf{r} = \{r^1, r^2, ..., r^n\} \in \mathbb{R}^n. \tag{3.61}$$

Both definitions are, therefore, equivalent.

Finally, we have a somewhat more sophisticated, but equivalent, definition of a frame in terms of 1-jets. Let $\phi$ be a local diffeomorphism from an open neighbourhood of $O \in \mathbb{R}^n$ to an open neighbourhood of $b \in \mathcal{B}$, such that $\phi(O) = b$. Consider the 1-jet $f_\phi = j_O^1 \phi$. As we know, a 1-jet consists of two elements: the first is just the value of the function at the source point, which is just the target point $b = \phi(O)$. The second element of the 1-jet is a linear map between the tangent space at the source point, namely $T_O\mathbb{R}^n$, and the tangent space at the target point, namely $T_b\mathcal{B}$. In view that, in our case, $\phi$ has been assumed to be a diffeomorphism, this linear map is actually an isomorphism. But $T_O\mathbb{R}^n$ is canonically isomorphic to $\mathbb{R}^n$ itself, since $\mathbb{R}^n$ is not just a manifold but also a vector space. Therefore, $f_\phi = j_O^1\phi$ provides us with a specific linear isomorphism from $\mathbb{R}^n$ to $T_b\mathcal{B}$, resulting thus in a frame in the sense of Equation (3.58)[5]. Conversely, given a basis $f$ at $b$, we can construct a local diffeomorphism in the following way. Let $X^I$ be a local coordinate system and let $\mathbf{f}_\alpha = f_\alpha^J \frac{\partial}{\partial X^J}$. In this coordinate system we define $f_\phi$ by the coordinate expressions:

$$X^J = f_\alpha^J r^\alpha. \tag{3.62}$$

In this way, we can convince ourselves that all three definitions are essentially equivalent, and when we talk about a frame $f$ at a point $b \in \mathcal{B}$, we will take the liberty of using the same notation, regardless of which one of the three equivalent interpretations is being invoked.

**The iterated frame bundle**

We recall that attaching to each point $b$ of $\mathcal{B}$ the collection of all the frames at $b$, we obtain the *principal frame bundle* $F\mathcal{B}$, whose bundle projection we denote by $\pi$. The structure group of $F\mathcal{B}$ is the general linear group $GL(n;\mathbb{R})$. A direct way to put this group in evidence is to consider the fibre at $O \in \mathbb{R}^n$ of the *standard frame bundle* $F\mathbb{R}^n$. Given a coordinate chart $X^I$ in an open subset $\mathcal{U} \subset \mathcal{B}$, the field of natural bases $\frac{\partial}{\partial x^I}$ can be used to express any frame $f$ uniquely as $\mathbf{f}_\alpha = f_\alpha^I \frac{\partial}{\partial X^I}$. Thus, the $n + n^2$ numbers $x^I, f_\alpha^I$ serve as coordinates for $F\mathcal{B}$ over the open subset $F\mathcal{U} = \pi^{-1}(\mathcal{U}) \subset F\mathcal{B}$.

From the above considerations, it follows that $F\mathcal{B}$ is a differentiable manifold of dimension $n + n^2$. As such, it makes sense to consider its principal frame bundle $FF\mathcal{B}$, which is a differentiable manifold of dimension $(n + n^2) + (n + n^2)^2$. The elements of this manifold, which we call the *iterated frame bundle* of $\mathcal{B}$, are not easy to visualize. Poetically, we may choose to call them *frames of frames*, but their usefulness is limited when defined in the generality we have envisaged thus far. We will soon see that certain subclasses of these frames of frames are of great practical importance for our application and, moreover, that our third definition of frame (namely, the one based upon 1-jets of local diffeomorphisms) turns out to be the easiest means to define these subclasses. Be that as it may, it is interesting at this point to repeat the procedure we used for $F\mathcal{B}$ so as to reveal how a coordinate chart of $FF\mathcal{B}$ induced by a coordinate chart of $\mathcal{B}$ looks. We have seen that on the open subset $F\mathcal{U}$ the coordinates induced by the chart $X^I$ in $\mathcal{U} \subset \mathcal{B}$ are given by:

$$f = (b, \{f_1, ..., f_n\}) \mapsto \{X^I, f_\alpha^i\}, \tag{3.63}$$

where, with some overlap of notation, $f$ denotes a generic element of $F\mathcal{B}$ (consisting, naturally, of a point $b$ and a basis therein). The local natural basis corresponding to this coordinate system of the $(n + n^2)$-dimensional manifold $F\mathcal{B}$ consists of the following the $n + n^2$ linearly independent vectors:

---

[5] It is not difficult to see that, regarding the inverse of the local diffeomorphism $\phi^{-1}$ as a chart of $\mathcal{B}$ with coordinates $r^\alpha$, the basis that corresponds to $j_O^1 \phi$ is precisely the natural basis of this coordinate system: $\frac{\partial}{\partial r^\alpha}$ at $b$.

$$\{\frac{\partial}{\partial X^I}, \frac{\partial}{\partial f^I_\alpha}\} \tag{3.64}$$

Any other local basis $g$ will also consist of $n+n^2$ linearly independent vectors $\mathbf{g}_\alpha$ and $\mathbf{g}_\alpha{}^\rho$. Each of these vectors can be expressed uniquely in terms of the coordinate-induced base vectors, namely:

$$\mathbf{g}_\alpha = g^I{}_\alpha \frac{\partial}{\partial X^I} + g^I{}_{\beta\alpha} \frac{\partial}{\partial f^I_\beta}, \tag{3.65}$$

and

$$\mathbf{g}_\alpha{}^\rho = g^J{}_\alpha{}^\rho \frac{\partial}{\partial X^J} + g^J{}_{\beta\alpha}{}^\rho \frac{\partial}{\partial f^J_\beta}. \tag{3.66}$$

In mixed index and block-matrix notation, these equations can be written together as follows:

$$\begin{Bmatrix} \mathbf{g}_\alpha \\ \mathbf{g}_\alpha{}^\rho \end{Bmatrix} = \begin{bmatrix} g^I{}_\alpha & g^I{}_{\beta\alpha} \\ g^I{}_\alpha{}^\rho & g^I{}_{\beta\alpha}{}^\rho \end{bmatrix} \begin{Bmatrix} \frac{\partial}{\partial X^I} \\ \frac{\partial}{\partial f^I_\beta} \end{Bmatrix}. \tag{3.67}$$

Consequently, and always following the same procedure as we used above for the manifold $F\mathcal{B}$ mutatis mutandis, an element $g$ of the iterated frame bundle $FF\mathcal{B}$ can be assigned the following local coordinates:

$$g \mapsto \{X^I, f^I_\alpha, g^I{}_\alpha, g^I{}_\alpha{}^\rho, g^I{}_{\alpha\beta}, g^I{}_{\alpha\beta}{}^\rho\}. \tag{3.68}$$

Considered as a principal bundle over the base manifold $F\mathcal{B}$, the structure group of $FF\mathcal{B}$ is $GL(n+n^2; \mathbb{R})$. A typical element of this group is conveniently represented as:

$$A = \{A^\alpha{}_\beta, A^\alpha{}_\beta{}^\sigma, A^\alpha{}_{\rho\beta}, A^\alpha{}_{\rho\beta}{}^\sigma\}, \tag{3.69}$$

whereby the group multiplication is given by:

$$\begin{aligned} AB &= \{(AB)^\alpha{}_\beta, (AB)^\alpha{}_\beta{}^\sigma, (AB)^\alpha{}_{\rho\beta}, (AB)^\alpha{}_{\rho\beta}{}^\sigma\} \\ &= \{A^\alpha{}_\gamma B^\gamma{}_\beta + A^\alpha{}_\gamma{}^\lambda B^\gamma{}_{\lambda\beta},\ A^\alpha{}_\gamma B^\gamma{}_\beta{}^\sigma + A^\alpha{}_\gamma{}^\lambda B^\gamma{}_{\lambda\beta}{}^\sigma, \\ &\quad A^\alpha{}_{\rho\gamma} B^\gamma{}_\beta + A^\alpha{}_{\rho\gamma}{}^\lambda B^\gamma{}_{\lambda\beta},\ A^\alpha{}_{\rho\gamma} B^\gamma{}_\beta{}^\sigma + A^\alpha{}_{\rho\gamma}{}^\lambda B^\gamma{}_{\lambda\beta}{}^\sigma\}. \end{aligned} \tag{3.70}$$

Or, in matrix notation:

$$\begin{bmatrix} (AB)^\alpha{}_\beta & (AB)^\alpha{}_\beta{}^\sigma \\ (AB)^\alpha{}_{\rho\beta} & (AB)^\alpha{}_{\rho\beta}{}^\sigma \end{bmatrix} = \begin{bmatrix} A^\alpha{}_\gamma & A^\alpha{}_\gamma{}^\lambda \\ A^\alpha{}_{\rho\gamma} & A^\alpha{}_{\rho\gamma}{}^\lambda \end{bmatrix} \begin{bmatrix} B^\gamma{}_\beta & B^\gamma{}_\beta{}^\sigma \\ B^\gamma{}_{\lambda\beta} & B^\gamma{}_{\lambda\beta}{}^\sigma \end{bmatrix}. \tag{3.71}$$

This group acts to the right on $FF\mathcal{B}$ according to the formula:

$$\mathbf{g}A = \langle (\mathbf{g}A)_\alpha\ (\mathbf{g}A)_\alpha^\rho \rangle = \langle \mathbf{g}_\gamma\ \mathbf{g}_\gamma{}^\sigma \rangle \begin{bmatrix} A^\alpha{}_\gamma & A^\gamma{}_\alpha{}^\rho \\ A^\gamma{}_{\sigma\alpha} & A^\gamma{}_{\sigma\alpha}{}^\rho \end{bmatrix}. \tag{3.72}$$

3.5 The Cosserat material $G$-structures and groupoid        85

### 3.5.2 Non-holonomic, semi-holonomic and holonomic frames

### The bundle of non-holonomic frames $\bar{F}^2\mathcal{B}$

We have already remarked that the iterated bundle $FF\mathcal{B}$ is too large to be of practical use in the theory of inhomogeneities. To clarify this point further, we note that the frames of $F\mathcal{B}$ that our construction of $FF\mathcal{B}$ allows do not fully take into consideration the fact that $F\mathcal{B}$ is a fibre bundle. Indeed, our construction of $FF\mathcal{B}$ considers $F\mathcal{B}$ as a general manifold of dimension $n+n^2$, without due regard to the fact that in each tangent space of $F\mathcal{B}$ there are clearly privileged vectors, namely, *vertical vectors* (tangent, by definition, to the fibre of $F\mathcal{B}$). Although it might certainly be possible to pursue a heuristic construction of a subset of $FF\mathcal{B}$ that takes this fact (and perhaps other facts as well) into consideration, our third definition of frame (using 1-jets of local diffeomorphisms) will allow us to achieve this goal rigorously, systematically and more or less automatically by simply restricting the class of local diffeomorphisms permitted.

It is useful at this point to bring back into the scene the standard frame bundle $F\mathbb{R}^n$. Due to the fact that the base manifold $\mathbb{R}^n$ has lots of extra structure, usually absent from a generic manifold $\mathcal{B}$ of dimension $n$, we have the luxury of finding distinguished elements, not available otherwise. In particular, there exists a distinguished frame $e$ in $F\mathbb{R}^n$ given by:

$$e = j^1_O \, id_{\mathbb{R}^n}, \qquad (3.73)$$

where $O$ denotes the canonical origin of $\mathbb{R}^n$ (another luxury). Here, $id_{\mathbb{R}^n}$ denotes the identity map of $\mathbb{R}^n$. In fact, we have already encountered this privileged frame in the guise of Equation (3.59). Consider now an open neighbourhood $\mathcal{U}$ of $O \in \mathbb{R}^n$. Denoting by $\pi$ the bundle projection of $F\mathbb{R}^n$, the set $\pi^{-1}(\mathcal{U})$ can be identified with the frame bundle $F\mathcal{U}$, a sub-bundle of $F\mathcal{B}$. Let:

$$\Phi : F\mathcal{U} \longrightarrow F\mathcal{B} \qquad (3.74)$$

be a principal-bundle morphism. Among other things (since fibres are mapped to fibres), this implies the existence of a well defined map:

$$\phi : \mathcal{U} \longrightarrow \phi(\mathcal{U}) = \mathcal{V} \subset \mathcal{B}. \qquad (3.75)$$

Thus, $\Phi$ is a principal-bundle isomorphism between $\pi_R^{-1}(\mathcal{U})$ and $\pi^{-1}(\mathcal{V})$, where $\pi_R$ and $\pi$ denote, respectively, the bundle projections of $F\mathbb{R}^n$ and $F\mathcal{B}$. Clearly, $e$ belongs to $\pi_R^{-1}(\mathcal{U})$ (see Figure 3.3).

Let $(b, f) = \Phi(0, e)$. We claim that the 1-jet:

$$g = j^1_{(O,e)} \Phi \qquad (3.76)$$

is a frame of $F\mathcal{B}$ at $(b, f)$. To convince ourselves that this statement is true, we need only realize that, since $F\mathbb{R}^n$ is a manifold of dimension $n+n^2$, there exist

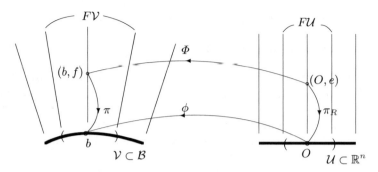

**Fig. 3.3.** A local principal-bundle isomorphism

local charts that map a neighbourhood of $(O, e)$ diffeomorphically to a neighbourhood of the origin of $\mathbb{R}^{n+n^2}$, and that one such chart can be canonically constructed on the basis of the standard coordinates in $\mathbb{R}^n$. Consequently, the local principal-bundle isomorphism $\Phi$ can also be viewed as a local isomorphism between an open set in $\mathbb{R}^{n+n^2}$ containing the origin, namely the set $\pi^{-1}(\mathcal{U}) \subset \mathbb{R}^{n+n^2}$, and a neighbourhood of $(b, f)$. Therefore, $g$ is indeed an element of $FF\mathcal{B}$, albeit not an arbitrary element, since $\Phi$ is a special map, a bundle morphism, that respects the bundle structure of the manifolds involved. This kind of frame of $F\mathcal{B}$ is called a *non-holonomic frame (of the second order)* of $\mathcal{B}$ at $b$.

The collection $\bar{F}^2\mathcal{B}$ of all non-holonomic frames of a manifold $\mathcal{B}$ has two different canonical principal-bundle structures[6], corresponding, respectively, to the following two canonical projections:

$$\bar{\pi}^2 : \bar{F}^2\mathcal{B} \longrightarrow \mathcal{B}$$
$$j^1_{(O,e)}\Phi \mapsto \phi(O) = \pi \circ \Phi(O, e), \qquad (3.77)$$

and

$$\bar{\pi}^2_1 : \bar{F}^2\mathcal{B} \longrightarrow F\mathcal{B}$$
$$j^1_{(O,e)}\Phi \mapsto \Phi(O, e), \qquad (3.78)$$

We will denote the structure groups of these two bundles $\bar{G}^2(n)$ and $\bar{G}^2_1(n)$, respectively. The group operation in each case can be retrieved by composition of the appropriate jets. We will now give a useful local-coordinate description of each of these bundles and of the corresponding group elements. As before, we denote the natural coordinates of $\mathbb{R}^n$ by $r^\alpha$ ($\alpha = 1, ..., n$) and a system of local coordinates in $\mathcal{B}$ by $X^I$ ($I = 1, ..., n$). The coordinates induced by these systems in the frame bundles $F\mathbb{R}^n$ and $F\mathcal{B}$ are denoted, respectively, by $s^\alpha_\beta$

---

[6] There exists, in fact, a third canonical principal-bundle structure corresponding to the "flipping" around of the two linear maps $f^I_\alpha$ and $g^I_\alpha$.

and $f^I_\beta$. All the considerations that follow are strictly local, that is, they are valid only in the respective coordinate patches. The coordinate patch of $F\mathbb{R}^n$ is assumed to contain the canonical frame $(O, e)$. An arbitrary differentiable map $\Phi$, not necessarily a principal-bundle morphism, from $F\mathbb{R}^n$ to $F\mathcal{B}$:

$$\Psi : F\mathbb{R}^n \longrightarrow F\mathcal{B}, \tag{3.79}$$

is represented in the given local coordinates by $n+n^2$ differentiable functions, as follows:

$$X^I = X^I(s^\alpha, s^\alpha_\beta), \tag{3.80}$$

and

$$f^I_\gamma = f^I_\gamma(s^\alpha, s^\alpha_\beta). \tag{3.81}$$

These functions, as given, do not necessarily map fibres into fibres, let alone linearly. It follows, therefore, that the 1-jet $j^1_{O,e}\Phi$ should produce an arbitrary element of $FF\mathcal{B}$. Let us verify that this is indeed the case, since we have constructed the elements of $FF\mathcal{B}$ in a different way. Recall that to obtain a coordinate representation of the 1-jet of a smooth function between manifolds, we just need to supplement the value of the function itself at the source point with the values of all the first derivatives of the coordinate expressions of the function at that point. That being the case, the 1-jet $g$ of $\Phi$ is given in coordinates by the following quantities:

$$\{X^I|_{(0,\delta^\alpha_\beta)},\ f^I_\gamma|_{(0,\delta^\alpha_\beta)},\ \left[\frac{\partial X^I}{\partial s^\gamma}\right]_{(0,\delta^\alpha_\beta)},\ \left[\frac{\partial X^I}{\partial s^\gamma_\rho}\right]_{(0,\delta^\alpha_\beta)},\ \left[\frac{\partial f^I_\sigma}{\partial s^\gamma}\right]_{(0,\delta^\alpha_\beta)},\ \left[\frac{\partial f^I_\sigma}{\partial s^\gamma_\rho}\right]_{(0,\delta^\alpha_\beta)}\} \tag{3.82}$$

A comparison with Equation (3.68) reveals that, except for notation, the results are identical.

Let us now, however, restrict the function to a local principal-bundle isomorphism, as we have required in the definition of a non-holonomic frame, namely, an element of $\bar{F}^2\mathcal{B}$. In this case, the coordinate expressions (3.80) and (3.81) are to be replaced by:

$$X^I = X^I(s^\alpha), \tag{3.83}$$

and

$$f^I_\gamma = \hat{f}^I_\rho(s^\alpha)\, s^\rho_\gamma. \tag{3.84}$$

Equation (3.84) is nothing but the coordinate representation of the subordinate function $\phi : \mathbb{R}^n \longrightarrow \mathcal{B}$. The fact that the argument $s^\alpha_\beta$ does not appear is precisely the mathematical expression of this fact. Equation (3.75), on the other hand, is the mathematical expression of two properties required by a principal-bundle morphism: that fibre are mapped into fibres and that the fibre maps preserve the right action of the structure group, $GL(n;\mathbb{R})$. For this particular form of map $\Phi$, Equation (3.82) becomes:

88     3 Uniformity of Cosserat media

$$g \mapsto \{X^I, f^I_\gamma, \left[\frac{\partial X^I}{\partial s^\gamma}\right], 0, \left[\frac{\partial \hat{f}^I_\sigma}{\partial s^\gamma}\right], f^I_\gamma \delta^\rho_\sigma\}. \tag{3.85}$$

Since the last entry contains the same information as the second, we will express a generic element of $\bar{F}^2\mathcal{B}$, namely, a non-holonomic frame, in coordinates simply as:

$$g \mapsto \{X^I, f^I_\gamma, g^I_\gamma, g^I_{\sigma\alpha}\}. \tag{3.86}$$

It should not be surprising by now that this expression is identical in form with a material implant operation. In other words, an implant results again in the specification of a frame (of the appropriate type) at a point.

We have pointed out that the total space $\bar{F}^2\mathcal{B}$ has two different natural structures as a principal bundle, with base manifolds $\mathcal{B}$ and $F\mathcal{B}$, respectively. In coordinates, the corresponding projections are given by:

$$\pi^2 : \bar{F}^2\mathcal{B} \longrightarrow \mathcal{B}$$
$$\{X^I, f^I_\gamma, g^I_\gamma, g^I_{\sigma\alpha}\} \mapsto \{X^I\}, \tag{3.87}$$

and

$$\pi^2_1 : \bar{F}^2\mathcal{B} \longrightarrow \mathcal{B}$$
$$\{X^I, f^I_\gamma, g^I_\gamma, g^I_{\sigma\alpha}\} \mapsto \{X^I, f^I_\gamma\}. \tag{3.88}$$

We still need to disclose the nature of the structure groups of these two principal bundles. We will denote these groups by $\bar{G}^2(n)$ and $\bar{G}^2_1(n)$, respectively. Since these are principal bundles, the structure group looks just like a fibre. Accordingly, an element $A$ of $\bar{G}^2(n)$ is given by:

$$A = \{A^\alpha_\beta, a^\alpha_\beta, a^\alpha_{\beta\gamma}\}, \tag{3.89}$$

while a typical element $a$ of the group $\bar{G}^2_1(n)$ looks as follows:

$$a = \{a^\alpha_\beta, a^\alpha_{\beta\gamma}\}. \tag{3.90}$$

These groups consist, in fact, of 1-jets of local principal-bundle automorphisms of $F\mathbb{R}^n$ evaluated at $(O, e)$. To reveal the group operation, we need just to compose two such automorphisms and calculate the 1-jet of the composite map in terms of the 1-jets of the factors. The only difference between these groups is that in the case of $\bar{G}^2(n)$ we require that the automorphisms preserve the origin $O \in \mathbb{R}^n$, while in the case of $\bar{G}^2_1(n)$ we require that they also preserve the standard frame $e$ at $O$. The results are

$$(AB) = \{(AB)^\alpha_\beta, (ab)^\alpha_\beta, (ab)^\alpha_{\beta\gamma}\} = \{A^\alpha_\rho B^\rho_\beta, a^\alpha_\rho b^\rho_\beta, A^\alpha_\rho b^\rho_{\beta\gamma} + a^\alpha_{\rho\sigma} B^\rho_\beta b^\sigma_\gamma\}. \tag{3.91}$$

and

$$(ab) = \{(ab)^\alpha_\beta, (ab)^\alpha_{\beta\gamma}\} = \{a^\alpha_\rho b^\rho_\beta, b^\alpha_{\beta\gamma} + a^\alpha_{\beta\sigma} b^\sigma_\gamma\}. \tag{3.92}$$

*Remark 3.5.* As a constructive exercise, one may want to obtain the composition laws (3.91) and (3.92) by following the procedure outlined, namely, the compositions of jets of local automorphisms of $F\mathbb{R}^n$.

## The bundle of semi-holonomic frames $\hat{F}^2\mathcal{B}$

The passage from the iterated frame bundle $FF\mathcal{B}$ to the bundle of non-holonomic frames $\bar{F}\mathcal{B}$ was achieved by restricting the class of admissible local diffeomorphisms. Indeed, the general local diffeomorphisms between $\mathbb{R}^{n+n^2}$ and $F\mathcal{B}$ as mere differentiable manifolds were restricted to those local diffeomorphisms that preserve the bundle structure, namely, local principal-bundle morphisms. We can restrict this class even further and thus obtain sub-bundles of the bundle of non-holonomic frames of potential interest in applications.

We have already remarked that, since it maps fibres to fibres, a bundle morphism $\Phi$ implies the existence of a uniquely defined subordinate diffeomorphism $\phi$ between the base manifolds, as depicted in Figure 3.3. The converse is not true in general, as far as uniqueness is concerned. Given a local diffeomorphism $\phi$ between the base manifolds of two fibre bundles, there exist many local bundle morphisms $\Phi$ whose subordinate map is the given diffeomorphism. In the case of the principal bundle of frames (or its associated bundles), however, it is possible to select canonically, among all the possible local principal-bundle morphisms that subsume $\phi$, a particular principal-bundle isomorphism, which we denote by $F\phi$, that can be seen as the "lift" of $\phi$ by differentiation. We have encountered this kind of lift in other contexts. If $\phi : \mathcal{B} \longrightarrow \mathcal{C}$ is a diffeomorphisms between the manifolds $\mathcal{B}$ and $\mathcal{C}$, then the principal-bundle isomorphism $F\phi : F\mathcal{B} \longrightarrow F\mathcal{C}$ is defined by:

$$F\phi(j_O^1\beta) = j_O^1(\phi \circ \beta). \tag{3.93}$$

Here, $\beta$ is a local diffeomorphism from an open neighbourhood of the origin of $\mathbb{R}^n$ to an open neighbourhood in $\mathcal{B}$. Its 1-jet, therefore, is appropriately a (first-order) frame at $\beta(O) \in \mathcal{B}$. Clearly, all elements of $F\mathcal{B}$ can be obtained in this way. The frame assigned by $F\phi$ to $j_O^1\beta$ is obtained by composing a representative (such as $\beta$) of the source frame with $\phi$ and then calculating its 1-jet. The result is the same as saying that, since the derivative map $\phi_*$ of $\phi$ maps tangent vectors into tangent vectors, $F\phi$ applied to any basis in $\mathcal{B}$ produces the basis in $\mathcal{C}$ consisting of the images by $\phi_*$ of the vectors of the source basis.

The first way to further restrict the principal-bundle isomorphisms used to construct a non-holonomic basis consists of requiring that given $\Phi$, its value at $(O, e)$ (but not necessarily elsewhere) coincide with the value of $F\phi$ at $(O, e)$, namely:

$$\Phi(O, e) = (F\phi)(O, e). \tag{3.94}$$

A non-holonomic frame obtained as the 1-jet of a local principal-bundle morphism satisfying this condition is called a *semi-holonomic frame*. Since the canonical frame $e$ of $\mathbb{R}^n$ is obtained by setting $\beta = id_{\mathbb{R}^n}$, Equation (3.93) implies that a frame is semi-holonomic if, and only if, the local fibre-bundle isomorphism $\Phi$ used to define it satisfies:

$$\Phi(O, e) = j_O^1\phi. \tag{3.95}$$

The collection $\hat{F}^2\mathcal{B}$ of all semi-holonomic frames at all points of a manifold $\mathcal{B}$ has two canonical structures as a principal bundle, one over $\mathcal{B}$ and the other over $F\mathcal{B}$, just as in the case of the more general non-holonomic frames. In terms of components and group actions, these bundles are obtained from their non-holonomic counterparts by setting $f^i_\alpha = g^i_\alpha$ in Equation (3.86). That is, a semi-holonomic frame $h$ is given in coordinates by:

$$h \mapsto \{X^I, h^I_\gamma, h^I_{\sigma\alpha}\}. \tag{3.96}$$

Similar identifications apply to the group elements and their respective operations. Semi-holonomic frames are not as useful as one might expect in Continuum Mechanics applications. The reason for this is that semi-holonomicity of a frame may be lost under a change of reference configuration and, therefore, it does not seem to convey an intrinsic material property.

*Remark 3.6.* It appears that a semi-holonomic Cosserat medium can have a legitimate physical meaning only if the macromedium has no bearing on the constitutive response.

### The bundle of holonomic frames $F^2\mathcal{B}$

The final, and strongest, restriction that we can impose on the local principal-bundle morphisms to generate frames consists of imposing the agreement of $\Phi$ with $F\phi$ not just at $(O, e)$, but everywhere in an open neighbourhood of this point of $F\mathbb{R}^n$. In so doing, we are identifying those elements of $\bar{F}^2(n)$ that happen to be 2-jets of a local diffeomorphism between $\mathbb{R}^n$ and $\mathcal{B}$. In short, we recover the concept of second-order frame-bundle $F^2\mathcal{B}$, fully dealt with in Chapter 2.

### 3.5.3 The Cosserat material $G$-structures

A Cosserat archetype is a non-holonomic second-order frame at a point of a uniform Cosserat body. Its material symmetries constitute a subgroup $\bar{\mathcal{G}}$ of $\bar{G}^2(3)$. The collection of material implants from the archetype to a point $\mathbf{X} \in \mathcal{B}$ constitutes a subset $\mathcal{P}_X$ of the set of all possible non-holonomic frames as $\mathbf{X}$. That is, at each point of $\mathcal{B}$, the archetype is mapped by all possible material isomorphisms to only a portion of the fibre (at that point) of the non-holonomic frame bundle $\bar{F}^2\mathcal{B}$. This portion is governed by the symmetry group of the archetype. In other words, the collection of all the sub-fibres $\mathcal{P}_X$ is closed under the right action of $\bar{\mathcal{G}} \subset \bar{G}^2(3)$. We have, therefore, a *reduction* of the principal bundle of non-holonomic frames $\bar{F}^2\mathcal{B}$ to another principal bundle with structure group $\bar{\mathcal{G}} \subset \bar{G}^2(3)$. This object is a (second-order) *non-holonomic G-structure*. Because it is completely determined by the material uniformity property of the body we call it a *Cosserat material G-structure*. A different archetype results in a different $G$-structure, but these different $G$-structures are conjugate of each other.

### 3.5.4 The Cosserat material groupoid

Given the principal frame bundle $F\mathcal{B}$ and two points $\mathbf{X}, \mathbf{Y} \in \mathcal{B}$, we may consider the collection of the 1-jets of all the local principal-bundle isomorphisms between a neighbourhood of $\mathbf{X}$ and a neighbourhood of $\mathbf{Y}$. In this way, by repeating this process for all pairs of points of $\mathcal{B}$, we construct a groupoid $\bar{\Pi}^2\mathcal{B}$, whose structure group is $\bar{G}^2(3)$. Given a constitutive law for a Cosserat body, we consider the subgroupoid of $\bar{\Pi}^2\mathcal{B}$ obtained by considering only those local principal-bundle isomorphisms whose 1-jets happen to be material isomorphisms. This subgroupoid is the *Cosserat material groupoid* of $\mathcal{B}$. If it is a transitive groupoid, then the body is materially uniform. If, in addition, this transitive groupoid is a Lie groupoid, we recover the notion of smooth material uniformity.

## 3.6 Homogeneity, flatness and integrable prolongations

In trying to characterize material homogeneity in terms of the notion of a property of a material $G$-structure, we found that, for the case of simple materials, local homogeneity is equivalent to the flatness of one, and hence any, of the conjugate material $G$-structures of the body. The case of second-grade bodies, however, necessitated some extra care, since, as we have discussed in Chapter 2, the flatness of one second-order $G$-structure does not imply the flatness of its conjugate $G$-structures. In resolving this question, we noted that one may interpret the flatness of a specific $G$-structure as the requirement that there exist a reference configuration such that all the points in a neighbourhood be in a specific state, namely, that of the chosen material archetype. Thus, the notion of flatness of a given $G$-structure was given a clear physical meaning. The flatness of the material groupoid, on the other hand, was found to be equivalent to the flatness of some material $G$-structure, a case that, again, has a clear physical meaning, namely, that there be a reference configuration such that all the points in a neighbourhood are in the same state, not necessarily a pre-specified one.

For the case of Cosserat bodies, of which second-grade bodies are a particular case, we naturally expect to have a repetition of the appearance of these two somewhat different notions of local homogeneity: with respect to a particular archetype (for instance, a stress-free one) or with respect to some archetype. And this is indeed the case. It turns out, however, that the situation in the case of Cosserat bodies is more complicated. Indeed, as we will presently show, the notion of flatness (or integrability) of a material $G$-structure is far too strong to characterize material homogeneity of any kind. When working in terms of components (as we have done in Sections 3.3 and 3.4), these geometrical subtleties are obscured by the mechanistic manipulation of the symbols. We happened to observe, for instance, that some indexed quantities vary, under coordinate transformation, as Christoffel symbols of a

## 3 Uniformity of Cosserat media

linear connection. We then proceeded to subtract them from another set of Christoffel symbols, thus obtaining tensorial quantities, and so on. Our aim in this section is to provide a somewhat more geometrical justification of the validity of such operations. We will see how a local section of the bundle of non-holonomic frames actually generates a linear connection by inducing a horizontal distribution in the frame bundle. We will then construct the notion of integrable prolongation and proceed to show how this idea must be invoked if one wants to obtain a property of a $G$-structure that reflects faithfully the physical concept of local homogeneity.

### 3.6.1 Sections of $\bar{F}^2\mathcal{B}$

The space of non-holonomic frames $\bar{F}^2\mathcal{B}$ acts as the total space for two different canonical principal bundles, as described in Equations (3.77) and (3.78) or, equivalently, in Equations (3.87) and (3.88), with projections $\bar{\pi}^2$ and $\bar{\pi}_1^2$, respectively. Recall that a second-order non-holonomic frame (being the 2-jet of a local diffeomorphism) contains canonically both a point at the base manifold and a linear frame therein. Accordingly, a section $\sigma : \mathcal{B} \longrightarrow \bar{F}^2\mathcal{B}$ of the first bundle, $(\bar{F}^2\mathcal{B}, \bar{\pi}^2)$, assigns to each point of $\mathcal{B}$ both a linear (i.e., first-order) frame and a (second-order) non-holonomic frame. On the other hand, a section $\gamma : F\mathcal{B} \longrightarrow \bar{F}^2\mathcal{B}$ of the second bundle, $(\bar{F}^2\mathcal{B}, \bar{\pi}_1^2)$, assigns to each point of $\mathcal{B}$ and to each linear frame therein a second-order non-holonomic frame. In this sense, we may somewhat imprecisely say that the section $\sigma$ is a "smaller" object than the section $\gamma$, and we ask the question as to whether there exists a canonical way to extend the former section so as to obtain a section of the bundle $(\bar{F}^2\mathcal{B}, \bar{\pi}_1^2)$.

Let us represent the section $\sigma$ in the following way:

$$b \mapsto \sigma(b) = \{b, f, g\}, \qquad (3.97)$$

where $b \in \mathcal{B}$, $f$ is a linear frame at $b$, and $g$ is the second-order non-holonomic frame proper. Let $\hat{f}$ be another linear frame at $b$. Since the general linear group $GL(n; \mathbb{R})$ acts transitively on $F\mathcal{B}$, there exists a unique $a \in GL(n; \mathbb{R})$ such that $\hat{f} = fa$. Now, the section $\sigma$ says nothing about assigning a non-holonomic frame to $\{b, \hat{f}\}$, and we are looking for a canonical way to do precisely that. Recall, that the structure group of $(\bar{F}^2\mathcal{B}, \bar{\pi}^2)$ is $\bar{G}^2(n)$. Therefore, if we only could, starting from the element $a \in GL(n\mathbb{R})$, canonically construct an element $A \in \bar{G}^2(n)$ consistent with it, we would be done! Indeed, we would simply have to act with $A$ to the right on $\{b, f, g\}$. Such an element does exist. It is given by the canonical inclusion:

$$\begin{aligned} j : GL(n; \mathbb{R}) &\longrightarrow \bar{G}^2(n) \\ a &\mapsto A = j(a) = \{a, a, 0\}. \end{aligned} \qquad (3.98)$$

More specifically, if the section $\sigma$ assigns to $b \in \mathcal{B}$ the element $\{b, f, g\}$, then the induced section $\gamma_\sigma : F\mathcal{B} \longrightarrow \bar{F}^2\mathcal{B}$ assigns to the pair $\{b, fa\}$ the element $\{b, \{f, g\}A\}$. In components, we have that if:

### 3.6 Homogeneity, flatness and integrable prolongations

$$\sigma(X^I) = \{X^I, f^I_\alpha, g^I_\alpha, g^I_{\alpha\beta}\}, \tag{3.99}$$

then:

$$\gamma_\sigma(\{X^I, f^I_\rho a^\rho_\alpha\}) = \{X^I,\ f^I_\rho a^\rho_\alpha,\ g^I_\rho a^\rho_\alpha,\ g^I_{\rho\tau} a^\rho_\alpha a^\tau_\beta\}. \tag{3.100}$$

In this way, we have extended the section $\sigma$ of $(\bar{F}^2\mathcal{B}, \hat{\pi}^2)$ to a section $\gamma_\sigma$ of $(\bar{F}^2\mathcal{B}\hat{\pi}_1^2))$. This section, however, is of a very special kind, known as an *invariant section*, namely a section with the property that:

$$\gamma(za) = \gamma(z)j(a), \quad \forall\, z = \{b, f\} \in F\mathcal{B}. \tag{3.101}$$

### 3.6.2 Invariant sections and linear connections

In section 3.3.4 we noted casually that the indexed quantities in (3.38), constructed as a combination of components of a given uniformity field of a Cosserat body, transform like Christoffel symbols under coordinate transformations. From this observation we can perhaps infer that there is a linear connection at play, but we have not shown what that connection is. We want now to demonstrate how a smooth local invariant section of $(\bar{F}^2\mathcal{B}, \bar{\pi}_1^2)$, such as the one generated by a Cosserat uniformity field, indeed gives rise to a local horizontal distribution[7] in the linear frame bundle $F\mathcal{B}$.

Let $\gamma$ be an invariant section, namely, a section of $(\bar{F}^2\mathcal{B}, \bar{\pi}_1^2)$ satisfying Equation (3.101). Therefore, at each point $z \in F\mathcal{B}$ we have some local diffeomorphism:

$$\Phi: F\mathbb{R}^n \longrightarrow F\mathcal{B}, \tag{3.102}$$

with $\Phi(O, e) = z$ and such that

$$\gamma(z) = j^1_{(O,e)}\Phi. \tag{3.103}$$

This map being a principal-bundle isomorphism whose underlying group isomorphism is assumed to be the identity, it must satisfy the following condition:

$$\Psi(za) = \Psi(z)a \quad \forall\, a \in GL(n;\mathbb{R}). \tag{3.104}$$

What this means is that the map between the fibre at $r \in \mathbb{R}^n$ and the corresponding fibre at $\psi(r) \in \mathcal{B}$ is completely characterized by the value $f(r) = \Psi(r, e)$. We have thus a well-defined smooth map:

$$f: \mathbb{R}^n \longrightarrow F\mathcal{B}. \tag{3.105}$$

The derivative of this map at $O \in \mathbb{R}^n$:

$$f_*(O): T_O\mathbb{R}^n \longrightarrow T_{f(O)}F\mathcal{B}, \tag{3.106}$$

being of rank $n$, has as a range an $n$-dimensional subspace of $T_zF\mathcal{B}$, where $z = f(O) = \psi(O)$. By construction, this subspace is not vertical. It is, therefore, a

---
[7] As described in Chapter 9.

*horizontal space* at $z$, which we denote by $H_z$. It is clear that this construction of a (local) horizontal distribution out of the information contained in a (local) section $\gamma$ does not depend on $\gamma$ being an invariant section. The invariant character of the section, however, needs to be invoked in order to prove that this horizontal distribution is consistent with the group action on $F\mathcal{B}$, a proof that is left to the reader. As a consequence, the horizontal distribution just obtained can be interpreted as a (local) linear connection.

*Remark 3.7.* As a useful exercise, one can write the above steps in terms of a coordinate chart, obtain the Christoffel symbols of the connection and then compare the result with Equation (3.38).

### 3.6.3 Prolongations

Recall that a smooth (local) section of the frame bundle $F\mathcal{B}$ can be regarded as a (local) parallelism, which we will now refer to as a (local) *linear parallelism*. By extension, a smooth (local) section of the second-order non-holonomic frame bundle $(\bar{F}^2\mathcal{B}, \bar{\pi}^2)$ will be called a (local) *second-order non-holonomic parallelism*. It establishes a choice of a particular second-order non-holonomic frame at each point of its domain. We will use the terminology of sections interchangeably with that of parallelisms, according to context.

Given a smooth non-holonomic parallelism on $\mathcal{U} \subset \mathcal{B}$, that is, a section of $(\bar{F}^2\mathcal{U}, \bar{\pi}^2)$:

$$\sigma : \mathcal{U} \longrightarrow \bar{F}^2\mathcal{B}, \qquad (3.107)$$

we obtain canonically two, in general different, smooth linear parallelisms on $\mathcal{U}$. The first one is obtained via the canonical projection $\bar{\pi}_1^2 : \bar{F}^2\mathcal{B} \longrightarrow F\mathcal{B}$ in terms of the following section of $F\mathcal{B}$:

$$q = \bar{\pi}_1^2 \circ \sigma. \qquad (3.108)$$

To obtain the second canonical linear parallelism on $\mathcal{U}$ induced by $\sigma$, we note that there exists also another canonical projection:

$$\tilde{\pi}_1^2 : \bar{F}^2\mathcal{B} \longrightarrow F\mathcal{B}, \qquad (3.109)$$

defined as:

$$\tilde{\pi}_1^2(j^1_{(O,e)}\Psi) = j^1_O\psi. \qquad (3.110)$$

Using this projection, we define the following section of $F\mathcal{U}$:

$$p = \tilde{\pi}_1^2 \circ \sigma. \qquad (3.111)$$

In terms of components in a local coordinates system, if $\sigma$ is given by:

$$\sigma(X^I) = \{X^J, q^I_\alpha(X^J), p^I_\alpha(X^J), r^I_{\alpha\beta}(X^J)\}, \qquad (3.112)$$

then the linear parallelisms $q$ and $p$ are represented, respectively, by:

$$q(X^I) = \{X^J,\ q^I_\alpha(X^J)\}, \qquad (3.113)$$

and

$$p(X^I) = \{X^J,\ p^I_\alpha(X^J)\}. \qquad (3.114)$$

We now ask the inverse question: given two linear parallelisms $q$ and $p$, is it possible to construct a non-holonomic parallelism $\sigma$, such that $q$ and $p$ are induced by $\sigma$ in the manner just shown? The objective is to construct at each point of $\mathcal{U}$ a second-order non-holonomic frame using just the information contained in $q$ and $p$. Recall that a a non-holonomic frame is primarily a basis of a tangent space of $F\mathcal{B}$, though not a general basis (as we have seen in Section 3.5.2), but one that is completely determined by a basis of an $n$-dimensional horizontal subspace. This is due to the fact that in constructing a non-holonomic frame we use the 1-jet of a local frame-bundle isomorphisms (rather than an arbitrary diffeomorphism). Consider the derivative of the map $q$ at a point $b \in \mathcal{U}$, namely:

$$q_*(b) : T_b\mathcal{U} \longrightarrow T_{q(b)}F\mathcal{B}. \qquad (3.115)$$

The section $p$, on the other hand, assigns at point $b$ a linear frame, which can be regarded as a basis $\mathbf{p}_\alpha$ ($\alpha = 1, ..., n$) of the tangent space $T_b\mathcal{U}$. Therefore, applying the map $q_*(b)$ to each vector of this basis, we obtain a basis $q_*(b)[\mathbf{p}_\alpha]$ ($\alpha = 1, ..., n$) for an $n$-dimensional horizontal subspace of $T_{q(b)}F\mathcal{B}$. In this way, we have constructed the desired section $\sigma$. To see how this looks in a coordinate chart $X^I$ ($I = 1, ..., n$), we note that the general coordinate expression of the derivative of Equation (3.113), evaluated at a point with coordinates $X^I$ and at a vector with components $a^I$ in the natural coordinate basis is:

$$q_*(X^I, a^I) = \{X^J,\ q^J_\alpha,\ a^J,\ \frac{\partial q^J_\alpha}{\partial X^K} a^K\}. \qquad (3.116)$$

Applying this expression to each of the vectors $\mathbf{p}_\alpha$ (with components $p^i{}_\alpha$, we obtain the non-holonomic section $\sigma$ as:

$$\sigma(X^I) = \{X^J,\ q^J_\alpha,\ p^J_\alpha,\ \frac{\partial q^J_\alpha}{\partial X^K} p^K_\beta\}. \qquad (3.117)$$

We call this non-holonomic parallelism a *prolongation*[8] of the linear parallelisms $q$ and $p$. A non-holonomic parallelism $\sigma(X^I) = \{X^J,\ q^J_\alpha,\ p^J_\alpha,\ r^I_{\alpha\beta}\}$ is a prolongation if, and only if:

$$r^I_{\alpha\beta} = \frac{\partial q^I_\alpha}{\partial X^K} p^K_\beta. \qquad (3.118)$$

According to Section 3.6.2, a non-holonomic parallelism, viewed as a section of $(\bar{F}^2\mathcal{B}, \bar{\pi}^2)$, can be canonically extended to an invariant section of

---

[8] A more thorough treatment of prolongations can be found in Chapter 10.

$(\bar{F}^2\mathcal{B}, \bar{\pi}_1^2)$, thus giving rise to a (local) linear connection $\Lambda$ on $\mathcal{U}$. In a coordinate system the Christoffel symbols of this connection are (see Remark 3.7):

$$\Lambda^I_{JK} = -r^I{}_{\alpha\beta}\, q^{-\alpha}_J\, p^{-\beta}_K. \tag{3.119}$$

In particular, if $\sigma$ happens to be a prolongation of the linear parallelisms $q$ and $p$, we have:

$$\Lambda^I_{JK} = -\frac{\partial q^I_\alpha}{\partial X^K}\, q^{-\alpha}_J. \tag{3.120}$$

Moreover, the section $\gamma$ (whether or not a prolongation) defines two (local) linear parallelisms $q$ and $p$ and, hence, two curvature-free local connections whose Christoffel symbols are, respectively:

$$\Delta^I_{JK} = -q^I{}_{\alpha,K}\, q^{-\alpha}_J, \tag{3.121}$$

and

$$\Gamma^I_{JK} = -p^I{}_{\alpha,K}\, p^{-\alpha}_J, \tag{3.122}$$

where commas are used for partial derivatives with respect to the coordinates.

A comparison of Equations (3.120) and (3.122) shows that a second-order non-holonomic parallelism is a prolongation if, and only if, the connections $\Delta$ and $\Lambda$ coincide. Equivalently, we may define the tensor:

$$D = \Delta - \Lambda, \tag{3.123}$$

and state that: A second-order non-holonomic parallelism is a prolongation if, and only if, the tensor $D$ vanishes identically.

A prolongation of $q$ and $p$ is said to be an *integrable prolongation* if the linear parallelism $p$ is integrable, namely, if the torsion of the $\Gamma$-connection vanishes identically. We may now assert that: A second-order non-holonomic parallelism is an integrable prolongation if, and only if, both the tensor $D$ and the torsion $\tau$ of the connection $\Gamma$ vanish identically.

We can easily extend the notion of integrable prolongation to a $G$-structure obtained as a reduction of the bundle $\bar{F}^2\mathcal{B}$ to a subgroup of $\bar{G}^2(n)$. We say that such a $G$-structure is an integrable prolongation if for every point of $\mathcal{B}$ there exists an adapted local section which is a non-holonomic integrable prolongation. As for the case of second-grade bodies, specific instances of homogeneity conditions for Cosserat media can be found in [39].

*Remark 3.8.* Beyond Cosserat media, there exist many other frameworks in which similar ideas can be used to develop specialized theories. Some instances are provided by structural beams and shells [90, 5, 38, 40], liquid crystals [42], electromagnetic deformable media [69], general relativity [34] and non-local media [36, 33].

# 4

# Functionally graded bodies

A functionally graded body (FGB) is, by definition and by actual industrial realization, a non-uniform body. Its material properties are intentionally made to vary smoothly from point to point. Can it, however, be homogeneous? The surprisingly positive answer to this question can be obtained by extending the notion of material isomorphism [41] so as to allow for the material comparison of points made of different materials as long as they have the same type of material symmetry. A common production technique of functionally graded bodies consists of embedding, in an underlying isotropic solid matrix, smoothly varying quantities of randomly oriented metal powder, thus preserving the isotropy of the matrix, albeit changing its elastic constants. Thus, a basis for material comparison of distant points is established and the theory of continuous distributions of dislocations in functionally graded bodies can get off the ground and fly.

## 4.1 The extended notion of material isomorphism

The notion of material isomorphism was introduced in Chapter 1 to represent the physical idea of material identity between two points of a body. The main mathematical construct to emerge from this notion manifested itself in the existence of certain non-singular linear maps between the tangent spaces of the two points involved. The availability of these maps was controlled exclusively by the material symmetry group of either point. When we reflect upon the derivation of the notions of material parallelism, material connection, dislocation density, material groupoid and material G-structure, we realize that nothing else beyond the existence of these linear maps and their smooth dependence on position was needed. In other words, even if a body is not materially uniform, as long as there exists a rational basis for establishing physically meaningful non-singular linear maps between different points, the possibility exists of replicating the same geometrical constructions and attribute to them similar physical meaning as in the case of uniform bodies. To highlight

98     4 Functionally graded bodies

the similarities, we will adopt a consistent terminology that is reminiscent of, though not identical to, that of uniform bodies. For specificity, we will deal only with simple (i.e., first-grade) bodies.

We say that two points, $\mathbf{X}_1$ and $\mathbf{X}_2$, of a body $\mathcal{B}$ are *symmetry-isomorphic* if their material symmetry groups, respectively $\mathcal{G}_1$ and $\mathcal{G}_2$, are conjugate to each other. Any linear map $\mathbf{A}_{12}$ between their tangent spaces, viz.:

$$\mathbf{A}_{12} : T_{\mathbf{X}_1}\mathcal{B} \longrightarrow T_{\mathbf{X}_1}\mathcal{B}, \tag{4.1}$$

such that:

$$\mathcal{G}_2 = \mathbf{A}_{12}\, \mathcal{G}_1\, \mathbf{A}_{12}^{-1}, \tag{4.2}$$

is, accordingly, called a *symmetry-isomorphism* from $\mathbf{X}_1$ (the *source*) to $\mathbf{X}_2$ (the *target*). We will denote by $\mathcal{A}_{12}$ the collection of all symmetry-isomorphisms between $\mathbf{X}_1$ and $\mathbf{X}_2$. Physically speaking, a symmetry isomorphism represents how a small (first-order) neighbourhood of one point is to be deformed so that its symmetry group coincides with that of another point. Notice that the two points need not be made of the same material, but only have the same kind of symmetry (isotropy, transverse isotropy, orthotropy, and so on).

A material body $\mathcal{B}$ is said to be *unisymmetric* if all its points are mutually symmetry-isomorphic. If the symmetry-isomorphisms can be chosen smoothly in a neighbourhood of each point, the body is said to be *smoothly unisymmetric*. The notions of symmetry-isomorphism and unisymmetry are the counterparts, respectively, of the notions of material isomorphism and uniformity introduced in Chapter 1.

## 4.2 Non-uniqueness of symmetry isomorphisms

We say that a symmetry-isomorphism is a *symmetry-automorphism* if its source and target coincide. The collection of all symmetry automorphisms of a point $\mathbf{X}$ is denoted by $\mathcal{A}_X$. How big is this set? To answer this question, we rewrite the defining condition (4.2), identifying now the source and target, as:

$$\mathcal{G}_X\, \mathbf{A} = \mathbf{A}\, \mathcal{G}_X. \tag{4.3}$$

This last equation is a necessary and sufficient condition for a non-singular linear map $\mathbf{A} : T_X\mathcal{B} \longrightarrow T_X\mathcal{B}$ to belong to $\mathcal{A}_X$. We conclude that the set $\mathcal{A}$ consists exactly of those automorphisms (of the tangent space at $\mathbf{X}$) that commute with the symmetry group.

In group theory, given a subgroup $\mathcal{G}$ of a group $\mathcal{H}$, the set of all the elements of $\mathcal{H}$ that commute with $\mathcal{G}$ constitutes the *normalizer* of $\mathcal{G}$ in $\mathcal{H}$, denoted by $\mathcal{N}(\mathcal{G})$. We can, therefore, say that the set $\mathcal{A}_X$ of symmetry-automorphisms at a point $\mathbf{X} \in \mathcal{B}$ is nothing but the normalizer $\mathcal{N}_X = \mathcal{N}(\mathcal{G}_X)$ of the symmetry group $\mathcal{G}_X$ within the general linear group $GL(T_X\mathcal{B})$ of transformations of the tangent space at **X**. We should be careful to interpret this definition, expressed by Equation (4.3), as stating that for each element $\mathbf{G} \in \mathcal{G}_X$ there exists some element $\mathbf{G}' \in \mathcal{G}_X$, not necessarily equal to **G**, such that $\mathbf{GA} = \mathbf{AG}'$. The more stringent requirement that $\mathbf{G} = \mathbf{G}'$ corresponds to the group theoretical notion of *centralizer*. The centralizer $\mathcal{C}(\mathcal{G})$ of a subgroup $\mathcal{G}$ of a group $\mathcal{H}$ is the collection of all the elements of $\mathcal{H}$ that commute *with every* element of $\mathcal{G}$. One can verify that both the normalizer $\mathcal{N}(\mathcal{G})$ and the centralizer $\mathcal{C}(\mathcal{G})$ of a subgroup $\mathcal{G} \subset \mathcal{H}$ are themselves subgroups of the original group $\mathcal{H}$. Moreover, both the subgroup $\mathcal{G}$ and its centralizer $\mathcal{C}(\mathcal{G})$ are subsets of the normalizer $\mathcal{N}(\mathcal{G})$.

We have just learned that the collection of symmetry-automorphisms at a point is in general larger than the symmetry group. Correspondingly, as one might expect, the degree of freedom in the choice of symmetry-isomorphisms between two points of a unisymmetric body is larger than its counterpart in the theory of uniform bodies. In a manner completely analogous to that used in Chapter 1, one can prove that the totality $\mathcal{A}_{12}$ of symmetry-isomorphisms between two points $\mathbf{X}_1$ and $\mathbf{X}_2$ is generated from any particular symmetry-isomorphism $\mathbf{A}_{12}$ by any of the expressions:

$$\mathcal{A}_{12} = \mathbf{A}_{12} \, \mathcal{N}_1 = \mathcal{N}_2 \, \mathbf{A}_{12} = \mathcal{N}_2 \, \mathbf{A}_{12} \, \mathcal{N}_1. \tag{4.4}$$

These equations are the counterparts of Equation (1.16) for uniform bodies.

*Remark 4.1.* A *spherical dilatation* is an automorphism $\mathbf{S} = \alpha \mathbf{I}$, where $\alpha$ is a positive constant and $\mathbf{I}$ is the identity transformation. Since $\mathbf{I}$ commutes with every group element, the normalizer $\mathcal{A}_X$ automatically contains all spherical dilatations at **X**.

*Example 4.2.* Let the symmetry group be trivial, namely: $\mathcal{G}_X = \{\mathbf{I}\}$. Then both the normalizer and the centralizer coincide with the general linear group. In particular: $\mathcal{A}_X = GL(T_X \mathcal{B})$, and $\mathcal{A}_{12}$ is the set of all non-singular maps between $T_{X_1}\mathcal{B}$ and $T_{X_2}\mathcal{B}$.

## 4.3 The material $N$-structure

Let a basis be chosen at some point $\mathbf{X}_0 \in \mathcal{B}$ of a smoothly unisymmetric body. Applying to this basis all the possible symmetry-isomorphisms from $\mathbf{X}_0$ to every point **X** in the body, we obtain at each point **X** a collection of frames, which we shall denote by $\mathcal{A}(\mathbf{X})$ (leaving the dependence on the initial choice of frame not explicitly indicated). Attaching to each **X** its corresponding set $\mathcal{A}(\mathbf{X})$ we obtain a fibre bundle $\mathcal{A}_B$ which we will name the *material fibre bundle of the unisymmetric body*. This construction is identical to the one that led us to the concept of a material fibre bundle $\mathcal{P}_B$ of a uniform body. Again, the fibre bundle $\mathcal{A}_B$ is a principal fibre bundle obtained as a reduction of the principal frame bundle to a subgroup of the general linear group.

The only difference between the present construction and that of Chapter 1 is that, whereas the reduction to $\mathcal{P}_B$ was governed by the symmetry group $\mathcal{G}$ of the material, in the case of $\mathcal{A}_B$ the reduction is governed by the normalizer $\mathcal{N}(\mathcal{G})$, a bigger subgroup of the general linear group. To emphasize this difference, we will refer to the unisymmetry G-structure as the *N-structure* of the functionally graded body $\mathcal{B}$.

Except for the difference, as just described, in their respective structure groups, all the geometrical constructions of Chapter 1 remain in place, mutatis mutandis. We are referring, in particular to the notions of local cross sections, their overlaps, the transition maps thereof, parallelisms, their associated linear connections and, finally, the idea of a transitive groupoid associated with the unisymmetric nature of the body.

## 4.4 Homosymmetry

A smooth local section of the material $N$-structure has, as before, the meaning of a smooth frame field over an open set $\mathcal{U} \subset \mathcal{B}$. Since the passage from the frame at one point to the frame at another is achieved now by a symmetry-diffeomorphism between these points, we will call such a local frame field a *unisymmetry field* of bases. If there exists a local coordinate system on $\mathcal{U}$ such that its natural basis coincides at each point with a given frame field, we have referred to this frame field as holonomic. We say that a smoothly unisymmetric body $\mathcal{B}$ is *locally homosymmetric* if, for every point $\mathbf{X} \in \mathcal{B}$ there exists a neighbourhood $\mathcal{U}_X$ in which a holonomic unisymmetry field of basis can be found. If the neighbourhood can be made as large as the whole body, we say that the unisymmetric body is *globally homosymmetric*. All the remarks made in Chapter 1 apply again now as far as expressing local homosymmetry in terms of existence of material connections with vanishing torsion and curvature.

In physical terms, if a functionally graded body is unisymmetric and locally homosymmetric, there exist special configurations, which we shall call *homosymmetric configurations*, such that entire chunks of the body (and possibly the whole body) have identical (not just conjugate) symmetry groups. If the only information available concerning the mechanical response of a functionally graded body is that it is unisymmetric, then not much more can be said beyond what has already been stated: either the body is locally homosymmetric or it isn't. In the last case, we can say that there exist non-removable symmetry-inhomogeneities, which can be the source, for example, of residual stresses within the body. The presence of these inhomogeneities is measured by the non-existence of a local parallelism giving rise to a torsionless connection. We remark that, because of the fact that the normalizer of any subgroup always contains the subgroup of spherical dilatations, the unisymmetric material connection is never unique, thus possibly leading to a rather vague characterization of homosymmetry or the absence thereof. If, however, the body points are also known to exhibit preferred states (such as in the case of elastic solids), a much sharper characterization of this kind of inhomogeneities can be devised, as we shall presently see, which comes remarkably close to that of uniform materials.

## 4.5 Unisymmetric homogeneity of elastic solids

If, at a point $\mathbf{X} \in \mathcal{B}$ a frame is chosen (that is, a basis of its tangent space), the symmetry group is represented by a subgroup of the general linear group $GL(3; \mathbb{R})$. A simple elastic material point is said to be *solid* if, for some choice of frame, the symmetry group is represented by a subgroup of the orthogonal group. This property can also be rephrased in terms of reference configurations. Indeed, since a reference configuration is tantamount to a coordinate chart, and since a chart induces at each point of its domain a natural basis for the tangent space, we can say that a solid point is characterized by the existence of a reference configuration in which its symmetry group is a subgroup of the orthogonal group. Such a configuration (or, more precisely, the 1-jet thereof) is called an *undistorted state*. Real elastic solids are endowed with a special subclass of undistorted states, called *natural states*, whereby the stress vanishes. Given one natural state, all the other natural states are generated by applying to it an arbitrary orthogonal transformation.

We will assume henceforth that our elastic solid body points have natural states. By virtue of this property, we can uniquely define an inner product in the tangent space of each point of the body in the following way: the inner product of two tangent vectors at a point is, by definition, equal to the ordinary (Cartesian) inner product of their images in one (and, therefore, every) natural state of that point. As a result, we conclude that an elastic simple solid body, regardless of whether or not it is uniform or even unisymmetric, is endowed with a natural inner product at each point. If this inner product varies smoothly, we obtain a unique *Riemannian structure* determined by the natural states. If the *curvature* of this Riemannian structure happens to vanish identically, the body can be (at least by chunks) be brought to configurations in which each point is in a natural state. Such configurations will be called *relaxed* or *natural* configurations. We emphasize that this notion depends only on solidity and smoothness, and not on uniformity or unisymmetry. A solid body whose natural Riemannian structure has an identically vanishing curvature will be called (locally) *relaxable*.

If, in addition to being an elastic simple solid, the body is also unisymmetric, we have at our disposal two independent differential geometric structures: the (unique) natural Riemannian structure induced by the solidity, and the $N$-structure induced by the unisymmetry. The interplay between these two structures gives rise to various combinations, the most stringent of which is contained in the following definition: An elastic solid simple body is said to be (locally) *unisymmetrically homogeneous* if it is both relaxable and homosymmetric, and if each point has a neighbourhood for which the natural configurations are also homosymmetric configurations. In other words, unisymmetrical homogeneity corresponds to the mutual compatibility of the two (integrable) geometric structures.

*Example 4.3.* A unisymmetric elastic simple body $B$ is said to be *triclinic* if its structural (symmetry) group is the trivial group. This symmetry poverty renders $B$ automatically homosymmetric, since the (trivial) symmetry groups of all points are equal to each other in every configuration. Therefore, if the body is relaxable, it is automatically unisymmetrically homogeneous. Note, however, that the body, even if it happens to be uniform, will in general not be homogeneous. This example clearly shows how diverse the notions of homogeneity, on the one hand, and unisymmetric homogeneity, on the other, can be.

*Example 4.4.* Consider now the other extreme, namely, an *isotropic* unisymmetric simple solid, namely, one for which the structure group is the full orthogonal group. Assume that this body is relaxable, and consider a natural configuration of a neighbourhood. Clearly, in such a configuration, the symmetry group of each and every point of the neighbourhood is the full orthogonal group. The body is, therefore, automatically homosymmetric and, hence, unisymmetrically homogeneous. But there is more. In this case, we have the degree of freedom furnished by the full orthogonal group to rotate the frames at each point at will. In particular, we can choose all the frames in the neighbourhood to be parallel (in the Cartesian sense of the relaxed configuration). In other words, if the body happens to be uniform, it is also (locally) homogenenous. This example shows that, for fully isotropic elastic solids, relaxability is synonymous with homogeneity (whether of the old or the new vintage).

*Example 4.5.* A *transversely isotropic* solid point has, as its symmetry group in a natural configuration, the group of rotations around a fixed axis, called the isotropy axis. Let a unisymmetric body be made of such points. If the body is relaxable, it may happen that at one (and therefore every) relaxed configuration the axes of isotropy are not parallel. The body is unisymmetrically inhomogeneous and, even if not uniform, can be said to contain a continuous distribution of dislocations. This example, among many others, justifies the extension of the theory of inhomogeneities to non uniform bodies.

The preceding examples point to the need of finding a unified geometric structure whose integrability or lack thereof will serve to assess whether or not a unisymmetric body is unisymetrically homogeneous.

## 4.6 The reduced $N$-structure

### 4.6.1 Algebraic preliminaries

Let $V$ and $W$ be two inner-product spaces of the same finite dimension. By virtue of the existence of a distinguished inner product, we can identify each of the vector spaces with its dual, an identification that we will implicitly assume in what follows. Thus, for example, if $\mathbf{A} : V \longrightarrow W$ is a linear map, its transpose can be seen as a linear map $\mathbf{A}^T : W \longrightarrow V$ between the spaces themselves, rather than between their dual spaces. A linear map $\mathbf{R} : V \longrightarrow W$ is said to be *orthogonal* if the following two conditions are satisfied:

$$\mathbf{R}\mathbf{R}^T = id_W, \quad \mathbf{R}^T\mathbf{R} = id_V, \tag{4.5}$$

where $id$ stands for the identity map of the subscript space.

The *polar decomposition theorem* asserts that if $\mathbf{A} : V \longrightarrow W$ is an isomorphism (i.e., a non-singular linear map), there exists a unique orthogonal map $\mathbf{Q} : V \longrightarrow W$ and unique symmetric positive-definite automorphisms: $\mathbf{S} : V \longrightarrow V$ and $\mathbf{T} : W \longrightarrow W$ such that:

$$\mathbf{A} = \mathbf{QS} = \mathbf{TQ}. \tag{4.6}$$

By symmetric we mean: $\mathbf{S} = \mathbf{S}^T$ and $\mathbf{T} = \mathbf{T}^T$, expressions that are meaningful thanks to the inner product structure. By positive definite we mean that for each non-zero $\mathbf{m} \in V$ and $\mathbf{n} \in W$, we must have: $\mathbf{m}.\mathbf{Sm} > 0$ and $\mathbf{n}.\mathbf{Tn} > 0$, where we are using the same notation (i.e., an interposed dot) for both inner products.

We now introduce an equivalence relation $\sim$ in the set $\mathcal{L}(V,W)$ of all non-singular linear maps from $V$ to $W$, by means of the following identification: two linear maps are said to be equivalent if they have the same orthogonal component in their polar decompositions. The corresponding set of equivalence classes is denoted by $\mathcal{L}(V,W)/\sim$. Let $\mathcal{G}_V$ and $\mathcal{G}_W$ be orthogonal subgroups, respectively, of $\mathcal{L}(V,V)$ and $\mathcal{L}(W,W)$, and let $\mathbf{A} \in \mathcal{L}(V,W)$ be a conjugation-like group isomorphism, viz:

$$\mathcal{G}_W = \mathbf{A}\mathcal{G}_V\mathbf{A}^{-1}. \tag{4.7}$$

It follows that if $\mathbf{Q}$ is in $\mathcal{G}_V$ the product $\mathbf{Q}' = \mathbf{A}\mathbf{Q}\mathbf{A}^{-1} \in \mathcal{G}_W$ is orthogonal. We have, therefore:

$$\mathbf{AQ} = \mathbf{Q}'\mathbf{A}. \tag{4.8}$$

Now, let $\mathbf{A} = \mathbf{RS}$ be the polar decomposition of $\mathbf{A}$, where $\mathbf{R}$ is orthogonal (as per Equation (4.5)) and $\mathbf{S}$ is symmetric positive-definite. Accordingly, we can rewrite (4.8) as follows:

$$\mathbf{SQ} = (\mathbf{R}^T\mathbf{Q}'\mathbf{R})\mathbf{S}. \tag{4.9}$$

But, by the uniqueness of the rotation factor in both polar decompositions (4.6), we obtain[1] :

$$\mathbf{Q} = \mathbf{R}^T \mathbf{Q}' \mathbf{R}. \tag{4.10}$$

We conclude that, in the case of orthogonal subgroups, all conjugation-like isomorphisms between the subgroups can be achieved exclusively by means of orthogonal transformations. Moreover, since these orthogonal transformations have been obtained from the polar decomposition, all elements of a given equivalence class give rise, if any, to the same isomorphism.

If $\mathcal{A}$ is the set of conjugation-like isomorphisms between $\mathcal{G}_V$ and $\mathcal{G}_W$, we denote by $\tilde{\mathcal{A}} = \mathcal{A}/\sim$ the set of all conjugation-like isomorphisms between $\mathcal{G}_V$ and $\mathcal{G}_W$ within the quotient space $\mathcal{L}(V,W)/\sim$. Roughly speaking, this set is the collection of all orthogonal conjugations between these two orthogonal subgroups. To learn how big this set is, we identify the two inner product spaces, namely $V = W$, and conclude that the size of $\mathcal{A}/\sim$ is controlled by the normalizer of $\mathcal{G}_V$ *within the orthogonal group* at $V$. We denote this normalizer by $\tilde{\mathcal{N}}_V = \mathcal{N}(\mathcal{G}_V)/\sim$. In complete analogy with Equation (4.4), we now have:

$$\tilde{\mathcal{A}} = \mathbf{R}\,\tilde{\mathcal{N}}_V = \tilde{\mathcal{N}}_W\,\mathbf{R} = \tilde{\mathcal{N}}_W\,\mathbf{R}\,\tilde{\mathcal{N}}_V. \tag{4.11}$$

where $\mathbf{R}$ is (the orthogonal representative of an equivalence class) in $\tilde{\mathcal{A}}$.

### 4.6.2 The $\tilde{N}$-structure of a solid functionally-graded unisymmetric body

On the basis of the preceding considerations, we are now in a position of defining a geometric structure whose integrability will encompass together the notions of relaxability and homosymmetry, namely, the combined notion of unisymmetric homogeneity. As we have done in the case of ordinary uniformity, we obtain this structure by attaching to each point of the body a set representing the "degrees of freedom" in the choice of pertinent isomorphisms. In this case, these isomorphisms are orthogonal transformations (relative to the metric induced by the natural states at each point) representing conjugation-like isomorphisms of the symmetry groups. In other words, in a notation inspired by the preceding algebraic considerations, the degrees of freedom alluded to above are precisely represented at each point $\mathbf{X} \in \mathcal{B}$ by the set $\tilde{\mathcal{N}}_X$. If we choose an orthogonal frame at some point $\mathbf{X}_0$, the maps $\mathbf{A} \in \tilde{\mathcal{A}}_{X_0 X}$ generate at each point $\mathbf{X}$ of the body a set of orthogonal frames related to each other by the right action of the group $\tilde{\mathcal{N}}_X$. Attaching to each point its corresponding set of frames thus generated, we obtain a G-structure which, for obvious reasons, we will call a *material $\tilde{N}$-structure* of the unisymmetric functionally-graded solid body. Clearly, a material groupoid can be constructed in a similar way.

---

[1] The proof just sketched is due to Coleman and Noll [8].

Assume now that the $\tilde{N}$-structure is integrable. This means that for each point there exists a smooth local section that can be obtained from a chart (in other words, the natural basis of the chart coincides at each point with a frame in the $\tilde{N}$-structure). Since the $\tilde{N}$-structure is clearly contained in the original $N$-structure (that is, the standard structure that doesn't take into consideration the extra benefits of solidity), we conclude that the body is locally homosymmetric. Moreover, if we interpret the given chart, or a smooth extension thereof, as a (reference) configuration, in this configuration all points within the domain of the chart will be in a natural state. It follows that the body is also locally relaxable. This proves, therefore, that integrability of the $\tilde{N}$-structure implies local unisymmetric homogeneity. Conversely, let the body be locally unisymmetrically homogeneous. This means that there exists configurations in which, within a neighbourhood, the points are in a natural state and the Euclidean parallelisms act as unisymmetry transformations. Adopting this configuration as a coordinate chart, we conclude that the $\tilde{N}$-structure is integrable. Summarizing: the integrability of the $\tilde{N}$-structure of a unisymmetric solid body is equivalent to the condition of unisymmetric homogeneity.

*Remark 4.6.* Unlike the normalizer $\mathcal{N}(\mathcal{G})$, which is always continuous, the reduced normalizer $\tilde{\mathcal{N}}(\mathcal{G})$ may turn out to be discrete, as we shall see in the examples. It follows that, in some important cases for the applications, the assessment of unisymmetric homogeneity, or lack thereof, will be straightforward and similar, if not identical, to that of ordinary homogeneity of a uniform body.

*Remark 4.7.* Even if a unisymmetric body is not a solid, if it were possible, for some physical reason, to choose for each material point a preferential density, then another (larger) groupoid could be constructed, whose structure group is the unimodular part of the normalizer. Its integrability would measure the possibility of achieving a configuration in which each point is at its preferential density while the symmetry groups are identical within a neighbourhood. The mathematical possibility exists too of defining an intermediate groupoid based upon the centralizer, rather than the normalizer, of the symmetry group. Its physical significance, as well as that of its integrability, are open to interpretation.

*Remark 4.8.* Although unrelated to the problem of functionally graded bodies, it is interesting to point out that the theory of uniformity and homogeneity of material bodies with internal constraints [19], leads to the appearance of the *stabilizer* of a subset of the general linear group. Thus, the G-structures and the groupoid based exclusively on the analysis of the constraint have that stabilizer as their structure group.

## 4.7 Examples

### 4.7.1 The isotropic solid

We briefly revisit Example 4.4. The normalizer of the full orthogonal group within the general linear group can be shown to consist of all (commutative) products of spherical dilatations and orthogonal transformations. The normalizer within the orthogonal group coincides, therefore, with the orthogonal group itself. This example shows clearly the merit of using the reduced normalizer, since it gets rid, so to speak, of the "undesirable" dilatations. From the physical point of view, the fact that the reduced normalizer delivers again the symmetry group, means that there is no difference between ordinary (i.e., uniformity-based) homogeneity and unisymmetrical homogeneity. In other words, were we to have established that a functionally graded body is unisymmetrically homogeneous, the awareness that the body is actually uniform would not add any extra information as far as the presence of continuous distributions of inhomogneities (dislocations, say) is concerned.

### 4.7.2 The transversely isotropic solid

The normalizer of the group of rotations about a fixed axis consists of (commutative) products of arbitrary spherical dilatations, dilatations along the axis of isotropy and rotations about this axis. The normalizer within the orthogonal group coincides, therefore, with the original group of symmetry. We obtain the remarkable result that in the two technologically important cases of full and transverse isotropy, there is no need for the solid body to be made of the same material at all points in order to define and determine the presence of continuous distributions of inhomogeneities. In mathematical terms, they are governed by isomorphic $G$-structures, so that the criteria of integrability are identical in both cases.

### 4.7.3 The $n$-agonal solids

We will denote by *n-agonal* an elastic body whose points are endowed with a symmetry group which, in a natural state, consists of the rotations generated by successive applications of a rotation of magnitude $2\pi/n$ (for some integer $n > 1$) about a fixed axis. Although the normalizer within the general linear group has a different form for $n = 2$ than for $n > 2$, it turns out that the normalizer within the orthogonal group is, in all cases, equal to the group of *all* rotations about the given axis. In other words, the orthogonal normalizer coincides with that of the transversely isotropic body. The physical consequence of the fact that the orthogonal normalizer is larger than the symmetry group is the following: If, having determined that the functionally graded body is unisymmetrically homogeneous, we become aware that the body is actually uniform, then the possibility exists that, although the axis of symmetry are all parallel in a relaxed configuration, a further adjustment is necessary to render the symmetry isomorphisms material isomorphisms. This adjustment can only consist of rotations about the axis at each point and will, in general lead to the development of stress. In other words, for these bodies, the conditions of homogeneity and unisymmetric homogeneity are different, the former being more stringent.

### 4.7.4 Orthotropic materials

The *rhombic* subgroup of the general linear group is the discrete Abelian group consisting of the unit matrix and the three diagonal matrices that contain a n entry equal to 1 and two entries equal to $-1$. These matrices represent rotations of $\pi$ around each of three mutually perpendicular axes. The normalizer of this group consists of all diagonal matrices (whether spherical or not). The orthogonal normalizer coincides with the original group itself, which is discrete. So, for the important class of orthotropic materials, there is no difference between the two varieties of homogeneity. Moreover, the condition of local homogeneity boils down to the vanishing of the torsion of the locally unique curvature-free connection.

## 4.8 Summary

A general classification of non-uniform bodies has not been attempted[2]. Nevertheless, if a body enjoys the rather weak property of unisymmetry (same "type" of material at all points), such as many functionally graded bodies do, then already a geometric object exists for its description: a $G$-structure (and a groupoid) with structure group $\mathcal{N}(\mathcal{G})$, the normalizer of the typical symmetry group $\mathcal{G}$. This is a rather large structure group, allowing for a considerable freedom in the choice of admissible frames. The integrability of this object, guaranteeing the existence of configurations whereby the groups become identical at all points of a neighbourhood, would allow for the comparison of points by a wide class of deformations, including dilatations. For special materials, such as elastic solids, some of these deformations may be deemed "undesirable", since they would imply the coexistence of points in natural and stressed states even in the privileged configurations guaranteed by the integrability condition. For elastic solids, therefore, a $G$-substructure (and a subgroupoid) can be constructed with structure group $\tilde{\mathcal{N}}(\mathcal{G})$, the normalizer of $\mathcal{G}$ within the orthogonal group. The integrability of this new object eliminates, at least in some instances of practical importance, the unwanted situations, since it ensures the existence of fully relaxed configurations in which the groups coincide at all points. The non-integrability can thus be seen as an indication of the existence of distributed dislocations (or other defects possibly causing

---

[2] For an early attempt at a geometric characterization of non-uniformity see [18].

residual stresses) *in a non-uniform body*. Finally, if the body happens to be uniform, the classical notion of homogeneity is recovered as the integrability of the smaller $G$-structure (and groupoid) with structure group $\mathcal{G}$. As far as this smaller structure is concerned, some of the previous isomorphisms permitted by $\tilde{\mathcal{G}}$ may be expected to be inadmissible. This situation will happen if there exist orthogonal automorphisms of a natural state which do not belong to the symmetry group. In other words, the criterion of two points having just the same symmetry group may not be as fine as the criterion of having the same constitutive equation. It is quite remarkable, though, that for many types of solids this is not the case, and both criteria give rise to the *same* measure of inhomogeneity.

The following diagram summarizes the relation between the various concepts discussed in this and earlier chapters:

| Concept: | Uniformity $\rightarrow$ | Elastic Solid Unisymmetry | $\rightarrow$ Unisymmetry | $\rightarrow$ | General Non-uniformity |
|---|---|---|---|---|---|
| Structural Group of the G-structures: | $\mathcal{G}$ $\subset$ | $\tilde{\mathcal{N}}(\mathcal{G})$ | $\subset$ $\mathcal{N}(\mathcal{G})$ | | (non-transitive groupoid) |
| Integrability: | Homogeneity $\rightarrow$ | Unisymmetrical Homogeneity | $\rightarrow$ Homosymmetry | | |

# Part II

# Material Evolution

# 5

# On energy, Cauchy stress and Eshelby stress

Material evolution can be regarded as the time-like counterpart of material uniformity. More specifically, rather than comparing the material responses of two different body points at one instant of time, the theory of material evolution concerns itself with the comparison of the material response of one and the same point at different times. When, in a precise sense, it can be asserted that this response has suffered a modification, we say that a process of material evolution has taken place at that particular material point. If the body happens to be uniform, one can interpret its material evolution as a temporal change in the inhomogeneity pattern, as most dramatically exhibited in the phenomenon of plasticity. A thermodynamical analysis reveals that the agents driving certain processes of material evolution can be identified with the so-called configurational or material forces. The idea of obtaining such forces by evaluating the change in free energy required to effect the motion of an isolated dislocation can be traced back to the pioneering work of Eshelby [51], in whose honour the Eshelby stress has been named. It was later found that certain integrals used in fracture mechanics to calculate the energy released during the propagation of a crack are intimately related to the Eshelby stress. It seems appropriate, therefore, to start our treatment of the theory of material evolution by devoting this chapter to these and other related concepts. It is only in the next chapter that we shall devote our full attention to the theory of material evolution proper.

## 5.1 Preliminary considerations

Our definition of material isomorphism, Equation (1.8), was formulated in terms of a putative elastic energy $\psi$ per unit spatial volume. On the other hand, the various stress tensors (Cauchy, Piola, Kirchhoff, Eshelby) used in Continuum Mechanics consist of sundry combinations of $\psi$, the deformation gradient $\mathbf{F}$, and the first derivative of $\psi$ with respect to the latter. Properly speaking, therefore, if we insist on using the energy itself as a measure of material response, we should incorporate within the definition of material isomorphism a *gauge*, since two material points may have exactly the same Cauchy stress for each and every value of **F** and still have different values of the energy (as allowed by the nature of this gauge). In the case of solid materials having natural states (namely, states of zero stress), it is possible to choose the value of the gauge equal to zero at one (and, therefore, every) natural state to normalize the energy constitutive equation before testing for uniformity. Be that as it may, our treatment is entirely based on the existence of a set of linear maps **P**, whose nature and meaning is not altered by these considerations.

The primary aim of this chapter is to explicitly reveal the relation between various energy densities and, in particular, the stress tensors of Cauchy and Eshelby. The latter, which will play a prominent role as a driving force for the evolution of inhomogeneities, is a purely referential quantity usually expressed in terms of the elastic energy density $\psi_R$ per unit volume induced by a reference configuration as:

$$b_I^J = \psi_R \delta_I^J - F_I^i T_i^J, \tag{5.1}$$

$T_i^J$ being the components of the Piola (or first Piola-Kirchhoff) stress, defined by:

$$T_i^J = \frac{\partial \psi_R}{\partial F_J^i}. \tag{5.2}$$

Since the value of the Eshelby stress depends, as it were, on the zero energy level adopted, a fact that seems to contradict conventional wisdom, a need to resolve this apparent paradox arises. In the process of explaining the origin of this paradox, we will clarify the different gauges to be expected in the definition of material isomorphism according to which measure of energy density is used.

## 5.2 The Cauchy stress revisited

Cauchy's perspective is that of a spatial entity busy recording an energy density $\psi$ per unit *spatial* volume of the present configuration. He then decides to construct his stress measure by means of a first-order differential operator. But he realizes that he needs to satisfy an additional criterion, which he formulates as follows: my stress measure should be indifferent to a constant additive change of energy within a *material* volume element. This criterion arises naturally from thermodynamic considerations, or from experience with the concept of conservative forces in Classical Mechanics. The only novelty is that one has to pin down the same particles when claiming that an addition of energy does not affect the stress, since it is the energy contained in a fixed ensemble of particles that makes physical sense. That is why the material, rather than the instantaneous spatial, volume is chosen.

## 5.2 The Cauchy stress revisited

To achieve his objective, Cauchy calculates:

$$\psi_R = J\psi, \tag{5.3}$$

where $J$ is the determinant of $\mathbf{F}$. Then he demands that if $\psi_R = \psi_R(\mathbf{F})$ is the constitutive equation of a material point, the constitutive law $\psi'_R(\mathbf{F}) = \psi_R(\mathbf{F}) + C$ must give rise to the same stress, regardless of the value of the added constant $C$. We may say that, as far as the energy density per unit referential volume is concerned, Cauchy's gauge is represented by an additive constant. In view of Equation (5.3), we can incorporate the gauge that corresponds to the energy density $\psi$ per unit spatial volume as follows:

$$\psi' = \psi + J^{-1}C, \tag{5.4}$$

so that the gauge in this case is an arbitrary additive multiple of the determinant of the deformation gradient. Differentiating (5.4) with respect to the deformation gradient yields:

$$\frac{\partial \psi'}{\partial F^i_I} = \frac{\partial \psi}{\partial F^i_I} - CJ^{-1}(F^{-1})^I_i. \tag{5.5}$$

Reading off $C$ from Equation (5.4) and plugging the result into Equation (5.5), we obtain:

$$\psi'(F^{-1})^I_i + \frac{\partial \psi'}{\partial F^i_I} = \psi(F^{-1})^I_i + \frac{\partial \psi}{\partial F^i_I}. \tag{5.6}$$

Multiplying through by $F^j_I$, this result can be expressed by saying that the spatial tensor with components:

$$t^j_i = \psi \delta^j_i + \frac{\partial \psi}{\partial F^i_I} F^j_I, \tag{5.7}$$

satisfies completely Cauchy's aspirations: (i) It is of first order in the *spatial* energy density; (ii) It is indifferent to the zero level of the *referential* (or material) energy density. At first sight, this formula does not seem to coincide with the usual formula for Cauchy's stress. Nevertheless, we can easily verify that:

$$t^j_i = \psi \delta^j_i + \frac{\partial \psi}{\partial F^i_I} F^j_I = J^{-1} \frac{\partial (J\psi)}{\partial F^i_I} F^j_I = J^{-1} \frac{\partial \psi_R}{\partial F^i_I} F^j_I, \tag{5.8}$$

which can be written as:

$$t^j_i = J^{-1} F^j_I T^I_i, \tag{5.9}$$

which is the standard formula giving the Cauchy stress in terms of the Piola stress of Equation (5.2). The particular perspective used in the present derivation can be used to motivate a similar reasoning for Eshelby's stress.

## 5.3 Eshelby's tensor as Cauchy's dual

Eshelby's perspective is in every respect complementary to Cauchy's. Situated at a fixed position in a reference configuration, he believes that he is immersed in some kind of "pseudo-space" and that, on the contrary, it is the fixed positions in physical space that have identities as "pseudo-particles". This fortunate "delusion", resulting (as in some optical illusions) from an exchange of roles between matter and space, gives rise to Eshelby's stress in exactly the same way as Cauchy generated his own stress.

Eshelby, then, records the energy density $\psi_R$ per unit *referential* volume. The fictitious motion recorded by Eshelby for his pseudo-particles is the so-called *inverse motion*, namely:

$$X^I = X^I(x^i, t). \tag{5.10}$$

According to this apparent motion he can, for example, check which "particles" come and go from a given referential control volume. Now he resolves to build a stress à la Cauchy, but from the other side of Alice's looking glass, as it were. So, he wants a first-order operator and the following additional criterion: my stress measure should be indifferent to a constant additive change of energy in my *pseudo-material* volume element (i.e., what Cauchy would regard as a fixed spatial volume element). To achieve this aim he calculates:

$$\psi = J^{-1}\psi_R, \tag{5.11}$$

and then he demands that $\psi(\mathbf{F})$ and $\psi(\mathbf{F}) + c$ give rise to the same stress, regardless of the value of the arbitrary constant $c$. Writing:

$$\psi'_R = \psi_R + Jc, \tag{5.12}$$

and then differentiating with respect to the deformation gradient, we conclude that the referential tensor with components:

$$b_I^J = \psi_R \delta_I^J - F_I^i T_i^J, \tag{5.13}$$

satisfies Eshelby's ambitions: (i) It is of first order in the *referential* energy density; (ii) It is indifferent to the zero level of the *spatial* energy density. Equation (5.13) is the standard form of Eshelby's stress. Nevertheless, following Cauchy's analogy, Eshelby stumbles upon the new formula:

$$b_I^J = -J\frac{\partial \psi}{\partial F_J^i}F_I^i, \tag{5.14}$$

a formula that hides the zero level "problem" as effectively as the usual formula (5.9) hides it for Cauchy's stress.

In hindsight, the definition of the Eshelby stress should have been obtained automatically from the standard formula for the Cauchy stress, via the concept of inverse motion, by the trivial exchange:

$$\psi_R \longrightarrow \psi, \quad \mathbf{F} \longrightarrow \mathbf{F}^{-1}. \tag{5.15}$$

But, unlike nature, history seldom chooses the shortest path.

## 5.4 Complete expressions of hyperelastic uniformity

On the basis of the considerations just presented, one can obtain the general expression for the energy gauges that affect our primitive expression (1.8) of a material isomorphism and, consequently, the uniformity condition (1.20). Indeed, the constitutive equation of a hyperelastic body can be equivalently expressed in terms of the energy density $\psi$ per unit spatial volume:

$$\psi = \psi(\mathbf{F}, \mathbf{X}), \tag{5.16}$$

or the energy density $\psi_R$ per unit referential volume:

$$\psi_R = \psi_R(\mathbf{F}, \mathbf{X}), \tag{5.17}$$

or a stress measure, such as the Cauchy stress $\mathbf{t}$:

$$\mathbf{t} = \mathbf{t}(\mathbf{F}, \mathbf{X}). \tag{5.18}$$

From the physical perspective, a material isomorphism between two body points, $\mathbf{X}_1$ and $\mathbf{X}_2$, should stipulate that the Cauchy stress must be the same for both points for all values of the deformation gradient $\mathbf{F}$ modulo a linear map $\mathbf{P}_{12}$ between their tangent spaces, namely:

$$\mathbf{t}(\mathbf{F}, \mathbf{X}_2) = \mathbf{t}(\mathbf{F}\mathbf{P}_{12}, \mathbf{X}_1). \tag{5.19}$$

Since, as we have seen, the Cauchy stress is indifferent to a constant added to $\psi_R$ (or, equivalently, a constant times the determinant of $\mathbf{F}$, added to $\psi$), we conclude that equation (1.8) can be extended to:

$$\psi(\mathbf{F}, \mathbf{X}_2) + J^{-1}C = \psi(\mathbf{F}\mathbf{P}_{12}, \mathbf{X}_1) + J^{-1}J_{P_{12}}^{-1}D, \tag{5.20}$$

where $C$ and $D$ are constants, without altering condition (5.19). Since this identity must hold true for all non-singular $\mathbf{F}$, we may set $\mathbf{F} = \mathbf{P}_{12}^{-1}$ to read off the added terms and write:

$$\psi(\mathbf{F}, \mathbf{X}_2) = \psi(\mathbf{F}\mathbf{P}_{12}, \mathbf{X}_1) + J^{-1}J_{P_{12}}^{-1}\left(\psi(\mathbf{P}_{12}^{-1}, \mathbf{X}_2) - \psi(\mathbf{I}, \mathbf{X}_1)\right). \tag{5.21}$$

If we now identify in this equation $\mathbf{X}_1$ with a material archetype (with constitutive law $\bar{\psi}(\mathbf{F})$) and $\mathbf{X}_2$ with a variable point $\mathbf{X}$ in the body, the uniformity condition (1.20) becomes extended to:

$$\psi(\mathbf{F}, \mathbf{X}) = \bar{\psi}(\mathbf{F}\mathbf{P}) + J_{FP}^{-1}\left(\psi(\mathbf{P}^{-1}, \mathbf{X}) - \bar{\psi}(\mathbf{I})\right), \tag{5.22}$$

where $\mathbf{P} = \mathbf{P}(\mathbf{X})$ is an implant field. This expression guarantees that the Cauchy stress at $\mathbf{X}$ due to an arbitrary deformation gradient $\mathbf{F}$, is identical to the one that the archetype would attain by composing the implant with the same deformation gradient.

Similarly, in terms of the energy density $\psi_R$ per unit referential volume, the extended uniformity condition is:

$$\psi_R(\mathbf{F},\mathbf{X}) = J_P^{-1}\bar\psi_R(\mathbf{FP}) + \psi_R(\mathbf{P}^{-1},\mathbf{X}) - J_P^{-1}\bar\psi_R(\mathbf{I}). \qquad (5.23)$$

These modifications do not have any effect upon the theory of inhomogeneities as developed so far, except that they allow to identify more bodies as being uniform than the more restricted condition (1.20) would. As already pointed out, however, for solid materials it is easy to dispose of the degree of freedom afforded by the gauge by assigning a value of zero energy to the natural states. Thus, for example, if the archetype is in a natural state, we have that $\bar\psi(\mathbf{I}) = 0$. Moreover, since $\mathbf{P}^{-1}(\mathbf{X})$ brings point $\mathbf{X}$ to a natural state, we also have that $\psi(\mathbf{P}^{-1},\mathbf{X}) = 0$, and equation (5.22) reverts to (1.20)[1].

## 5.5 The Eshelby and Mandel Stresses in the Context of Material Uniformity

A fresh look at the uniformity condition (1.20) that will turn out to be useful later on when we deal with the concept of material evolution is the following: We may regard the energy density, according to Equation (1.20), as depending not just on the variable $\mathbf{F}$, but also on the variable $\mathbf{P}$. In other words, we may consider that these two variables are, in principle, independent and both can vary in time. The first variable, $\mathbf{F}$, represents the changes taking place in the neighbourhood of a particle as a result of the kinematic evolution of this neighbourhood in physical space. The second variable, $\mathbf{P}$, on the other hand, represent changes that take place in the material neighbourhood of the body itself. These changes can be seen as a "remodelling" taking place in the body and governed by some physically valid principle (for example, dislocation movement, plasticity, damage, growth, etc.). The expenditure of energy necessary to accomplish, or resulting from, each of these two types of processes (i.e., spatial and material) gives rise to two correspondingly different types of "stress". From this point of view, we will now see that the Eshelby stress corresponds (and is dual) to the material kinematics, its driving force, as it were, in the same way as the classical Cauchy or Piola stresses are the driving forces of the spatial kinematics.

Written in terms of the energy density $\psi_R$ per unit referential volume, the uniformity condition (1.20) reads:

$$\psi_R(F_I^i,\mathbf{X}) = J_P^{-1}\bar\psi_R(F_I^i P_\alpha^I(\mathbf{X})), \qquad (5.24)$$

where, for simplicity, we have eliminated the gauge freedom, assuming that it has been adjusted by means of a natural state. If we differentiate with respect to $\mathbf{F}$ we obtain the Piola stress, namely:

---

[1] In the framework of a comparison between the symmetry group of the Cauchy stress and that of the strain-energy function, more rigorous considerations of this type can be found in [87], Section 85.

## 5.5 The Eshelby and Mandel Stresses in the Context of Material Uniformity

$$T_i^I = \frac{\partial \psi_R}{\partial F_I^i} = \frac{\partial \bar{\psi}_R}{\partial (F_K^i P_\alpha^K)} P_\alpha^I. \tag{5.25}$$

We now differentiate, instead, with respect to **P** to assess the energy cost brought about by remodelling, and obtain:

$$-b_I^\alpha = \frac{\partial \psi_R}{\partial P_\alpha^I} = -J_P^{-1}(P^{-1})_I^\alpha + J_P^{-1} \frac{\partial \bar{\psi}_R}{\partial (F_K^j P_\alpha^K)} F_I^j. \tag{5.26}$$

Combining Equations (5.24), (5.25) and (5.26), yields the following result:

$$b_I^\alpha = \left[ \psi_R \delta_I^J - T_i^J F_I^i \right] (P^{-1})_J^\alpha. \tag{5.27}$$

The quantity within the square brackets, namely:

$$b_I^J = \psi_R \delta_I^J - T_i^J F_I^i, \tag{5.28}$$

is precisely the Eshelby stress, as defined in Equation (5.13).

Had we started from the uniformity condition as given in Equation (1.20) in terms of the energy density $\psi$ per unit spatial volume, the result would have been almost identical:

$$\frac{\partial \psi}{\partial P_\alpha^I} = (-\psi \delta_i^j + t_i^j)(F^{-1})_j^K (P^{-1})_K^\alpha F_I^i = J^{-1} b_I^J (P^{-1})_J^\alpha. \tag{5.29}$$

These calculations clearly show the physical meaning of the Eshelby tensor, as intended by Eshelby himself, as measuring the energy changes associated with the motion of inhomogeneities.

It remains to repeat these calculations using our third option, namely, the energy density $\psi_\rho$ per unit mass. If we assume that the implant maps satisfy the mass consistency condition, the uniformity condition (1.20) can be written as:

$$\psi_\rho(F_I^i, \mathbf{X}) = \bar{\psi}_\rho(F_I^i P_\alpha^I(\mathbf{X})). \tag{5.30}$$

Notice that the Piola stress can now be evaluated as:

$$T_i^I = \frac{1}{\rho_R} \frac{\partial \psi_\rho}{\partial F_I^i}. \tag{5.31}$$

Carrying out calculations similar to the ones that led to the Eshelby stress, we obtain:

$$m_I^\alpha = \frac{1}{\rho_R} \frac{\partial \psi_\rho}{\partial P_\alpha^I} = \left[ T_i^J F_I^i \right] (P^{-1})_J^\alpha. \tag{5.32}$$

The quantity within the square brackets:

$$\beta_I^J = T_i^J F_I^i, \tag{5.33}$$

is the component expression of the *Mandel stress* $\boldsymbol{\beta}$. It differs from the Eshelby stress (apart from sign) in the absence of the spherical part proportional to the energy and, thus, it does not arouse any paradoxical suspicions. It tends to appear naturally in formulations of material evolution equations, as we shall see later.

## 5.6 Eshelby-stress identities

### 5.6.1 Consequences of balance of angular momentum

Recall that the balance of angular momentum implies the symmetry of the Cauchy stress, which in turn implies the symmetry of the product $\mathbf{TF}^T$, where $\mathbf{T}$ is the Piola stress. The Eshelby stress:

$$\mathbf{b} = \psi_R \mathbf{I} - \mathbf{F}^T \mathbf{T}, \tag{5.34}$$

when multiplied by the right Cauchy-Green tensor $\mathbf{C} = \mathbf{F}^T \mathbf{F}$, yields therefore:

$$\mathbf{bC} = \psi_R \mathbf{C} - \mathbf{F}^T \mathbf{T} \mathbf{F}^T \mathbf{F} = \psi_R \mathbf{C} - \mathbf{F}^T \mathbf{F} \mathbf{T}^T \mathbf{F} = (\mathbf{bC})^T, \tag{5.35}$$

or, in components:

$$b_I^J C_{JK} = b_K^J C_{JI}. \tag{5.36}$$

In other words, the Eshelby stress is symmetric with respect to the spatial metric (pulled back to the body).

### 5.6.2 Consequences of a continuous symmetry group

If the material has a continuous symmetry group (such as for a fully or transversely isotropic solid), the Eshelby stress performs no "work" on an infinitesimal remodelling which belongs to the Lie algebra of the symmetry group. Although this statement is almost self-evident, we provide a derivation in components that will prove useful in manipulations. Let $\mathbf{G}(\lambda)$ be a one-parameter subgroup of the symmetry group of the archetype, with $\mathbf{G}(0) = \mathbf{I}$. We have then:

$$J_P^{-1} \bar{\psi}_R (F_I^i P_\alpha^I) = J_{PG}^{-1} \bar{\psi}_R (F_I^i P_\alpha^I G_\beta^\alpha), \tag{5.37}$$

for all values of $\lambda$. Taking derivatives of both sides with respect to $\lambda$, we obtain:

$$0 = \left( -J_{PG}^{-1} G_\nu^\mu \bar{\psi}_R (F_I^i P_\alpha^I G_\beta^\alpha) + J_{PG}^{-1} \frac{\partial \bar{\psi}_R}{\partial F_I^i P_\alpha^I G_\nu^\alpha} F_J^i P_\mu^J \right) \frac{dG_\nu^\mu}{d\lambda}, \tag{5.38}$$

which can be written as:

$$0 = \left( b_I^J \ (P^{-1})_J^\alpha \ P_\mu^I \ (G^{-1})_\alpha^\nu \right) \frac{dG_\nu^\mu}{d\lambda}. \tag{5.39}$$

For $\mathbf{P} = \mathbf{G} = \mathbf{I}$ (that is, identifying the point with the archetype and moving within the Lie algebra of the group), we verify our assertion. As an example, consider a fully isotropic solid archetype in a natural state. Its symmetry group is the full orthogonal group. The Lie algebra of this group consists of all skew-symmetric matrices. We conclude that, for full isotropy only, the product $\mathbf{bD}$ is symmetric, where $\mathbf{D} = (\mathbf{PP}^T)^{-1}$ is the metric induced by the implants.

### 5.6.3 Consequences of the balance of linear momentum

The Eshelby stress is a well-defined referential quantity whether or not the body is uniform. In the case of uniformity, however, the law of balance of linear momentum implies a particularly attractive identity. We consider here the case of Statics in the absence of body forces[2]. The equilibrium equation in a Cartesian reference reads:

$$T^I_{i\ ,I} = 0. \tag{5.40}$$

We start by noting that the referential gradient of the energy density $\psi_R$ in a uniform material can be calculated as:

$$\frac{\partial \psi_R}{\partial X^I} = \frac{\partial (J_P^{-1} \bar{\psi}_R)}{\partial X^I} = T^J_i F^i_{J,I} + T^K_i F^i_J P^{-\alpha}_K P^J_{\alpha,I} - \psi_R P^{-\alpha}_J P^J_{\alpha,I}, \tag{5.41}$$

where we have made use of the chain rule in a manner similar to that of Equations (5.25) and (5.26). Invoking the definition of the Eshelby stress and Equation (1.32) for the Christoffel symbols of the local parallelism induced by the implants $\mathbf{P}(\mathbf{X})$, we can write Equation (5.41) as:

$$\frac{\partial \psi_R}{\partial X^I} = T^J_i F^i_{J,I} + b^K_J \Gamma^J_{KI}. \tag{5.42}$$

We now evaluate the referential divergence of the Eshelby stress as:

$$b^J_{I\ ,J} = \left( \psi_R \delta^J_I - F^i_I T^J_i \right)_{,J} = b^K_J \Gamma^J_{KI}, \tag{5.43}$$

where we have made use of Equations (5.40) and (5.42)[3]. This remarkable identity [43, 44] can be put in a better invariant form by introducing a modified Eshelby stress via the re-scaling:

$$\tilde{\mathbf{b}} = J_P \mathbf{b}. \tag{5.44}$$

If we now indicate by means of a semicolon the covariant derivative with respect to the material connection $\Gamma$, it is not difficult to verify that Equation (5.43) can be recast in the form:

$$\tilde{b}^J_{I\ ;J} = \tilde{b}^J_K \tau^K_{JI} + \tilde{b}^J_I \tau^K_{JK}, \tag{5.45}$$

where $\tau$ is the torsion of the material connection as defined in Equation (1.60).

---

[2] For a full dynamical formulation in terms of the so-called *balance of pseudo-momentum*, see [68]. See also [95]

[3] If curvilinear coordinates are used in a reference configuration, the equation retains its tensorial character in the sense that in the right-hand side would appear the differences between the Christoffel symbols of the material connection and those of the coordinate system.

It may be argued that for the case of a continuous symmetry group Equation (5.43) cannot be true, since the left-hand side is clearly independent of the material connection used, whereas the right-hand side isn't. But let us check in which way can two material connections differ. Given a fixed archetype, two implant fields $\mathbf{P}(\mathbf{X})$ and $\hat{\mathbf{P}}(\mathbf{X})$ must be related by:

$$\hat{\mathbf{P}}(\mathbf{X}) = \mathbf{P}(\mathbf{X})\,\mathbf{G}(\mathbf{X}), \tag{5.46}$$

in which $\mathbf{G}(\mathbf{X})$ is a smooth assignment to each point in a neighbourhood of an element in the symmetry group of the archetype. Evaluating the respective Christoffel symbols, we obtain:

$$\hat{\Gamma}^I_{JK} = \Gamma^I_{JK} - G_\alpha^{-\beta}\, P_J^{-\alpha}\, P_\rho^I\, G^\rho_{\beta\,,K}. \tag{5.47}$$

It follows that the extra term is precisely of the form required by Equation (5.39) for the vanishing of its contraction with the Eshelby stress. On the other hand, it is not to be expected that the more elegant Equation (5.45) will be indifferent to a change of material connection.

For the case of a discrete symmetry group, Equation (5.45) can be neatly interpreted as follows. If the body is locally homogeneous, the torsion of the (locally unique) connection will vanish and, therefore, so will the divergence of the Eshelby stress with respect to the material connection. In other words, the absence of inhomogeneity translates itself into a conservation law of sorts.

### 5.6.4 Inhomogeneity with compact support and the J-integral

Suppose that a uniform body $\mathcal{B}$ is not locally homogeneous but has the following property: For every point outside a simply connected closed subset $\mathcal{C}$ of the body, there exists an open neighbourhood in which an integrable uniformity field can be found. Physically, this may correspond to a case in which the distribution of dislocations is limited to a small portion of the body, as could happen in a localized plastic process.

Under these conditions, it makes sense to consider a translation of this inhomogeneous blob en masse. Let $\mathbf{U}$ be a vector in the reference configuration representing the instantaneous material velocity of this translation at time $t = 0$ (Figure 5.1). The uniformity field, as a function of time, in a sufficiently small neighbourhood of $\mathcal{C}$, is given by:

$$\mathbf{P}(\mathbf{X}, t) = \mathbf{P}(\mathbf{X} - t\,\mathbf{U}, 0), \tag{5.48}$$

for sufficiently small times. The total energy content in the body is:

$$W(t) = \int_\mathcal{B} \psi_R(\mathbf{F}, \mathbf{X}) dV = \int_\mathcal{B} J_P^{-1} \bar{\psi}_R(\mathbf{F}\mathbf{P}(\mathbf{X}, t)) dV. \tag{5.49}$$

We now freeze $\mathbf{F}$ and calculate the power expended in moving the inhomogeneous blob. Thermodynamically speaking, this power will be dissipated in the process of moving the dislocations. We obtain:

## 5.6 Eshelby-stress identities

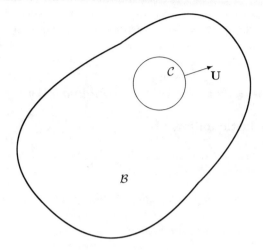

**Fig. 5.1.** Inhomogeneity with compact support

$$\frac{dW}{dt} = \int_{\mathcal{B}} \left( -J_P^{-1} P_I^{-\alpha} \dot{P}_\alpha^I \bar{\psi}_R + J_P^{-1} \frac{\partial \bar{\psi}_R}{\partial (F_K^i P_\alpha^K)} F_I^i \dot{P}_\alpha^I \right) dV. \tag{5.50}$$

But from (5.48) we calculate:

$$\dot{P}_\alpha^I = P_{\alpha,J}^I U^J. \tag{5.51}$$

Plugging this result into Equation (5.50) and using Equations (1.32), (5.26), (5.27) and (5.28) we get:

$$\frac{dW}{dt} = U^J \int_{\mathcal{B}} b_K^I \, \Gamma_{IJ}^K \, dV, \tag{5.52}$$

which, in view of Equation (5.43), can be written as:

$$\frac{dW}{dt} = U^I \int_{\mathcal{B}} b_{I,J}^J \, dV. \tag{5.53}$$

If we carry out this computation in a reference configuration which is homogeneous everywhere, except for the support of the inhomogeneity, the divergence of the Eshelby stress in the region $\mathcal{B} - \mathcal{C}$ vanishes, since so do the Christoffel symbols $\Gamma$. It follows, then, by a direct application of the divergence theorem that the integral:

$$J = U^I \oint_{\mathcal{S}} b_I^J \, N_J \, dA, \tag{5.54}$$

has a constant value on all closed surfaces $\mathcal{S}$ (with unit normal **N**) enclosing the inhomogeneity. It is this kind of reasoning that serves to establish a link with the classical results of Fracture Mechanics (such as the J-integral) and to justify the idea that the Eshelby stress (or its divergence) represents a kind of material or configurational force responsible for the motion of dislocations and other defects.

## 5.7 The Eshelby stress in thermoelasticity

### 5.7.1 Thermoelastic uniformity

A thermoelastic heat-conducting simple body is characterized by constitutive equations of the form:

$$T_i{}^I = T_i{}^I(F^j{}_J,\ \theta,\ \theta_{,J},\ X^J), \tag{5.55}$$

$$Q^I = Q^I(F^j{}_J,\ \theta,\ \theta_{,J},\ X^J), \tag{5.56}$$

$$\psi_R = \psi_R(F^j{}_J,\ \theta,\ \theta_{,J},\ X^J), \tag{5.57}$$

$$s_R = s_R(F^j{}_J,\ \theta,\ \theta_{,J},\ X^J), \tag{5.58}$$

where $\theta$ is the absolute temperature, $\mathbf{Q}$ is the referential heat-flux vector and $s_R$ is the entropy per unit reference volume. In this context, $\psi_R$ denotes the referential Helmholtz free energy, related to the referential internal energy $u_R$ by the equation: $\psi_R = u_R - \theta\, s_R$. Imposition of the second law of thermodynamics in the form of the Clausius-Duhem inequality, results in severe restrictions on the choice of possible constitutive functions. Specifically, these restrictions are:

$$\frac{\partial \psi_R}{\partial \theta_{,I}} = 0, \tag{5.59}$$

$$T_i{}^I = \frac{\partial \psi_R}{\partial F^i{}_I}, \tag{5.60}$$

$$s_R = -\frac{\partial \psi_R}{\partial \theta}, \tag{5.61}$$

and

$$Q^I\, \theta_{,I} \leq 0. \tag{5.62}$$

It follows that in reality a thermoelastic heat conductor is characterized by just two constitutive equations, namely:

$$\psi_R = \psi_R(F^j{}_J,\ \theta,\ X^J), \tag{5.63}$$

and

$$Q^I = Q^I(F^j{}_J,\ \theta,\ \theta_{,J},\ X^J). \tag{5.64}$$

When defining the notions of material isomorphism and uniformity within a thermoelastic context [28, 47], we must take into consideration both of these equations, so that both the mechanical and thermal responses are included in the comparison of material points. In terms of archetypal response functions $\bar\psi_R$ and $\bar Q^\alpha$, a uniform thermoelastic conductor must, therefore, satisfy the following two equations:

$$\psi_R(F^j{}_J,\ \theta,\ X^J) = J_P^{-1}\ \bar\psi_R(F^j{}_J P^J_\alpha,\ \theta), \tag{5.65}$$

and

$$Q^I(F^j{}_J,\ \theta,\ \theta_{,J},\ X^J) = J_P^{-1}\ P^I_\alpha\ \bar Q^\alpha(F^j{}_J P^J_\alpha,\ \theta,\ \theta_{,I} P^I_\alpha), \tag{5.66}$$

for some smooth implant maps $P^I_\alpha = P^I_\alpha(X^J)$.

Since the geometric structure associated with the notions of uniformity and homogeneity rests only upon the nature of the material isomorphisms, we observe that the fact that thermal effects have been taken into consideration is of little relevance (beyond the fact that the supply of implant fields **P** is now governed by more constitutive equations and variables)[4]. Material connections, in particular, are based upon the availability of these implant fields, but their definition remains the same as before.

### 5.7.2 The Eshelby stress identity

Having eliminated, by virtue of the Clausius-Duhem inequality, the dependence of the free energy on the temperature gradient, the procedure entailed by Equations (5.24-5.26) to assess the energy expenditure brought about by a change in the uniformity field will yield identical results to those of the purely elastic case. The Eshelby stress, in particular, is still given by:

$$b^J_I = \psi_R\,\delta^J_I - F^i{}_I T_i{}^J. \tag{5.67}$$

In the thermoelastic case, however, the referential divergence of this stress will be affected by the presence of the temperature as an argument of the free energy density. For the static case in the absence of body forces, a direct calculation along the lines of Equations (5.40-5.43) yields the following result:

$$b^J_{I,J} = b^K_J \Gamma^J_{KI} - s_R\,\theta_{,I}. \tag{5.68}$$

What is surprising about this result is the fact that even if the body is perfectly homogeneous and is placed in a homogeneous reference configuration, whereby

---

[4] To clarify this point further, note that we can recover the purely elastic model from the thermoelastic one by restricting attention to isothermal processes only. In this case, the heat flux vector vanishes and the free energy can be interpreted as the stored elastic energy at a given temperature. Clearly, the symmetry group can only increase or, at most, remain the same since there are fewer conditions to be satisfied. In other words, the supply of material isomorphisms may be more limited in the thermoelastic case, since every thermoelastic isomorphism is also an elastic one, but not viceversa.

the Christoffel symbols vanish, the divergence of the Eshelby stress will in general not vanish, unless the temperature is uniform. On further reflection, we notice that the definition of reference configuration in Continuum Mechanics is purely geometric: it does not involve any thermal reference. The presence of the term $s_R\,\theta_{,I}$ expresses the fact that there is more to an actual configuration than meets the eye. An observer bent on looking at the situation as if it were purely mechanical will end up feeling an unexpected material force, as if there were some kind of invisible dislocation.[5]

### 5.7.3 Thermal stresses

To further clarify the preceding remarks, it may be instructive to revisit the naive theory of thermal stresses. By this we mean the purely mechanical theory in which the temperature plays just the role of a parameter not affecting the mechanical response except for a "shift" of the "resting strain" at which the stress vanishes. The so-called coefficient of thermal expansion makes its appearance within this framework in the linearized theory. This basic idea has proven to be very fruitful in other contexts as well, such as the description of the neural activation of skeletal muscle, in which a "contractile element", whose length can be varied at no cost, is incorporated within classical rheological models of springs and dashpots.

The main concept to be introduced in the theory of thermal stresses, and the one that distinguishes it from the more general theory of thermoelasticity, is the notion of *thermal strain* or, more appropriately for the nonlinear context, *thermal deformation gradient*. The justification of this simplification of the general theory rests upon the fact that many materials in nature can be quite accurately described, as far as their thermoelastic response is concerned, in this way. Our point of departure is an elastic material point, or material archetype, with constitutive law:

$$\mathbf{t} = \mathbf{f}(\mathbf{F}), \tag{5.69}$$

where $\mathbf{t}$ denotes the Cauchy stress and $\mathbf{f}$ is its functional dependence on the deformation gradient $\mathbf{F}$. But now we assign to this law a specific temperature $\theta_0$, which we will call the *reference temperature*. Assume that we proceed to change the temperature to another value $\theta \neq \theta_0$, while fixing the deformation gradient at its base value $\mathbf{F} = \mathbf{I}$. As we know from experience, even if at the reference temperature the stress happened to vanish (in other words, even if at the reference temperature the point happened to be in a natural state), at the new value of the temperature the stress will no longer vanish. We denote by $\mathbf{H}(\theta)$ a value of the deformation gradient $\mathbf{F}$ which, when substituted *in the original constitutive law* (5.69), delivers this value of the stress. We write this condition in the following way:

---

[5] Kroener [60] remarked that, in fact, "any state of eigenstresses can be described formally in terms of dislocation theory".

## 5.7 The Eshelby stress in thermoelasticity

$$\mathbf{t}(\mathbf{I}, \theta) = \mathbf{f}(\mathbf{H}(\theta)), \qquad (5.70)$$

with an obvious notation.

So far, we have described the state of affairs when the temperature is changed and the specimen is kept undeformed (namely, at $\mathbf{F} = \mathbf{I}$). What happens if we change not only the temperature but also the deformation gradient? At this point we make the fundamental simplifying assumption of the theory: The stress at a given temperature $\theta$ and at arbitrary values of the deformation gradient is governed by composition *with the same* $\mathbf{H}(\theta)$ as for $\mathbf{F} = \mathbf{I}$. This assumption is expressed by the following fundamental formula of the theory of thermal stresses:

$$\mathbf{t}(\mathbf{F}, \theta) = \mathbf{f}(\mathbf{F}\, \mathbf{H}(\theta)), \quad \forall \mathbf{F}. \qquad (5.71)$$

This fundamental formula appears to be of little use, since we have not addressed the question of uniqueness of the *thermal deformation gradient* $\mathbf{H}(\theta)$. Let, therefore, $\hat{\mathbf{H}}(\theta)$ be a different thermal deformation gradient, also satisfying Equation (5.71). We have:

$$\mathbf{f}(\mathbf{F}\, \mathbf{H}^{-1}(\theta)\, \hat{\mathbf{H}}(\theta)) = \mathbf{t}(\mathbf{F}\, \mathbf{H}^{-1}(\theta), \theta) = \mathbf{f}(\mathbf{F}), \quad \forall \mathbf{F}, \qquad (5.72)$$

which means that the product $\mathbf{G} = \mathbf{H}^{-1}(\theta)\, \hat{\mathbf{H}}(\theta)$ belongs necessarily to the symmetry group of the archetypal constitutive equation (5.69). This in turn means that the degree of freedom in the choice of the thermal strain at any given temperature is dictated exclusively by the symmetry group. Moreover, the symmetry group is preserved at all temperature levels.

Let $\mathcal{B}$ be a material body whose points $\mathbf{X}$ abide by a constitutive law of this type. If at the reference temperature $\theta_0$ the body is uniform with uniformity field $\mathbf{P}(\mathbf{X})$, it will remain uniform at all temperatures, since the constitutive law changes by composition to the right, just as the material isomorphisms do. In terms of the archetypal stored elastic energy $\bar{\psi}_R$ at the reference temperature, the uniformity condition for arbitrary temperatures reads:

$$\psi_R(\mathbf{F}, \theta, \mathbf{X}) = J_{PH}^{-1}\, \bar{\psi}_R(\mathbf{F}\, \mathbf{P}(\mathbf{X})\, \mathbf{H}(\theta)). \qquad (5.73)$$

As an example, suppose that at the reference temperature the body is in a globally homogeneous configuration, whereby all the implant maps are equal to $\mathbf{I}$. If we now specify a smooth temperature field $\theta = \theta(\mathbf{X})$, it follows from Equation (5.73) that the maps $\mathbf{H}(\theta(\mathbf{X}))$ play the role of new non-trivial material implants[6]. For an observer not equipped with a thermometer, it will appear that the body has suddenly and mysteriously become inhomogeneous. The corresponding material connection, according to Equation (1.32), is given by:

---

[6] Note that these implants will in general violate the mass consistency condition, since temperature changes are usually accompanied by changes in volume.

126    5 On energy, Cauchy stress and Eshelby stress

$$\Gamma_{IJ}^K = -H_I^{-\alpha} \frac{dH_\alpha^K}{d\theta} \theta_{,J}. \tag{5.74}$$

This connection is unique if the symmetry group is discrete. Otherwise, there will remain the degree of freedom afforded by the continuity of the group, just as in the general theory of uniformity. Be that as it may, our purely mechanical observer will calculate the divergence of the Eshelby stress and obtain, in accordance with Equations (5.43) and (5.74):

$$b_{I\phantom{,}}^{J}{}_{,J} = -b_I^K H_K^{-\alpha} \frac{dH_\alpha^J}{d\theta} \theta_{,I}. \tag{5.75}$$

In the preceding treatment, the temperature played merely a parametric role. From the more rigorous thermodynamic point of view, however, Equation (5.73) can be regarded as a perfectly legitimate, albeit simple, expression for the free-energy density of a uniform thermoelastic body, as a comparison with Equation (5.65) quickly reveals. But it is crucial to realize that, within this larger thermoelastic context, the implant maps are simply given by $\mathbf{P}(\mathbf{X})$, and not by $\mathbf{P}(\mathbf{X}) \mathbf{H}(\theta(\mathbf{X}))$. Indeed, in Equation (5.65) $\mathbf{H}$ plays no role whatsoever. Rather, it is included in the general dependence on $\theta$. In particular, for the example above, in which the $\mathbf{P}$-implants are all equal to $\mathbf{I}$, we conclude that the thermoelastic body is homogeneous (regardless of the temperature field). To reconcile both points of view, we recall that in the full thermoelastic context the governing identity for the Eshelby stress is given by Equation (5.68), which includes the term $s_R\, \theta_{,I}$. Recalling Equation (5.61) for the entropy density and using the chain rule in Equation (5.73) (with $\mathbf{P} \equiv \mathbf{I}$), we obtain:

$$s_R = -\frac{\partial \psi_R}{\partial \theta} = J_H^{-1} H_I^{-\alpha} \frac{dH_\alpha^I}{d\theta} \bar{\psi}_R - J_H^{-1} \frac{\partial \bar{\psi}_R}{\partial (F^i{}_J H_\alpha^J)} F^i{}_I \frac{dH^I}{d\theta}. \tag{5.76}$$

But since:

$$T_i^I = \frac{\partial \psi_R}{\partial F^i_I} = J_H^{-1} \frac{\partial \bar{\psi}_R}{\partial (F^i{}_J H_\alpha^J)} H_\alpha^I, \tag{5.77}$$

we can write Equation (5.76) as:

$$s_R = \left( \psi_R\, \delta_I^J - F^i{}_I\, T_i{}^J \right) H_J^{-\alpha} \frac{dH_\alpha^I}{d\theta} = b_I^J H_J^{-\alpha} \frac{dH_\alpha^I}{d\theta}, \tag{5.78}$$

where we have used Equation (5.67). Substituting this result in Equation (5.68), we finally obtain:

$$b_{I\phantom{,}}^{J}{}_{,J} = b_J^K \Gamma_{KI}^J - s_R\, \theta_{,I} = -b_L^J H_J^{-\alpha} \frac{dH_\alpha^L}{d\theta} \theta_{,I}, \tag{5.79}$$

where we have exploited the vanishing of the Christoffel symbols in our particular example. This result coincides with that of Equation (5.75), obtained by different (purely mechanical) means.

Having thus elucidated the computation of the material forces associated with thermal inhomogeneities in the case of the theory of thermal stresses, one cannot escape the conclusion that the simple expression $s_R\,\theta_{,I}$ is the generalization of the inhomogeneity force in the context of genuinely general thermoelastic materials, namely, those for which the material properties depend on the temperature in a completely general way (rather than via some thermal strain). This term should, therefore, appear in various path integrals in the mechanics of fracture of thermoelastic materials.

### 5.7.4 The material heat conduction tensor

Up to this point, the uniform constitutive equation (5.64) for the heat flux vector has played only the role of possibly limiting the supply of material isomorphisms. It has not otherwise appeared in any considerations involving the Eshelby stress, whose nature is linked to the free-energy density. Since the Clausius-Duhem inequality does not establish any connection between the free energy and the heat flux, the latter has remained conspicuously absent. Nevertheless, we may try to emulate the procedure followed in the derivation of the Eshelby stress so as to obtain a heat-flux equivalent of the Eshelby stress and to investigate the repercussions that the balance of energy (which we have not exploited so far) might have upon it. Unlike the case of the Eshelby stress, whose physical meaning was immediately palpable, we will proceed in a merely formal way, leaving the interpretation of the results to further speculation.

We will limit the treatment to a class of materials that admit a scalar potential $\phi_R = \phi_R(F^i_{,I}, \theta, \theta_{,I})$ for the heat flux, according to the prescription:

$$Q^I = \frac{\partial \phi_R}{\partial \theta_{,I}}. \tag{5.80}$$

Limited as this class of materials may appear to be, it is certainly general enough to at least include Fourier-like materials with anisotropic heat conductivities that may depend on temperature and deformation. For these materials the potential is given by:

$$\phi_R = -\frac{1}{2} K^{IJ} \theta_{,I} \theta_{,J}, \tag{5.81}$$

where $K^{IJ} = K^{IJ}(F^k_K, \theta)$ is the (symmetric) heat conductivity tensor.

In terms of the heat-flux potential, the uniformity condition (5.66) translates into the following equation:

$$\phi_R(F^i_{,I},\,\theta,\,\theta_{,I},\,X^J) \;=\; J_P^{-1}\,\bar{\phi}_R(F^i_I P^I_\alpha,\,\theta,\,\theta_{,I} P^I_\alpha), \tag{5.82}$$

where we continue to use a superposed bar to indicate quantities in the archetype. Following the lead of Equations (5.26-5.28), we define the *material (or Eshelby-like) heat-conduction tensor* as:

$$d_I^J = -\frac{\partial \phi_R}{\partial P_\alpha^I} P_\alpha^J. \tag{5.83}$$

This tensor is the (formal) heat-flux analogue of the Eshelby stress: it represents the expenditure of "heat potential" brought about by a change in the inhomogeneity pattern. Carrying out the computations, we obtain:

$$d_I^J = \phi_R \delta_I^J - \Phi_i^J F_I^i - Q^J \theta_{,I}, \tag{5.84}$$

with

$$\Phi_i^I \equiv \frac{\partial \phi_R}{\partial F_I^i}. \tag{5.85}$$

If the heat flux potential is independent of the deformation, the material heat conduction tensor reduces to:

$$d_I^J = \phi_R \delta_I^J - Q^J \theta_{,I}. \tag{5.86}$$

In a static steady-state condition and in the absence of radiation sources, the law of energy balance is:

$$Q^I{}_{,I} = 0. \tag{5.87}$$

For these conditions, the divergence of the material heat conduction tensor can be computed as:

$$d_I^J{}_{,J} = d_K^J \Gamma_{JI}^K + \frac{\partial \phi_R}{\partial \theta} \theta_{,I}, \tag{5.88}$$

in complete analogy with Equation (5.68). If, as is the case in the classical Fourier law, the potential (and, therefore, the conductivity tensor) is independent of temperature, Equation (5.88) reduces to its hyperelastic counterpart:

$$d_I^J{}_{,J} = d_K^J \Gamma_{JI}^K, \tag{5.89}$$

implying the existence of heat-flux related path-independent (or surface-independent) integrals around isolated inhomogeneities.

## 5.8 On stress, hyperstress and Eshelby stress in second-grade bodies

The equilibrium equation of a second-grade hyperelastic body in the absence of body forces or couples is:

$$T_{i,I}^I - T_{i,IJ}^{IJ} = 0, \tag{5.90}$$

where $T_i^I$ and $T_i^{IJ}$ are the Piola stress and hyperstress expressed, respectively, as:

## 5.8 On stress, hyperstress and Eshelby stress in second-grade bodies

$$T_i^I = \frac{\partial \psi_R}{\partial F_I^i}, \qquad (5.91)$$

and

$$T_i^{IJ} = \frac{\partial \psi_R}{\partial F_{IJ}^i}. \qquad (5.92)$$

In terms of the energy density $\psi_R$ per unit referential volume, the uniformity condition (2.22) can be written as:

$$\psi_R = \psi_R(F_I^i, F_{IJ}^i, \mathbf{X}) = J_P^{-1} \bar{\psi}_R(F_I^i P_\alpha^I, F_{IJ}^i P_\alpha^I P_\beta^J + F_I^i Q_{\alpha\beta}^I). \qquad (5.93)$$

For later use, we note the following identities, obtained by combining Equations (5.91) and (5.92) with Equation (5.93) and using the chain rule of differentiation:

$$\frac{\partial \bar{\psi}_R}{\partial (F_I^i P_\alpha^I)} = J_P P_J^{-\alpha} \left( T_i^J - T_i^{KL} P_K^{-\beta} P_L^{-\gamma} Q_{\beta\gamma}^J \right) \qquad (5.94)$$

and

$$\frac{\partial \bar{\psi}_R}{\partial (F_{IJ}^i P_\alpha^I P_\beta^J + F_I^i Q_{\alpha\beta}^I)} = J_P P_K^{-\alpha} P_L^{-\beta} T_i^{KL}. \qquad (5.95)$$

Following the cue of the first-grade case, we now take derivatives of Equation (5.93) with respect to both $\mathbf{P}$ and $\mathbf{Q}$. Using the identities just derived, we obtain:

$$-b_I^\alpha = \frac{\partial \psi_R}{\partial P_\alpha^I} = \left( -\psi_R \delta_I^J + T_i^J F_J^i - T_i^{KL} F_I^i P_K^{-\beta} P_L^{-\gamma} Q_{\beta\gamma}^J + 2T_i^{JL} F_{IL}^i \right) P_J^{-\alpha}, \qquad (5.96)$$

and

$$-b_I^{\alpha\beta} = \frac{\partial \psi_R}{\partial Q_{\alpha\beta}^I} = T_i^{KL} F_I^i P_K^{-\alpha} P_L^{-\beta}. \qquad (5.97)$$

Isolating the expressions that are independent of the archetype, we define the Eshelby stress and hyperstress, respectively, as:

$$b_M^N = b_M^\rho P_\rho^N + b_M^{\rho\sigma} Q_{\rho\sigma}^N, \qquad (5.98)$$

and

$$b_M^{NL} = b_M^{\rho\sigma} P_\rho^N P_\sigma^L. \qquad (5.99)$$

More explicitly:

$$b_M^N = \psi_R \delta_M^N - T_i^N F_M^i - 2T_i^{NL} F_{ML}^i. \qquad (5.100)$$

and

$$b_M^{NL} = -T_i^{NL} F_M^i. \qquad (5.101)$$

Note that, in the absence of second-grade effects, $b_M^N$ becomes the ordinary Eshelby stress, while $b_M^{NL}$ vanishes.

130     5 On energy, Cauchy stress and Eshelby stress

In order to derive a consequence of the equilibrium equation in terms of these material entities, we start by evaluating the (total) derivative of (5.93) with respect to **X** as:

$$\frac{\partial \psi_R}{\partial X^M} = T_i^I F_{IM}^i + t_i^{IJ} F_{IJ,M}^i - b_I^\alpha P_{\alpha,M}^I - b_I^{\alpha\beta} Q_{\alpha\beta,M}^I. \tag{5.102}$$

Next we calculate:

$$b_{M,N}^N - b_{M,NL}^{NL} = -b_I^\alpha P_{\alpha,M}^I - b_I^{\alpha\beta} Q_{\alpha\beta,M}^I, \tag{5.103}$$

where we have made use of (5.90), (5.100), (5.101) and (5.14). It is interesting to note that, just as was the case in first-grade materials, the right-hand side of this equation is equal to the derivative of the energy $\psi_R$ with respect to its *explicit* dependence on **X**. Be that as it may, using Equations (2.59), (2.60), (5.98) and (5.99), Equation (5.103) can be written as:

$$b_{M,N}^N - b_{M,NL}^{NL} = b_I^J \Gamma_{JM}^I + b_I^{NL} \Lambda_{NL;M}^I, \tag{5.104}$$

a semicolon indicating covariant derivative with respect to the first-grade material connection $\Gamma$. This equation is the generalization of (5.43) to second-grade bodies. We note that if the body is homogeneous and is placed in a homogeneous configuration, the right-hand side of Equation (5.104) vanishes.

## 5.9 On stress, microstress and Eshelby stress in Cosserat bodies

### 5.9.1 Equilibrium equations

For our purposes in this section, the equations of equilibrium of a Cosserat body in the absence of any body forces or couples will suffice. These equations can be obtained, for the hyperelastic case, as the conditions for rendering the stored elastic energy stationary. The total energy stored in the body is given, following Equation (3.14), by:

$$\Psi = \int_B \psi_R(K_I^i, F_I^i, K_{I,J}^i, X^K) \, dV, \tag{5.105}$$

where the integration extends over the body in a reference configuration with Cartesian volume element $dV$. By standard techniques of the calculus of variations, we obtain the following equilibrium equations:

$$T_i{}^I{}_{,I} = 0, \tag{5.106}$$

and

$$\left[S_i{}^I - S_i{}^{IJ}{}_{,J}\right]_{,I} = 0, \tag{5.107}$$

in terms of the (macromedium) Piola stress:

$$T_i{}^I = \frac{\partial \psi_R}{\partial F^i{}_I}, \qquad (5.108)$$

the *micromedium Piola stress*:

$$S_i{}^I = \frac{\partial \psi_R}{\partial K^i{}_I}, \qquad (5.109)$$

and the *micromedium Piola hyperstress*:

$$S_i{}^{IJ} = \frac{\partial \psi_R}{\partial K^i{}_{I,J}}. \qquad (5.110)$$

If one defines the *effective micromedium Piola stress* as:

$$\Sigma_i{}^I \equiv S_i{}^I - S_i{}^{IJ}{}_{,J}, \qquad (5.111)$$

the micromedium equilibrium equation (5.107) can be written as follows:

$$\Sigma_i{}^I{}_{,I} = 0, \qquad (5.112)$$

a formula that resembles its macromedium counterpart.

### 5.9.2 Eshelby stresses

Rewriting the uniformity condition (3.26) in terms of the energy per unit volume in the reference configuration as:

$$\psi_R(K^i{}_I, F^i{}_I, K^i{}_{I,J}, X^I)$$
$$= J_P^{-1} \bar{\psi}_R(K^i{}_I Q^I{}_\alpha, F^i{}_I P^I{}_\alpha, K^i{}_{I,J} P^J{}_\beta Q^I{}_\alpha + K^i{}_I R^I{}_{\alpha\beta}), \qquad (5.113)$$

where $\bar{\psi}_R$ denotes the energy per unit volume in the archetype (with respect to a chosen frame therein), and using the chain rule of differentiation together with the definitions (5.108-5.110), we obtain:

$$J_P^{-1} \frac{\partial \bar{\psi}_R}{\partial (F^i{}_L P^L{}_\alpha)} = T_i{}^I P_I{}^{-\alpha}, \qquad (5.114)$$

$$J_P^{-1} \frac{\partial \bar{\psi}_R}{\partial (K^i{}_{L,M} P^M{}_\beta Q^L{}_\alpha + K^i{}_L R^L{}_{\alpha\beta})} = S_i{}^{IJ} P_J{}^{-\beta} Q_I{}^{-\alpha}, \qquad (5.115)$$

and

$$J_P^{-1} \frac{\partial \bar{\psi}_R}{\partial (K^i{}_L Q^L{}_\alpha)} = [S_i{}^I - S_i{}^{LM} P_M{}^{-\sigma} Q_L{}^{-\rho} R^I{}_{\rho\sigma}] Q_I{}^{-\alpha}. \qquad (5.116)$$

We now compute the derivatives of the energy density with respect to the various components of the implant field as:

$$-b_I^\alpha = \frac{\partial \psi_R}{\partial P_\alpha^I} = \left[ -\psi_R \, \delta_I^J + F_I^i \, T_i^{\ J} + K_{L,I}^i \, S_i^{\ LJ} \right] P_J^{-\alpha}, \quad (5.117)$$

$$-c_I^\alpha = \frac{\partial \psi_R}{\partial Q_\alpha^I} = \left[ K_I^i \, S_i^{\ J} - K_I^i \, S_i^{\ LM} P_M^{-\sigma} Q_L^{-\rho} R_{\rho\sigma}^I + K_{I,L}^i \, S_i^{\ JL} \right] Q_J^{-\alpha}, \quad (5.118)$$

and

$$-c_I^{\alpha\beta} = \frac{\partial \psi_R}{\partial R_{\alpha\beta}^I} = K_I^i \, S_i^{\ LM} P_M^{-\beta} Q_L^{-\alpha}. \quad (5.119)$$

Following the same idea as in the case of second-grade materials, we now define the Eshelby macrostress as:

$$b_I^J = b_I^\alpha \, P_\alpha^J, \quad (5.120)$$

the micromedium Eshelby stress as:

$$c_I^J = c_I^\alpha \, Q_\alpha^J + c_I^{\alpha\beta} \, R_{\alpha\beta}^J, \quad (5.121)$$

and, finally, the micromedium Eshelby hyperstress as:

$$c_I^{JK} = c_I^{\alpha\beta} \, P_\beta^K \, Q_\alpha^J. \quad (5.122)$$

Explicitly, we have the following relations between the various Eshelby stresses just defined and the various Piola stresses, energy density and Cosserat deformation components:

$$b_I^J = \psi_R \, \delta_I^J - F_I^i \, T_i^{\ J} - K_{L,I}^i \, S_i^{\ LJ}, \quad (5.123)$$

$$c_I^J = -K_I^i \, S_i^{\ J} - K_{I,L}^i \, S_i^{\ JL}, \quad (5.124)$$

and

$$c_I^{JK} = -K_I^i \, S_i^{\ JK}. \quad (5.125)$$

We observe that Equations (5.123-5.125) can also be obtained directly from (5.117-5.119) by setting the local implant to $(\mathbf{I}, \mathbf{I}, \mathbf{0})$. In other words, the various Eshelby stresses represent the energy expenditure needed to produce a small material change measured from the present state of the reference configuration. This state will keep changing in time if the body undergoes a material evolution.

We now check whether an *effective micromedium Eshelby stress* is worth defining. By analogy with Equation (5.111), we define it as:

$$C_I^J \equiv c_I^J - c_I^{JK}{}_{,K}. \quad (5.126)$$

Carrying out the computations by means of Equations (5.124) and (5.125), we obtain the following nice result:

## 5.9 On stress, microstress and Eshelby stress in Cosserat bodies

$$C_I^{\ J} = -K^i{}_I \Sigma_i{}^J, \tag{5.127}$$

which shows that, as far as the micromedium is concerned, the effective Eshelby stress behaves like a Mandel-type stress. The absence of the spherical term is due, at least in our derivation, to the fact that the determinant of $\{K^i{}_I\}$ plays no role in the uniformity condition (5.113).

### 5.9.3 Eshelby stress identities

We now investigate the repercussions of the equilibrium equations of the Cosserat body on the referential divergence of the various Eshelby stresses. We start with the micromedium. Invoking the micromedium equilibrium equation (5.112), the divergence of Equation (5.127) yields:

$$C_I^{\ J}{}_{,J} = -K^i{}_{I,J} \Sigma_i{}^J. \tag{5.128}$$

Using Equation (5.127) once again, we arrive at the following result:

$$C_I^{\ J}{}_{,J} = \Pi^M_{IJ} C_M^{\ J}, \tag{5.129}$$

with $\Pi^I_{JK} \equiv K^{-I}_{\ \ ,i} K^i{}_{J,K}$.

Notice how this result resembles the identity for the Eshelby stress of a simple body, as expressed in Equation (5.43). Nevertheless, although the quantities $\Pi^M_{IJ}$ are indeed legitimate Christoffel symbols of a connection in the body, this connection is not material but deformational. It is the absence of the spherical term (the only difference between a referential Mandel stress and an Eshelby stress) that precludes the appearance of a true material connection.

Starting from Equation (5.123), we can calculate the divergence of the Eshelby macrostress. After rather drastic simplifications stemming from the various relations derived so far, we arrive at the following identity:

$$b_I^{\ J}{}_{,J} = b_J^K \Gamma^J_{KI} + c_J^K \Delta^J_{KI} - C_J^K \Pi^J_{KI} + c_L^{KJ} \Lambda^L_{KJ|I}, \tag{5.130}$$

where a vertical bar indicates covariant differentiation with respect to the $\Delta$-connection. Note again the presence of the deformational connection $\Pi$, which in general will have a non-vanishing torsion (otherwise, we could invoke Equation (5.129) and replace the $C_J^K \Pi^J_{KI}$ with $C_I^{\ J}{}_{,J}$).

*Remark 5.1.* In addition to those presented in this chapter, the concept of material or Eshelby-like stresses is amenable to generalization to many other contexts, such as: electromagnetic deformable media [69, 45, 70], general relativity [34] and non-local media [49].

# 6

# An overview of the theory of material evolution

In our geometrical description of the theory of continuous distributions of inhomogeneities, time has played no role whatsoever. Our inhomogeneity patterns are frozen, as it were, inside the material and remain so during the process of deformation. On the other hand, we know that in many practical applications inhomogeneities, once present in the material, tend to move or otherwise evolve in time. The phenomenon of metal plasticity constitutes a perfect example of a massive motion of dislocations in solid materials, and the heat generated by the internal friction thus created can be felt in as mundane an experiment as the large torsion of a metal paper clip. Another important example is provided by the volumetric growth and remodelling of biological tissues, such as bone and muscle. As more material is added (growth) or removed (resorption), internal stresses develop to accommodate the new state of affairs. In these examples, all of which are encompassed by the theory of material evolution, the temporal change takes place only inasmuch as the material undergoes a process of re-accommodation or remodelling, always keeping its "chemical" identity. One can contemplate other types of phenomena, whereby changes of phase or even alchemic transmutation of elements occur. The general theory to be presented, however, is not as general as to include such phenomena.

## 6.1 What is material evolution?

Broadly speaking, material evolution is the temporal counterpart of the notion of material isomorphism. In the case of a material isomorphism, we consider two different material points at a fixed instant of time and compare their constitutive responses. If these responses can be made to match each other by composition of the deformation with a single local diffeomorphism of the body, we say that the two points are materially isomorphic. In the case of material evolution, on the other hand, we consider *one and the same material point* $\mathbf{X}$ *at two different instants of time*. If there exists a local automorphism at $\mathbf{X}$ such that the responses at the two different times can be made to match one another for all deformations, we can say that the point has retained the essential features of its constitutive behaviour. If this automorphism belongs to the original symmetry group at the initial time, however, we must conclude that the constitutive law has not changed, since this degree of freedom (namely, the symmetry group) is naturally available in any case. In other words, there has been no evolution. If, on the other hand, the automorphism doing the matching job does not belong to the original symmetry group, we find that the the material responses at the two instants of time are not identical, but merely materially isomorphic. In this case we assert that a material evolution has taken place. Although we have described the notion of material evolution by referring to two isolated instants of time, we are interested in the case in which material evolution takes place smoothly as time goes on within an interval of interest. Thus, material evolution in this sense is the temporal counterpart of material uniformity.

Material evolution, as just described, is a point-wise phenomenon. We may, therefore, have a non-uniform body undergoing material evolution point by point. That said, a case of obvious interest is furnished by the combination of uniformity with evolution. In this case, a single material archetype can be used to effect both the space and the time implants, imbuing the theory with a unified spirit and with greater elegance. These features are most prominent within a general relativistic context [34].

Whether in a point-wise or a uniform context, the Eshelby stress (or, rather, the Mandel stress) will emerge naturally as the thermodynamic dual, or "driving force", of the rate of material evolution. This fact should not be surprising, since we have already unravelled the meaning of the Eshelby stress as essentially the derivative of the free-energy density with respect to the material implant.

To render a particular theory of material evolution complete and self-contained, one needs to supply an *evolution law* prescribing the specific way in which the evolution takes place. This law usually takes the form of a system of first-order ordinary differential equations for the implants as a function of time. Thus, from a more general point of view, the theory of material evolution can be seen as a theory of materials with internal state variables, the internal variables being precisely the implant fields. What distinguishes the theory of evolution from the general case is the particular way in which these internal variables are enmeshed in the formulation. Instead of being just an extra set of independent constitutive variables, they enter the constitutive laws via a *right composition* with the variables affected by the deformation. This very special feature of the evolution variables is ultimately responsible for the relevance of the Eshelby stress in the formulation. Indeed, the Eshelby stresses turn out to be nothing else but the forces associated with this special kind of internal variables.

Evolution laws cannot be completely arbitrary, but must satisfy certain formal requirements. In this sense, they are similar to constitutive laws, which are also subject to certain formal limitations. In the case of the evolution laws, however, these requirements do not arise merely from the principle of material frame indifference, since a frame is an essentially *spatial* notion, whereas evolution is a fundamentally *material* phenomenon. Rather, we will impose other safeguards on the evolution laws to guarantee that they are in some sense independent of any particular reference configuration, consistent with the symmetry of the material and representative of a true evolution. The more delicate issue of thermodynamical restrictions will be considered as well.

## 6.2 A geometric picture

For the case of the material evolution of a uniform body (rather than the point-wise evolution of a non-uniform one) we have a situation whereby the body remains uniform as time goes on, but the pattern of distribution of inhomogoneities in its midst keeps changing. A common case in everyday engineering practice is provided by the bending of an initially homogeneous steel beam (Figure 6.1). As the load is gradually increased, it eventually crosses a threshold ("yield") value $F_Y$, whereupon the beam begins to exhibit larger and larger plastic regions. The fact that upon unloading the material still "remembers" its original modulus of elasticity is witness to the validity of the proposed model inasmuch as it retains the original elastic archetype, but renders its implants into the body point- and time-dependent.

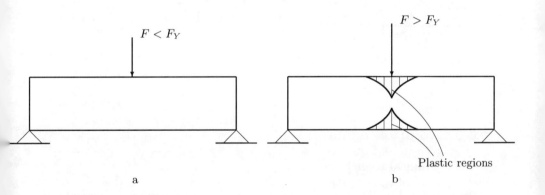

**Fig. 6.1.** Metal beam: (a) Homogeneous; (b) Inhomogeneous

This physical picture has a differential-geometric counterpart. Assume first that the material has no non-trivial symmetries. In that case, choosing a particular basis at one point of the body $\mathcal{B}$, the (trivial) $G$-structure associated with its uniformity consists of a fixed cross section of the principal frame bundle $F\mathcal{B}$. As the plastic process begins to develop, this section starts evolving leading us to other (different, but still trivial) $G$-structures, as shown in Figure (6.2). The new sections will in general not be obtainable as a lift of a coordinate system in the base manifold. If, on the other hand, the material has a continuous symmetry group, the material $G$-structure is no longer a single cross section. Instead, each fibre consists of a continuous (and closed) subset of the set of local frames. A uniformity field is now a locally smooth section cutting through these subsets. As plasticity develops, the $G$-structure moves to a counterpart with no common cross sections with the original $G$-structure. In other words, although an evolution law may be specified in terms of a particular cross section, one has to make sure that, for a true evolution, the section constantly moves outwards of the present $G$-structure. Note that the representation in Figure (6.2) is merely schematic: it couldn't be otherwise, since the true picture is 12-dimensional! Nevertheless, this kind of representation can be very useful to trigger the right geometrical ideas.

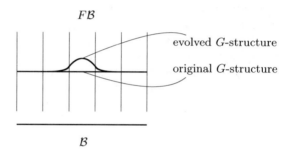

**Fig. 6.2.** $G$-structure evolution

## 6.3 Evolution equations

### 6.3.1 General form

In an evolving material process (or *remodelling process*), we will observe that, for a given archetype, the material implants at each point will be functions of time. For the case of a simple material, we will have:

$$\mathbf{P} = \mathbf{P}(\mathbf{X}, t). \tag{6.1}$$

It is possible to imagine some devices (consisting perhaps of nano-scale sensors and motors attached at a sufficiently large number of body points) that take care of enforcing any desired material remodelling. Under such hypothetical conditions we cannot properly speak of evolution equations. What we mean by an evolution equation is a differential equation which the material somehow "solves" in "real time" simultaneously with all the other field and constitutive equations. Thus, an expression such as (6.1) will be the final outcome of a solution process that furnishes also the deformation and temperature fields. The most common form of an evolution law at a point $\mathbf{X} \in \mathcal{B}$ is provided by a first-order differential equation:

$$\dot{\mathbf{P}} = \mathbf{f}(\mathbf{P}, \mathbf{a}, \mathbf{X}), \tag{6.2}$$

where $\mathbf{f}$ is a tensor-valued function of the present value of the implant $\mathbf{P}$ and of a list of other arguments which we have collectively denoted by $\mathbf{a}$. A superposed dot denotes the time derivative. The list of arguments usually will include the Eshelby stress $\mathbf{b}$, and it may also include other variables such as the deformation gradient. We assume, however, that the time variable ($t$) will never appear explicitly, since the evolution law describes essentially a material behaviour, and intrinsic material properties are independent of absolute time.

As we have already pointed out, the evolution function $\mathbf{f}$ cannot be completely arbitrary. In the sequel, we will motivate and derive a number of restrictions to be imposed on this function.

### 6.3.2 Reduction to the archetype

The particular mathematical expression of the evolution law (6.2) depends on a choice of archetype and on a choice of reference configuration. Leaving the archetype fixed, we are interested in investigating the effect that a change of reference configuration has on the expression of the evolution law. Figure (6.3) shows the basic scheme, where we use a "hat" to distinguish quantities pertaining to the second reference configuration.

The change of reference configuration itself is denoted by $\lambda$ and its gradient by $\nabla \lambda$. The evolution law in the second reference configuration is given by:

$$\dot{\hat{\mathbf{P}}} = \hat{\mathbf{f}}(\hat{\mathbf{P}}, \hat{\mathbf{a}}_{\nabla \lambda}, \lambda(\mathbf{X})), \tag{6.3}$$

where:

$$\hat{\mathbf{P}} = \nabla \lambda \, \mathbf{P}, \tag{6.4}$$

and

$$\dot{\hat{\mathbf{P}}} = \nabla \lambda \, \dot{\mathbf{P}}. \tag{6.5}$$

As far as the argument $\mathbf{a}$ is concerned, its transformed version $\hat{\mathbf{a}}_{\nabla \lambda}$ depends on the nature of this argument. As an example, the Eshelby stress transforms according to the formula:

$$\hat{\mathbf{b}}_{\nabla \lambda} = J_{\nabla \lambda}^{-1} \, \nabla \lambda^{-T} \, \mathbf{b} \, \nabla \lambda^{T}. \tag{6.6}$$

Combining all these results, we can write:

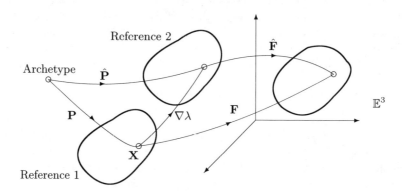

**Fig. 6.3.** Change of reference configuration

$$\dot{\mathbf{P}} = \nabla\lambda^{-1}\,\hat{\mathbf{f}}(\nabla\lambda\,\mathbf{P},\,\hat{\mathbf{a}}_{\nabla\lambda},\,\lambda(\mathbf{X})). \tag{6.7}$$

If we choose a change of reference configuration that instantaneously brings the point $\mathbf{X}$ into coincidence with the archetype, namely: $\nabla\lambda = \mathbf{P}^{-1}$, Equation (6.7) yields:

$$\dot{\mathbf{P}} = \mathbf{P}\,\hat{\mathbf{f}}(\mathbf{I},\,\hat{\mathbf{a}}_{\mathbf{P}^{-1}}). \tag{6.8}$$

In this equation, the function $\hat{\mathbf{f}}$ represents how a material point which coincides instantaneously with the archetype reacts to the instantaneous application of any value of the argument $\mathbf{a}$ measured therein. This is clearly a material property that can be determined once and for all. Renaming it $\bar{\mathbf{f}}$, we can write the general form of the evolution law as:

$$\dot{\mathbf{P}} = \mathbf{f}(\mathbf{P},\,\mathbf{a},\,\mathbf{X}) = \mathbf{P}\,\bar{\mathbf{f}}(\hat{\mathbf{a}}_{\mathbf{P}^{-1}}). \tag{6.9}$$

We observe that this equation, being based as it is on a single archetype, already incorporates the assumed uniformity of the body. For the specific case in which the argument list $\mathbf{a}$ consists of just the Eshelby stress $\mathbf{b}$, we obtain:

$$\dot{\mathbf{P}} = \mathbf{P}\,\bar{\mathbf{f}}(J_P\,\mathbf{P}^T\,\mathbf{b}\,\mathbf{P}^{-T}). \tag{6.10}$$

Defining the *inhomogeneity velocity gradient at the reference configuration* as:

$$\mathbf{L}_P = \dot{\mathbf{P}}\mathbf{P}^{-1}, \tag{6.11}$$

we can write Equation (6.10) as:

$$\mathbf{L}_P = \mathbf{P}\,\bar{\mathbf{f}}(J_P\,\mathbf{P}^T\,\mathbf{b}\,\mathbf{P}^{-T})\,\mathbf{P}^{-1}. \tag{6.12}$$

We note that the terminology "velocity gradient" may convey the wrong impression that $\mathbf{L}_P$ is actually the gradient of a vector field, which it is not the case in general. The pull-back of the inhomogeneity velocity gradient to the archetype is given by:

$$\bar{\mathbf{L}}_P = \mathbf{P}^{-1} \mathbf{L}_P \mathbf{P} = \mathbf{P}^{-1} \dot{\mathbf{P}}. \tag{6.13}$$

We can, therefore, write Equation (6.12) also as:

$$\bar{\mathbf{L}}_P = \bar{\mathbf{f}}(J_P \mathbf{P}^T \mathbf{b} \mathbf{P}^{-T}), \tag{6.14}$$

or, in the more general case of an arbitrary list of arguments $\mathbf{a}$:

$$\bar{\mathbf{L}}_P = \bar{\mathbf{f}}(\hat{\mathbf{a}}_{\mathbf{P}^{-1}}). \tag{6.15}$$

We say that in this form the evolution law has been *reduced to the archetype*. In conclusion, the dependence of an evolution law on the implant $\mathbf{P}$ cannot be arbitrary, but must be of the explicit form shown in Equations (6.14) or (6.15). This drastic reduction should not be surprising, since the implant $\mathbf{P}$ does not have an absolute physical meaning, but represents a means of comparison between two states of a material point. To completely determine an evolution law, we need to know just how a given archetype responds instantaneously to arbitrary values of the list of arguments $\mathbf{a}$.

*Remark 6.1.* If we insist on imposing the mass consistency condition (1.11) for the material isomorphisms at a fixed point $\mathbf{X}$ at different times, and if there is no change in density at the reference configuration, then the determinant $J_P$ of $\mathbf{P}(\mathbf{X}, t)$ will remain constant in time. In that case:

$$0 = \frac{dJ_P}{dt} = J_P\, P_I^{-\alpha}\, \dot{P}^I_{\alpha} = J_P\, tr \bar{\mathbf{L}}_P. \tag{6.16}$$

So, the mass consistency condition translates in this case into a traceless evolution function $\bar{\mathbf{f}}$. The physical meaning of this condition, namely, the conservation of the local density, may be rendered as the absence of bulk growth.

### 6.3.3 The principle of actual evolution

When the material symmetry group is discrete, each fibre of the material $G$-structure consists of a discrete set of points. Consequently, any smooth change of the implant field will necessarily lead to points not belonging to the original $G$-structure. In other words, any non-trivial evolution law represents an actual process of material evolution, whereby a true material remodelling takes place. The material $G$-structure evolves into nearby conjugate $G$-structures.

The situation is quite different when the material symmetry group is continuous (i.e., a Lie group). For, in this case, each fibre of the $G$-structure consists of a differentiable manifold which, as we know, is diffeomorphic to the Lie group of symmetries of the archetype. It is, therefore, possible to take advantage of the degree of freedom afforded by the continuity of the group so as to change the implant field smoothly while always remaining in the original material $G$-structure. For this reason, when an evolution law is proposed, we must make sure that it instantaneously prescribes a change of the implant field leading outward of the present $G$-structure. We call this notion the *principle of actual evolution*. We remark once again that, although we are expressing these ideas in terms of the $G$-structure associated with a uniform body, they are in fact applicable in a point-wise sense. In particular, the principle of actual evolution expresses the fact that an evolution law must represent a time-dependent material automorphism that does not belong identically to the symmetry group of the point.

In order to obtain a precise mathematical statement of the principle of actual evolution, we record what the situation is that we are trying to prevent from happening. And that is an evolution of the form:

$$\mathbf{P}(\mathbf{X}, t) = \mathbf{P}(\mathbf{X}, 0) \, \mathbf{G}(t), \tag{6.17}$$

for $t > 0$, where $\mathbf{G}(t)$ is a time-dependent element of the symmetry group $\bar{\mathcal{G}}$ of the archetype and $t = 0$ is, say, the present time. Clearly, $\mathbf{G}(0) = \mathbf{I}$. Differentiating this equation with respect to time and setting $t = 0$ we get:

$$\dot{\mathbf{P}}(\mathbf{X}, 0) = \mathbf{P}(\mathbf{X}, 0) \left[ \frac{d\mathbf{G}(t)}{dt} \right]_{t=0}. \tag{6.18}$$

But the indicated derivative on the right-hand side of this equation is a vector tangent to the group $\bar{\mathcal{G}}$ at the unit element $\mathbf{I}$, namely, an element of the Lie algebra $\bar{\mathfrak{g}}$ of $\bar{\mathcal{G}}$ (or, equivalently, an infinitesimal generator of a one-parameter subgroup). We conclude that the situation we are trying to avoid is:

$$\bar{\mathbf{L}}_P \in \bar{\mathfrak{g}}. \tag{6.19}$$

Comparing this condition with Equation (6.15), we can formulate the principle of actual evolution as follows: For a true evolution to take place, the evolution function $\bar{\mathbf{f}}$ must yield values outside of the material Lie algebra of the archetype.

As an example, consider the case of an isotropic solid and assume the archetype to be in a natural state, in which we have chosen an orthonormal frame. Then, the symmetry group is the full orthogonal group, whose Lie algebra consists of all skew-symmetric matrices. The principle of actual evolution requires that $\bar{\mathbf{f}}$ must consistently produce matrices with a non-vanishing symmetric part.

As a result of the principle of actual evolution, two different evolution laws whose results differ by an element of the material Lie algebra must be considered equivalent. In the example of the isotropic solid this means that only the symmetric part of the resulting inhomogeneity velocity gradient matters as far as the evolution is concerned. Therefore, two evolution laws (for an isotropic solid) whose results differ by a skew-symmetric matrix are equivalent. For this statement to make sense not just instantly but for finite times, we must ensure that any evolution law (which is formulated in terms of particular sections of the $G$-structure) produces the same $G$-structure evolution, regardless of the particular section chosen at the initial time. This result will be guaranteed by a further requirement of consistency between the evolution law and the material symmetry group, a requirement to be discussed presently.

### 6.3.4 Material symmetry consistency

#### Change of archetype

Material evolution is a matter of comparison between the constitutive responses of a body point at two different times. The archetype is not involved in this process, but is only used as an auxiliary device to obtain a concrete expression of the form (6.15).

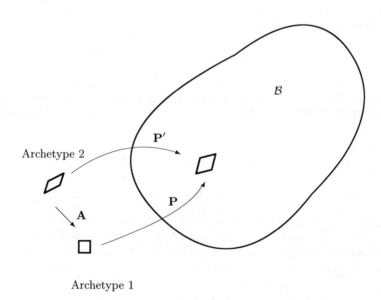

**Fig. 6.4.** Change of archetype: $\mathbf{P}' = \mathbf{P}\,\mathbf{A}$

Let two archetypes be given, related by the linear map $\mathbf{A}$, as shown in Figure 6.4. Indicating with primes the quantities associated with the second archetype, the corresponding evolution equation is:

$$\bar{\mathbf{L}}'_{P'} = \bar{\mathbf{f}}'(\hat{\mathbf{a}}_{\mathbf{P}'-1}). \qquad (6.20)$$

To establish the relationship between the functions $\bar{\mathbf{f}}$ and $\bar{\mathbf{f}}'$, we note that since $\mathbf{A}$ is independent of time we have, in addition to:

$$\mathbf{P}' = \mathbf{PA}, \qquad (6.21)$$

also:

$$\dot{\mathbf{P}}' = \dot{\mathbf{P}}A, \qquad (6.22)$$

whence, invoking Equation (6.13), we obtain:

$$\bar{\mathbf{L}}'_{P'} = \mathbf{A}^{-1}\bar{\mathbf{L}}_P\mathbf{A}. \qquad (6.23)$$

Using this result in conjunction with Equations (6.15) and (6.20), we can write:

$$\bar{\mathbf{f}}'(\hat{\mathbf{a}}_{A^{-1}P^{-1}}) = \mathbf{A}^{-1}\bar{\mathbf{f}}(\hat{\mathbf{a}}_{P^{-1}})\mathbf{A}. \qquad (6.24)$$

**Symmetry consistency**

A change $\mathbf{A}$ of archetype is a *symmetry of the evolution law* if the functions $\bar{\mathbf{f}}$ and $\bar{\mathbf{f}}'$ are one and the same function. In other words, $\mathbf{A}$ is a symmetry of the evolution law if, and only if, Equation (6.24) can be replaced by:

$$\bar{\mathbf{f}}(\hat{\mathbf{a}}_{A^{-1}P^{-1}}) = \mathbf{A}^{-1}\bar{\mathbf{f}}(\hat{\mathbf{a}}_{P^{-1}})\mathbf{A}. \qquad (6.25)$$

The collection $\bar{\mathcal{A}}$ of all the linear maps $\mathbf{A}$ satisfying this equation identically for all $\mathbf{P}$ is never empty, since the identity $\mathbf{I}$ is always a member of $\bar{\mathcal{A}}$. In fact, this collection is a multiplicative group, as can be verified directly. We call this group the *evolution symmetry group with respect to the given archetype*. Since an evolution equation is the reflection of an intrinsic material behaviour, we must demand that every constitutive symmetry be also a symmetry of the constitutive law. We call this requirement the *material symmetry consistency condition*. It is summarized in the following expression:

$$\bar{\mathcal{G}} \subset \bar{\mathcal{A}}, \qquad (6.26)$$

where (as, for instance, in Equation (1.21)), we denote by $\bar{\mathcal{G}}$ the material symmetry group of the archetype.

**A corollary**

An important corollary of symmetry consistency (so important, in fact, as to almost justify the consistency requirement itself), is the one already alluded to at the end of section 6.3.3. An evolution law is necessarily formulated in terms of the evolution of an implant map (from a given archetype). We know, however, that implant maps are not unique in general, their multiplicity being controlled by the material symmetry group of the archetype. The burning question is, therefore, the following: If $\mathbf{P} = \mathbf{P}(t)$ is a solution of Equation (6.15) for a given initial condition $\mathbf{P}(0) = \mathbf{P}_0$ and for any arbitrarily given $\mathbf{a} = \mathbf{a}(t)$, is the solution $\mathbf{P}'(t)$ of Equation (6.15) (for a small enough interval of time) with initial condition $\mathbf{P}'(0) = \mathbf{P}_0 \mathbf{G}$ and the same $\mathbf{a} = \mathbf{a}(t)$ related to $\mathbf{P}(t)$ by $\mathbf{P}'(t) = \mathbf{P}(t)\mathbf{G}(t)$? Here, $\mathbf{G}(t)$ is some, possibly time-dependent, element of the material symmetry group of the archetype. More pictorially, we can rephrase this question in the case of a uniform material as follows: Given two different sections of the initial material $G$-structure, does the evolution law move them, as time goes on, consistently to sections of the same, time-dependent, $G$-structure? The symmetry consistency condition guarantees that the answer to this question is positive. The verification of this statement is left as an exercise. In the course of carrying it out, we obtain the additional information that $\mathbf{G}(t) = \mathbf{G}(0)$, namely, that the "separation" between the two solutions remains constant in time. In terms of $G$-structures this can be described as the fact that the evolution law causes the cross sections to move while preserving their relative positions, like synchronized swimmers.

*Remark 6.2.* The following exercise is proposed to the reader: Reduce an evolution law of the implicit form

$$\phi(\mathbf{P}, \dot{\mathbf{P}}, \mathbf{b}, \mathbf{F}, \dot{\mathbf{F}}) = 0, \tag{6.27}$$

as much as possible. Assume the archetype to be an isotropic solid in a natural state. Use the following criteria: reduction to the archetype, actual evolution, symmetry consistency and material frame indifference (as expected because of the explicit appearance of the deformation gradient). Such more general evolution laws are often invoked in models of growth, whose mechanism is hypothesized to be possibly triggered by signals sent by biochemical sensors of strain or strain rate.

*Remark 6.3.* The combination of the principle of actual evolution with the principle of symmetry consistency can, somewhat surprisingly, lead to a relaxation of the latter. Indeed, if two evolution laws are to be considered identical when they differ by an element of the Lie algebra of the material symmetry group, we may extend Equation (6.25) by adding to the right-hand side an arbitrary element of this Lie algebra. As a result, more evolution laws will become acceptable than otherwise. An interesting example of this situation will be shown in Chapter 7, Section 7.5, within the framework of second-grade materials.

## 6.4 The field equations of remodelling and bulk growth

The balance equations that govern processes of remodelling and growth can be regarded as the result of focusing attention on a single component of a chemically reacting mixture of several substances. Mass is, therefore, not necessarily conserved and transfers of momentum, angular momentum, energy and entropy appear in the equations as extra contributors to the total balance. In a process of growth, these extra terms may be, at least in part, attributed to the very addition or subtraction of mass. At best, we may have a case of "compliant" growth, whereby the new mass happens to enter at the same velocity, specific energy and specific entropy as the local substratum. At worst, not only will the entering quantities be at a different state than the substratum, but the process of growth (or even just remodelling) itself may entail other sources of discrepancy.[1]

By bulk, or volumetric, growth we understand a process of addition or removal of mass while the body particles retain their identity. It is only the mass density that changes with time. In contradistinction with this situation it is possible, and certainly meaningful and practical, to consider growth by addition of mass at the boundaries of the body. Thus, for example, holes may be closed or created which change the topology of the original body. These more complicated processes are excluded from the present analysis.[2]

### 6.4.1 Balance of mass

While at some conceivable level of analysis (such as that of chemically reacting mixtures) the appearance or disappearance of mass of one species may be accounted for by concomitant losses or gains in other species, in a bulk growth model we accept the existence of sources and sinks of mass as part of the theory. For a fixed volume $\Omega$ in a reference configuration, we write:

$$\frac{D}{Dt} \int_\Omega \rho_R \, d\Omega = \int_\Omega \Pi \, d\Omega + \int_{\partial \Omega} M \, dS, \qquad (6.28)$$

where $\rho_R$ is the possibly time-varying mass density in the reference configuration, $\Pi$ is the (smooth) volumetric source of mass and $M$ is a possible mass flux through the boundary $\partial \Omega$, with exterior unit normal $\mathbf{N}$. The existence of a mass-flux vector $\mathbf{M}$ such that:

$$M = \mathbf{M} \cdot \mathbf{N} \qquad (6.29)$$

being guaranteed by Cauchy's theorem[3], we can write the local form of Equation (6.28) as:

---

[1] In [11] the "non-compliant" source of internal energy is lumped as a clearly indicated extra term in the balance equation, so that the compliant case can be easily recovered by eliminating this extra term. This practice, which will be followed here, was also adopted for the totality of balance laws in [48], where the compliant contributions were called "reversible". To avoid unnecessary confusion with the use of this terminology in the strictest thermodynamical sense, we prefer to call them "compliant".

[2] For some indication of the difficulties involved in modelling these processes, see [82].

[3] Or assumed a priori.

$$\frac{\partial \rho_R}{\partial t} = \Pi + \text{Div } \mathbf{M}. \tag{6.30}$$

In terms of the spatial counterparts of $\rho_R$, $\Pi$ and $\mathbf{M}$, given respectively by:

$$\rho = J_F^{-1} \rho_R, \tag{6.31}$$

$$\pi = J_F^{-1} \Pi, \tag{6.32}$$

and

$$\mathbf{m} = J_F^{-1} \mathbf{F} \mathbf{M} \tag{6.33}$$

with

$$m = \mathbf{m} \cdot \mathbf{n} \tag{6.34}$$

we obtain the following Eulerian balance law:

$$\frac{D\rho}{Dt} = \pi + \text{div } \mathbf{m} - \rho \, \text{div } \mathbf{v}, \tag{6.35}$$

where $\mathbf{v}$ is the velocity field and where we have distinguished between the spatial divergence operator ($div$) and its referential counterpart ($Div$).

### 6.4.2 Balance of linear momentum

Starting with the Eulerian formulation, the balance of linear momentum in a moving material volume $\omega$ in an inertial frame states is:

$$\frac{D}{Dt} \int_\omega \rho \, \mathbf{v} \, d\omega = \int_\omega \mathbf{f}_s \, d\omega + \int_{\partial\omega} \mathbf{t}_s \, ds + \int_\omega (\pi \, \mathbf{v} + \bar{\mathbf{p}}) \, d\omega + \int_{\partial\omega} (m \, \mathbf{v} + \bar{\mathbf{m}}) \, ds. \tag{6.36}$$

In this equation, $\mathbf{f}_S$ is the body force per unit spatial volume and $\mathbf{t}_S$ is the surface traction per unit spatial area. The terms $\bar{\mathbf{p}}$ and $\bar{\mathbf{m}}$ represent, respectively, the non-compliant sources of volumetric momentum and of momentum flux. We note that, on account of Equation (6.34), we can write:

$$m \, \mathbf{v} = (\mathbf{v} \otimes \mathbf{m}) \, \mathbf{n}. \tag{6.37}$$

Invoking now Cauchy's tetrahedron argument[4], we deduce the existence of a (Cauchy) stress tensor $\mathbf{t}$ such that:

$$\mathbf{t}_S + \bar{\mathbf{m}} = \mathbf{t} \, \mathbf{n}. \tag{6.38}$$

Taking into consideration Equations (6.37) and (6.38) and using the balance of mass in the form of Equation (6.35), we obtain the following local Eulerian version of the balance of momentum:

$$\rho \frac{Dv^i}{Dt} = f_S^i + \bar{p}^i + t^{ij}{}_{,j} + m^j \, v^i{}_{,j}. \tag{6.39}$$

---

[4] Under the usual assumptions.

The Lagrangian counterpart of this equation is:

$$\rho_R \frac{\partial v^i}{\partial t} = f_R^i + \bar{p}_R^i + T^{iI}{}_{,I} + M^I v^i{}_{,I}. \tag{6.40}$$

In the last two equations we have used (Cartesian) components to avoid any ambiguity of the notation. The relation between the Lagrangian and Eulerian densities is:

$$f_R^i = J_F f_S^i, \tag{6.41}$$

$$\bar{p}_R^i = J_F \bar{p}^i, \tag{6.42}$$

and

$$T^{iI} = J_F F^{-I}{}_j t^{ij}, \tag{6.43}$$

where $T^{iI}$ are the components of the Piola stress $\mathbf{T}$.

### 6.4.3 Balance of angular momentum

In the absence of body and surface couples, the global balance of angular momentum is expressed as:

$$\frac{D}{Dt} \int_\omega \mathbf{r} \times (\rho \mathbf{v}) \, d\omega = \int_\omega \mathbf{r} \times (\mathbf{f}_S + \pi \mathbf{v} + \bar{\mathbf{p}}) \, d\omega + \int_{\partial \omega} \mathbf{r} \times [(\mathbf{t} + \mathbf{v} \otimes \mathbf{m})\mathbf{n}] \, ds, \tag{6.44}$$

where $\mathbf{r}$ is the spatial position vector with respect to the inertial frame. Enforcing the balance of mass and linear momentum, the local form of this equation boils down to the symmetry of the following modified Cauchy stress:

$$\tilde{\mathbf{t}} \equiv \mathbf{t} + \mathbf{v} \otimes \mathbf{m}. \tag{6.45}$$

Note that, due to the momentum contribution of the incoming mass, the Cauchy stress itself is not symmetric even in the absence of the non-compliant extra mass flux $\bar{\mathbf{m}}$. Introducing the modified Piola stress as:

$$\tilde{\mathbf{T}} \equiv \mathbf{T} + \mathbf{v} \otimes \mathbf{M}, \tag{6.46}$$

and using Equation (6.43), the symmetry of $\tilde{\mathbf{t}}$ translates into the symmetry of $\tilde{\mathbf{T}} \mathbf{F}^T$, namely:

$$\tilde{T}^{iI} F^j{}_I = \tilde{T}^{jI} F^i{}_I. \tag{6.47}$$

### 6.4.4 Balance of energy

Denoting by $u_\rho$ the internal energy per unit mass, we can write the global Eulerian form of the energy balance as:

## 6.4 The field equations of remodelling and bulk growth

$$\frac{D}{Dt} \int_\omega \rho(u_\rho + \frac{1}{2}\mathbf{v}\cdot\mathbf{v})\, d\omega$$
$$= \int_\omega (\mathbf{f}_S\cdot\mathbf{v} + \rho r_\rho)\, d\omega + \int_\omega [\pi(u_\rho + \frac{1}{2}\mathbf{v}\cdot\mathbf{v}) + \bar{u} + \bar{\mathbf{p}}\cdot\mathbf{v}]\, d\omega$$
$$+ \int_{\partial\omega}\left((\mathbf{t}_S + \bar{\mathbf{m}}).\mathbf{v} + m(u_\rho + \frac{1}{2}\mathbf{v}\cdot\mathbf{v}) + h\right) ds, \tag{6.48}$$

where $\bar{u}$ is the non-compliant volumetric contribution to the internal energy[5] and where $r_\rho$ and $h$ represent, respectively, the rate of non-mechanical energy supply per unit mass ("radiation") and per unit area ("conduction"). The latter can be expressed (again, through the tetrahedron argument) in terms of a heat flux vector $\mathbf{q}$:

$$h = -\mathbf{q}\cdot\mathbf{n}. \tag{6.49}$$

Using Equations (6.34), (6.35), (6.38) and (6.39), the local form of Equation (6.48) is obtained as:

$$\rho\frac{Du_\rho}{Dt} = \rho r_\rho + \bar{u} + t^{ij}v_{i,j} + m^j u_{\rho,j} - q^j_{,j}. \tag{6.50}$$

Its Lagrangian counterpart is:

$$\rho_R\frac{\partial u_\rho}{\partial t} = \rho_R r_\rho + \bar{U} + T^{iI}v_{i,I} + M^I u_{\rho,I} - Q^I_{,I}, \tag{6.51}$$

with:

$$Q^I = J_F\, F^{-I}_{\ i}\, q^i, \tag{6.52}$$

and

$$\bar{U} = J_F\, \bar{u}. \tag{6.53}$$

### 6.4.5 The Clausius-Duhem inequality and its consequences

What does the Second Law of Thermodynamics have to say when one has come to terms with the assumption that, at least for modelling purposes, matter is continuous? A possible answer to this question is the one embodied in the Clausius-Duhem inequality. It is not our intention to either contest or defend the validity of this, or any other, particular form of the Second Law. And this is not for lack of space, but for lack of knowledge. We simply do not know how such a fundamental law of nature can be rendered compatible with such a bold, albeit manifestly useful, assumption. To further complicate matters, there is the issue of the use of internal state variables, which can be regarded as inexorably intertwined with any modelling of material evolution. Is the formulation of the Second Law for a continuous medium to be modified,

---

[5] In principle, this term could have been absorbed into $r_\rho$, with the understanding that it may eventually be specified constitutively, rather than just externally.

150    6 An overview of the theory of material evolution

perhaps augmented, in the presence of internal variables? Or must we recalcitrantly cling to the original form and live with the consequences? Be that as it may, we proceed to the formulation of the Clausius-Duhem inequality adding terms that are consistent with the philosophy used for the balance equations formulated so far.

**The Clausius-Duhem inequality**

Defining the entropy content per unit mass, $s_\rho$, we write the Eulerian version of the global Clausius-Duhem inequality as follows:

$$\frac{D}{Dt}\int_\omega \rho s_\rho \, d\omega \geq \int_\omega \left(\frac{\rho r_\rho + \bar{u} - \bar{h}}{\theta} + \pi s_\rho\right) d\omega + \int_{\partial \omega} \left(\frac{h}{\theta} + m s_\rho\right) ds, \tag{6.54}$$

$\theta$ being the absolute temperature. We note that we have added a further volumetric sink $\bar{h}$ of non-compliant entropy, which should be specified constitutively.[6] The resulting local form is:

$$\rho \frac{Ds_\rho}{Dt} \geq \frac{\rho r_\rho + \bar{u} - \bar{h}}{\theta} + m^i s_{\rho,i} - \left(\frac{q^i}{\theta}\right)_{,i}. \tag{6.55}$$

The Lagrangian version is:

$$\rho_R \frac{\partial s_\rho}{\partial t} \geq \frac{\rho_R r_\rho + \bar{U} - \bar{H}}{\theta} + M^I s_{\rho,I} - \left(\frac{Q^I}{\theta}\right)_{,I}, \tag{6.56}$$

with an obvious notation.

Introducing the Helmholtz free energy per unit mass:

$$\psi_\rho \equiv u_\rho - \theta s_\rho, \tag{6.57}$$

we can use the balance of energy to rewrite Equation (6.56) in the form:

$$\rho_R \dot{\psi}_\rho + \rho_R \dot{\theta} \, s_\rho - \bar{H} - M^I(\psi_{\rho,I} + \theta_{,I} s_\rho) - T_i^I v^i{}_{,I} + \frac{1}{\theta} Q^I \theta_{,I} \leq 0. \tag{6.58}$$

For later use, notice the subtle distinctions (even in the absence of mass flow) between this form of the Clausius-Duhem inequality, which does not presume mass conservation, and its usual counterpart, namely:

$$\dot{\psi}_R + s_R \dot{\theta} + \frac{1}{\theta} Q^I \theta_{,I} - T_i^I v^i{}_{,I} \leq 0, \tag{6.59}$$

where $\psi_R$ and $s_R$ are measured per unit referential *volume*. Notice also that the entropy sink $\bar{H}$ may be present even if growth does not occur. It may, indeed, be responsible for the process of remodelling.

---

[6] We follow here the idea of Cowin and Hegedus [11], with a slightly different notation.

## Is mass flux possible?

Properly speaking, the field equations we have presented are adequate for simple materials only. Had we wanted to include second-grade effects, we might have changed, for instance, the energy balance equation to include the power of the hyperstress. That said, we will proceed to assume that the second gradient of the deformation is included in the list of arguments of the constitutive equations. By the principle of equipresence, we will include it in all constitutive equations and let the Clausius-Duhem inequality eliminate this argument wherever it is inconsistent with the principle of dissipation. This is not a futile exercise, since we have introduced the new constitutive quantity **M** (the mass flux) and it is quite possible that the Clausius-Duhem inequality may allow the gradient of **F** to stay as an independent variable of **M**, much in the same way as it allows the gradient of $\theta$ to remain as an independent variable of **Q**. Let us assume, therefore, that the constitutive equations are as follows:

$$T_i{}^I = T_i{}^I(F^j{}_J,\ F^j{}_{J,K},\ \theta,\ \theta_{,J}), \tag{6.60}$$

$$Q^I = Q^I(F^j{}_J,\ F^j{}_{J,K},\ \theta,\ \theta_{,J}), \tag{6.61}$$

$$M^I = M^I(F^j{}_J,\ F^j{}_{J,K},\ \theta,\ \theta_{,J}), \tag{6.62}$$

$$\psi_R = \rho_R\,\psi_\rho = \psi_R(F^j{}_J,\ F^j{}_{J,K},\ \theta,\ \theta_{,J}), \tag{6.63}$$

and

$$s_R = \rho_R\,s_\rho = s_R(F^j{}_J,\ F^j{}_{J,K},\ \theta,\ \theta_{,J}). \tag{6.64}$$

The volumetric mass source $\Pi$ and the non-compliant sources of momentum $\bar{\mathbf{p}}_R$, of internal energy $\bar{U}$ and of entropy $-\bar{H}$ may be either given externally or constitutionally. We are now about to substitute the above expressions in the Clausius-Duhem inequality, where the time-derivative of the free energy (among other quantities) needs to be evaluated. Clearly, when calculating this time derivative, the fact that evolution is taking place plays a prominent role. In other words, we should express the constitutive equations in terms of (time-dependent) implants of an archetype. Nevertheless, in order to clarify the issue of the second gradient of the deformation, let us assume for a moment (as we certainly may) that this type of evolution has been frozen, leaving the more general case for later treatment. Substituting Equations (6.60-6.64) in Equation (6.58), we obtain:

$$\rho_R\left(\frac{\partial\psi_\rho}{\partial F^i{}_I}\dot{F}^i{}_I + \frac{\partial\psi_\rho}{\partial F^i{}_{I,J}}\dot{F}^i{}_{I,J} + \frac{\partial\psi_\rho}{\partial\theta}\dot\theta + \frac{\partial\psi_\rho}{\partial\theta_{,I}}\dot\theta_{,I}\right)$$
$$- M^K\left(\frac{\partial\psi_\rho}{\partial F^i{}_I}F^i{}_{I,K} + \frac{\partial\psi_\rho}{\partial F^i{}_{I,J}}F^i{}_{I,JK} + \frac{\partial\psi_\rho}{\partial\theta}\theta_{,K} + \frac{\partial\psi_\rho}{\partial\theta_{,I}}\theta_{,IK} + \theta_{,K}\,s_\rho\right)$$
$$+ \rho_R\,\dot\theta\,s_\rho - \bar{H} - T_i{}^I\,v^i{}_{,I} + \frac{1}{\theta}Q^I\theta_{,I} \le 0, \tag{6.65}$$

or, collecting some terms:

$$\left[\rho_R \frac{\partial \psi_\rho}{\partial F^i_I} - T_i{}^I\right] \dot{F}^i_I + \left[\rho_R \frac{\partial \psi_\rho}{\partial F^i_{I,J}}\right] \dot{F}^i_{I,J}$$

$$+ \left[\rho_R \frac{\partial \psi_\rho}{\partial \theta} + \rho_R s_\rho\right] \dot{\theta} + \left[\rho_R \frac{\partial \psi_\rho}{\partial \theta_{,I}}\right] \dot{\theta}_{,I}$$

$$- M^K \left(\frac{\partial \psi_\rho}{\partial F^i_I} F^i_{I,K} + \frac{\partial \psi_\rho}{\partial F^i_{I,J}} F^i_{I,JK} + \frac{\partial \psi_\rho}{\partial \theta} \theta_{,K} + \frac{\partial \psi_\rho}{\partial \theta_{,I}} \theta_{,IK} + \theta_{,K} s_\rho\right)$$

$$- \bar{H} + \frac{1}{\theta} Q^I \theta_{,I} \leq 0. \tag{6.66}$$

It follows that:

$$\rho_R \frac{\partial \psi_\rho}{\partial F^i_I} - T_i{}^I = 0, \tag{6.67}$$

$$\rho_R \frac{\partial \psi_\rho}{\partial F^i_{I,J}} = 0, \tag{6.68}$$

$$\rho_R \frac{\partial \psi_\rho}{\partial \theta} + \rho_R s_\rho = 0, \tag{6.69}$$

and

$$\rho_R \frac{\partial \psi_\rho}{\partial \theta_{,I}} = 0. \tag{6.70}$$

According to Equation (6.68) the free-energy density is independent of the second gradient of the deformation and, on account of Equations (6.67) and (6.69), so are the Piola stress and the entropy density. Using Equations (6.67-6.70), Equation (6.66) yields the following residual inequality:

$$-\frac{1}{\rho_R} M^K T_i{}^I F^i_{I,K} - \bar{H} + \frac{1}{\theta} Q^I \theta_{,I} \leq 0. \tag{6.71}$$

We remark once again that the dissipative evolution terms are still missing, a deficiency that we will soon remedy. Nevertheless, it is clear that the Clausius-Duhem inequality allows the second gradient of the deformation to remain as an independent variable of the mass flux, as expected in a diffusive phenomenon.

**Evolution**

Having established that mass flux is consistently tenable even in the absence of second-grade effects in the free energy density, it is clear, nevertheless, that a consistent formulation of the evolution of a material with mass flux must be carried out within the theory of second-grade materials. The reason for this is that the second gradient of the deformation appears in the constitutive

## 6.4 The field equations of remodelling and bulk growth

equation of the mass flux and, therefore, the time-dependent material isomorphisms must take this fact into consideration. Accordingly, we must postpone the treatment of evolution with mass flux to the next chapter, in which we deal with second-grade bodies. We will see there that it is possible to subordinate the second-grade evolution to the first-grade one in a precise sense, but even this subordination is not just a trivial cancelling of the evolution of the second-grade part of the implants (such a cancelling would not be invariant, for example, under a change of reference configuration). Leaving these issues to the next chapter, we will now continue with the legitimate first-grade case, whereby the dependence on the second gradient of the deformation disappears altogether, implying (according to the residual inequality (6.71)) that the mass flux vanishes identically. In other words, the bulk growth (if any) under consideration is due entirely to the volumetric contribution $\Pi$.

Having thus postponed the analysis of the second-grade case, we are left with the prescription of the time-evolution of the first-grade implants $\mathbf{P}(\mathbf{X}, t)$. Assuming the mass consistency condition to be satisfied, if we denote the constant density of the archetype by $\bar{\rho}_R$, the density at the reference configuration is given by:

$$\rho_R = \bar{\rho}_R\, J_P^{-1}, \tag{6.72}$$

whence:

$$\dot{\rho}_R = -\rho_R\, P_I^{-\alpha}\, \dot{P}_\alpha^I, \tag{6.73}$$

or:

$$tr(\bar{\mathbf{L}}_P) = -\frac{\dot{\rho}_R}{\rho_R}. \tag{6.74}$$

Comparing this result with Equation (6.30), we conclude that:

$$\Pi = -\left(\rho_R\, tr\bar{\mathbf{L}}_P + Div\mathbf{M}\right), \tag{6.75}$$

which in the present analysis reduces to:

$$\Pi = -\rho_R\, tr\bar{\mathbf{L}}_P. \tag{6.76}$$

This equation implies that (having assumed that the mass consistency condition is satisfied) no separate constitutive law needs to be given for the volumetric mass source $\Pi$, since it will be dictated by the evolution of the implants. It is worthwhile noting that the time evolution of the determinant of $\mathbf{P}$ tells us whether there is instantaneous growth (negative time-derivative) or resorption (positive time-derivative). In other words, growth is taking place if $tr\bar{\mathbf{L}}_P$ is negative, while the reduced (or increased) value of the determinant of $\mathbf{P}$ is a measure of accumulated growth (or resorption) from a given time origin. If $\mathbf{P}$ is non-spherical, there may also be rotations and distortions in addition to growth or resorption. Thus, the only difference between a theory of growth and a purely remodelling theory (such as the case of visco-elastoplasticity) resides in the fact that in the latter we require that the evolution function be traceless. In a theory of growth and remodelling it is precisely the trace of the inhomogeneity velocity gradient that will carry the information about the volumetric growth.

The Clausius-Duhem inequality (6.58) is not to be violated in any conceivable thermomechanical process, thus leading to restrictions on the possible constitutive and evolution laws of a given class of materials. A class of materials is defined by a choice of a class of constitutive functionals and of a list of independent arguments therein. In the case of bodies undergoing material evolution in the sense precisely defined in previous sections, an Equation such as (1.20) clearly displays the fact that the list of independent variables must include the time-dependent implant field $\mathbf{P}$, which acts, therefore, as a special kind of internal state variable[7]. What is special about this internal state variable is that it enters the constitutive law mainly in a multiplicative way, like some kind of "change of scale", only.

Assuming, for example, that the archetype consists of a simple thermoelastic heat conductor, the list of arguments of its constitutive law consists of $F^i_I$, $\theta$ and $\theta_{,I}$. Consequently,

$$\psi_R = J_P^{-1} \, \bar{\psi}_R(F^i_I P^I_\alpha, \, \theta, \, \theta_{,I} P^I_\alpha). \tag{6.77}$$

Moreover, we have:

$$\rho_R \dot{\psi}_\rho = \rho_R \frac{\partial}{\partial t}\left(\frac{\psi_R}{\rho_R}\right) = \dot{\psi}_R - \frac{\dot{\rho}_R}{\rho_R}\psi_R = \dot{\psi}_R + (tr\bar{\mathbf{L}}_P)\,\psi_R. \tag{6.78}$$

and

$$\dot{\psi}_R = -J_P^{-1}\, P_I^{-\alpha}\, \dot{P}^I_\alpha\, \bar{\psi}_R + J_P^{-1}\,\frac{\partial \bar{\psi}_R}{\partial F^i_I}\,\dot{F}^i_I$$
$$+ J_P^{-1}\,\frac{\partial \bar{\psi}_R}{\partial \theta}\,\dot{\theta} + J_P^{-1}\,\frac{\partial \bar{\psi}_R}{\partial \theta_{,I}}\,\dot{\theta}_{,I} + J_P^{-1}\,\frac{\partial \bar{\psi}_R}{\partial P^I_\alpha}\,\dot{P}^I_\alpha. \tag{6.79}$$

But, with some abuse of notation,

$$\frac{\partial \bar{\psi}_R}{\partial F^i_I} = \frac{\partial \bar{\psi}_R}{\partial (F^i_K P^K_\alpha)}\, P^I_\alpha, \tag{6.80}$$

$$\frac{\partial \bar{\psi}_R}{\partial P^I_\alpha} = \frac{\partial \bar{\psi}_R}{\partial (F^i_K P^K_\alpha)}\, F^i_I + \frac{\partial \bar{\psi}_R}{\partial (\theta_{,K} P^K_\alpha)}\, \theta_{,I}, \tag{6.81}$$

and

$$\frac{\partial \bar{\psi}_R}{\partial \theta_{,I}} = \frac{\partial \bar{\psi}_R}{\partial (\theta_{,K} P^K_\alpha)}\, P^I_\alpha. \tag{6.82}$$

Introducing these results in Equation (6.79) and collecting terms, we obtain:

---

[7] See, e.g., [7].

## 6.4 The field equations of remodelling and bulk growth

$$\psi_R \left[ J_P^{-1} \frac{\partial \bar{\psi}_R}{\partial (F^i_K P^K_\alpha)} P^I_\alpha - T_i^{\ I} \right] \dot{F}^i_I + \left[ J_P^{-1} \frac{\partial \bar{\psi}_R}{\partial \theta} + s_R \right] \dot{\theta}$$

$$+ \left[ J_P^{-1} \frac{\partial \bar{\psi}_R}{\partial (\theta_{,K} P^K_\alpha)} P^I_\alpha \right] \dot{\theta}_{,I}$$

$$+ \left[ J_P^{-1} \frac{\partial \bar{\psi}_R}{\partial (F^i_K P^K_\alpha)} F^i_I + \frac{\partial \bar{\psi}_R}{\partial (\theta_{,K} P^K_\alpha)} \theta_{,I} \right] \dot{P}^I_\alpha$$

$$- \bar{H} + \frac{1}{\theta} Q^I \theta_{R,I} \leq 0, \quad (6.83)$$

the last term in (6.78) having cancelled the term $J_P^{-1} P_I^{-\alpha} \bar{\psi}_R$, appearing as the first term in the right-hand side of (6.79). Expression (6.83) is linear in $\dot{F}^i_I$, $\dot{\theta}$ and $\dot{\theta}_{,I}$, since these quantities are not constitutively related to any other quantities appearing in the expression. Naturally, this statement is true provided that the list of arguments in the evolution law does not include the aforementioned time rates. It follows from this linearity that the first three expressions enclosed within square brackets must vanish identically, viz.:

$$J_P^{-1} \frac{\partial \bar{\psi}_R}{\partial (F^i_K P^K_\alpha)} P^I_\alpha = \frac{\partial \psi_R}{\partial F^i_I} = T_i^{\ I}, \quad (6.84)$$

$$J_P^{-1} \frac{\partial \bar{\psi}_R}{\partial \theta} = \frac{\partial \psi_R}{\partial \theta} = -s_R, \quad (6.85)$$

and

$$J_P^{-1} \frac{\partial \bar{\psi}_R}{\partial (\theta_{,K} P^K_\alpha)} P^I_\alpha = \frac{\partial \psi_R}{\partial \theta_{,I}} = 0. \quad (6.86)$$

These restrictions coincide, not surprisingly, with the constitutive restrictions of thermoelasticity. It is only the residual inequality which turns out to be augmented in the case of an evolving material. Notice that the fact that the term involving $\dot{\mathbf{P}}$ does not vanish is due to the relation imposed by the evolution law (6.2) between $\dot{\mathbf{P}}$ and at least some of the variables appearing in its bracketed coefficient in Equation (6.83). Using Equations (6.84) and (6.86) as well as the definition of the Eshelby stress as given in Equation (5.13) and the definition of the inhomogeneity velocity gradient in Equation (6.11), the residual inequality provided by the remaining two terms in Equation (6.83) can be expressed as:

$$tr(\mathbf{L}_P \boldsymbol{\beta}^T) - \bar{H} + \frac{1}{\theta} Q^I \theta_{,I} \leq 0. \quad (6.87)$$

where $\boldsymbol{\beta}$ is the Mandel stress (5.33), namely, (minus) the Eshelby stress devoid of its spherical part (proportional to the free energy). Notice the natural way in which the Mandel stress makes its appearance as the thermodynamic dual of the inhomogeneity velocity gradient. The contraction of the product of these two quantities provides the dissipation associated with the motion of dislocations or other kinds of inhomogeneity. Although it is conceivable to satisfy the residual inequality (6.87) by some delicate compensation between the mechanisms of dislocation motion and heat transfer, it seems more reasonable to assume that the conduction term $Q^I \theta_{,I}$ is in itself non-positive, leaving as an obviously sufficient condition for the inhomogeneity part the following restriction:

$$tr(\mathbf{L}_P \boldsymbol{\beta}^T) - \bar{H} \leq 0. \tag{6.88}$$

For the particular case in which the argument **a** of the evolution law (6.15) is just the Mandel stress $\boldsymbol{\beta}$, this restriction boils down to the following constraint on the evolution function:

$$tr[\bar{\mathbf{f}}(\mathbf{z})\,\mathbf{z}^T] - \bar{H} \leq 0, \tag{6.89}$$

for all values of the argument **z**. This formula follows directly from (6.10) and (6.88) and the fact that $J_P > 0$. If one sets $\bar{H} = 0$, a trivial way to satisfy this inequality is by choosing $\bar{f}(\mathbf{z}) = -k\mathbf{z}$, where $k$ is a positive material constant (analogous to a viscosity or a heat conduction coefficient). On the other hand, the presence of the extra term $\bar{H}$ makes, in principle, possible to sustain any evolution law by means of a finely adjusted control mechanism that systematically removes entropy from the system. In living organisms, it is quite possible that equilibrium is actively maintained in this way, just as one might hold a broom upside-down on the tip of a finger. The preceding calculations indicate that the thermodynamic dual of the inhomogeneity velocity gradient is not exactly the Eshelby stress, but rather the Mandel stress. In some sense this is a sobering and, at the same time, comforting remark, since it seems to eliminate the apparent paradox implied by the arbitrariness of the zero energy level.

## 6.5 An alternative approach

Our presentation of the fundamental equations of the mechanics and thermodynamics of evolving materials has been based on the consideration of the implant maps **P** as a set of passive internal state variables. We have seen, and we will illustrate it with concrete examples in the next sections, how this minimalist point of view leads to whole new vistas in terms of our ability to encompass a great variety of anelastic behaviours, including viscoelasticity, growth and remodelling. Many authors[8], on the other hand, have suggested that this framework can be considerably enlarged by advancing the assumption of extra balance laws to be satisfied by the material forces (such as the Eshelby stress) and by the addition of extra terms to the dissipation inequality. It seems somewhat futile to argue as to which is the right approach, since, as convincingly argued by Clifford Truesdell, Continuum Mechanics is, by its very nature, a *theory of models*. The philosophical boundary between balance

---

[8] See, for example, [54, 77, 12].

laws (supposed to apply to a very large universe of models) and constitutive laws (restricted to smaller classes) is at best fuzzy. The main requirement of any model is ultimately one of mathematical consistency. The value of a model is perhaps measured by its ability to withstand the vicissitudes of time, an ability determined to a great extent by its suitability to describe a large number of phenomena accurately, its elegance and simplicity and, to some extent, the authority of those propounding it. The present coexistence of several points of view (which, in our opinion, are not antagonistic but complementary) is a testimony of the richness of the subject and the vibrancy of Continuum Mechanics, both as a scientific discipline and as a veritable Weltanschauung, a vibrancy that seems to gainsay, time and time again, all the prophesies of its imminent doom and to cause it to rise cyclically from its proverbial ashes.

Forces are bounded linear operators on vector spaces of velocities (or virtual displacements). What this means is that once a kinematic substratum has been established for a continuum theory, the most general notion of force (as an entity producing virtual power on the space of virtual velocities) is completely determined and it encompasses both internal and external forces. This general approach can be pursued to the point of formulating completely global notions of stresses and of configurational (Eshelby-like) forces[9]. In the case of evolving materials of the kind we have been considering in this chapter, we may regard the implant fields $\mathbf{P}$ as added kinematic degrees of freedom on the same footing as the deformation gradient $\mathbf{F}$. At a given body point and at a given value of $\mathbf{F}$, we consider the collection of associated virtual "velocities" $\dot{\mathbf{F}}$ (or virtual "displacements" $\delta \mathbf{F}$). These are mixed body-spatial quantities. A local linear operator on the collection of virtual velocities at a point is known as a Piola (or first Piola-Kirchhoff) stress $\mathbf{T}$. The result of the linear operation, namely:

$$tr(\mathbf{T}\dot{\mathbf{F}}^T) = T_i{}^I \dot{F}^i_I, \qquad (6.90)$$

is known as the *virtual power* of the "force" (or stress) $\mathbf{T}$ on the virtual velocity $\dot{\mathbf{F}}$. A similar statement can be made in a purely static context in terms of virtual displacements, but we will continue using the language of velocities.

Having enlarged our kinematic outlook to include the time-dependent implant fields $\mathbf{P}$, we consider at a given body point and at a given value of the implant the collection of virtual implant velocities $\dot{\mathbf{P}}$, and we call a linear operator thereon a *material or configurational force* $\tilde{\mathbf{b}}$. Note that, when so introduced, this is a mixed tensor with one leg standing on the archetype and the other on the reference configuration (just as a Piola stress stands between the reference configuration and physical space). Moreover, although we are using the suggestive notation $\tilde{\mathbf{b}}$, this configurational force is yet to be connected with our previous notion of Eshelby stress. The evaluation of the linear operator $\tilde{\mathbf{b}}$ on a virtual implant velocity, namely:

---

[9] For this point of view in a general framework, see [81, 49].

$$tr(\tilde{\mathbf{b}}\dot{\mathbf{P}}^T) = \tilde{b}_I^\alpha \dot{P}_\alpha^I, \tag{6.91}$$

is called the *virtual power of the configurational force* on the given virtual implant velocity.

So far, we have just formulated definitions. Consider for a moment the classical case of a non-evolving material. We know that the mechanical field equations can be derived by postulating a *principle of virtual power*. Recall that to achieve this aim one introduces the virtual power (EVP) of the external forces as:

$$EVP \equiv \int_\Omega f_i v^i \, d\Omega + \int_{\partial\Omega} t_i v^i \, dS. \tag{6.92}$$

Here $f_i$ are the spatial components of the external body force per unit referential volume, $t_i$ are the spatial components of the surface traction per unit referential area of the unsupported part of the boundary, and $v^i$ are the components of a virtual ordinary velocity field. We are not including inertia effects, for the sake of simplicity. We now define the internal virtual power (IVP) as the integral of the expression (6.90), viz:

$$IVP \equiv \int_\Omega T_i^{\ I} \dot{F}^i_{\ I} \, d\Omega. \tag{6.93}$$

The principle of virtual power stipulates the satisfaction of the identity:

$$IVP = EVP, \tag{6.94}$$

for all virtual velocities. Now, in this classical case there is an understood extra compatibility condition between the virtual ordinary velocities ($\mathbf{v}$) and the virtual velocities of the deformation gradient ($\dot{\mathbf{F}}$). This condition establishes that the deformation gradient velocities must be derived from the (globally smooth) ordinary velocity field. That is, the identity (6.94) is enforced only under the condition that:

$$\dot{F}^i_{\ I} = v^i_{\ ,I}. \tag{6.95}$$

In this case, integration by parts and the arbitrariness of the virtual velocity field $\mathbf{v}$ are immediately seen to imply the field equation:

$$T_i^{\ I}{}_{,I} + f_i = 0, \tag{6.96}$$

and the natural boundary condition:

$$T_i^I N_I = t_i, \tag{6.97}$$

where $\mathbf{N}$ is the unit exterior normal to the (unsupported part of) boundary in the reference configuration. Thus, the "weak formulation" (6.94) delivers the same differential equations and boundary conditions as the usual "strong" formulation and, in fact, generalizes the latter for weaker conditions on the space of functionals.

## 6.5 An alternative approach

While the consistency between the weak and strong points of view in the classical case is well grounded on a tradition that goes as far back as d'Alembert, there is no a-priori reason to suppose that the principle of virtual power can be extended meaningfully to include remodelling phenomena or, for that matter, any case where internal state variables enter the physical picture. But it certainly can be done formally, as the following treatment shows. The first thing to be done is straightforward: the internal virtual power is to be supplemented with the integral of the local virtual power of the configurational force. Instead of Equation (6.93), we have now the augmented expression:

$$IVP \equiv \int_\Omega \left( T_i^{\ I} \dot{F}^i_{\ I} + \tilde{b}_I^{\ \alpha} \dot{P}^I_{\ \alpha} \right) d\Omega. \tag{6.98}$$

The second modification entails, as one would expect, a generalization of the external virtual power expression. It is here that an important difference between the classical case and the new formulation arises. For, whereas in the classical case we have at our disposal an ordinary velocity field (whose referential gradient delivers the velocity of $\mathbf{F}$), in the augmented formulation the implant field "velocity" $\dot{\mathbf{P}}$ is not necessarily integrable. In other words, there doesn't in general exist a vector field whose referential gradient is $\dot{\mathbf{P}}$. One way to come out of this impasse is to postulate the existence of an external material body force $\mathbf{B}$ which performs virtual power on *the same* virtual field $\dot{\mathbf{P}}$ as the internal material force $\tilde{\mathbf{b}}$. The nature of this new external force is left to be specified as part of each particular theory. It may very well vanish altogether or it may be meaningfully stipulated from physical considerations. At any rate, as an external body force, $\mathbf{B}$ is supposed to be given directly, rather than determined by any constitutive equation. The augmented external virtual power becomes:

$$EVP \equiv \int_\Omega \left( f_i v^i + B_I^{\ \alpha} \dot{P}^I_{\ \alpha} \right) d\Omega + \int_{\partial\Omega} t_i v^i \, dS. \tag{6.99}$$

Assuming the independence of the spatial and material virtual velocity fields, we obtain now the extra balance equation:

$$\tilde{b}_I^{\ \alpha} - B_I^{\ \alpha} = 0, \tag{6.100}$$

which doesn't seem much of a balance equation, but which can be intelligently exploited, as we shall presently see.

To complete the theory, a *dissipation principle* is postulated in the form of the following inequality:

$$\frac{D}{Dt} \int_\Omega \psi_R \, d\Omega \leq \int_\Omega \left( T_i^{\ I} \dot{F}^i_{\ I} + \tilde{b}_I^{\ \alpha} \dot{P}^I_{\ \alpha} \right) d\Omega, \tag{6.101}$$

where we have omitted the non-mechanical (heating) terms. Assuming the body to be uniform with a constitutive law given by Equation (5.24), namely:

$$\psi_R(F_I^i, \mathbf{X}) = J_P^{-1} \bar{\psi}_R(F_I^i P_\alpha^I(\mathbf{X})), \tag{6.102}$$

we obtain the local form of (6.101) as:

$$\left[ -J_P^{-1} P_I^{-\alpha} \bar{\psi}_R + J_P^{-1} \frac{\partial \bar{\psi}_R}{\partial F_K^i P_\alpha^K} F_I^i - \tilde{b}_I^\alpha \right] \dot{P}_\alpha^I$$

$$+ \left[ J_P^{-1} \frac{\partial \bar{\psi}_R}{\partial F_K^i P_\alpha^K} P_\alpha^I - T_i^I \right] \dot{F}_I^i \leq 0, \tag{6.103}$$

which, in view of the identity:

$$\frac{\partial \psi_R}{\partial F_I^i} = J_P^{-1} \frac{\partial \bar{\psi}_R}{\partial F_K^i P_\alpha^K} P_\alpha^I, \tag{6.104}$$

can be written as:

$$-\left[ \psi_R \delta_I^J - \frac{\partial \psi_R}{\partial F_J^i} F_I^i - \tilde{b}_I^J \right] \dot{P}_\alpha^I P_J^{-\alpha} + \left[ \frac{\partial \psi_R}{\partial F_I^i} - T_i^I \right] \dot{F}_I^i \leq 0, \tag{6.105}$$

with

$$\tilde{b}_I^J \equiv -\tilde{b}_I^\alpha P_\alpha^J. \tag{6.106}$$

Rather than hastily interpreting this as the identical vanishing of the quantities within the square brackets, one now introduces the *extra energetic responses*[10] $\hat{\mathbf{T}}$ and $\hat{\mathbf{b}}$ satisfying the residual inequality:

$$-\hat{T}_i^I \dot{F}_I^i + \hat{b}_I^J (L_P)_J^I \leq 0. \tag{6.107}$$

while making the identifications:

$$\psi_R \delta_I^J - \frac{\partial \psi_R}{\partial F_J^i} F_I^i - \tilde{b}_I^J = -\hat{b}_I^J, \tag{6.108}$$

and

$$\frac{\partial \psi_R}{\partial F_I^i} - T_i^I = -\hat{T}_i^I. \tag{6.109}$$

We are still left with the task of specifying the extra energetic quantities $\hat{b}$ and $\hat{T}$. Assume, for simplicity, that the latter vanishes (so that we recover the usual formula for the Piola stress in terms of the derivative of the referential free-energy density), but that the former does not. We see, then, that in this approach, the quantity $\tilde{b}_I^J$ is not quite the Eshelby stress, but rather the sum of the Eshelby stress and the extra energetic term $\hat{b}_I^J$. To satisfy the inequality (6.107), an evolution equation will have to be given in terms of the pull-back of $\hat{\mathbf{b}}$ (rather than that of $\tilde{\mathbf{b}}$) to the archetype.

Returning now to the extra balance equation (6.100), we see that, even if the external material body force $\mathbf{B}$ were to vanish, the repercussion will not

---

[10] We are following here the terminology of DiCarlo [13].

## 6.5 An alternative approach

be the vanishing of the Eshelby stress, but the vanishing of $\tilde{b}_I{}^J$. In that case, the extra energetic term $\hat{\mathbf{b}}$ will boil down to the classical Eshelby stress, and the evolution equation will coincide with that of the previous formulation, whereby no new balance law was postulated.

At the other extreme, the dissipation inequality $\hat{b}_I{}^J (L_P)^I{}_J \leq 0$ can always be trivially satisfied by setting $\hat{b}_I{}^J = 0$ identically, thus rendering all remodelling processes apparently "reversible". In this case, it follows from Equation (6.108) that the material force $\tilde{\mathbf{b}}$ coincides with the classical Eshelby stress. The extra balance equation (6.100), however, requires now that the external material body force $\mathbf{B}$ be also equal to the Eshelby stress. In this way, one can specify any evolution law whatsoever (for example, one of those we call below of the "self-driven" type) and always satisfy the dissipation inequality at the price of an external agent ($\mathbf{B}$) doing the job of carrying out the prescription of the evolution law by means of external sources of power. More to the point, if we substitute the extra balance law (6.100) into the dissipation principle (6.105), we obtain in general:

$$-\left[\psi_R \delta_I^J - \frac{\partial \psi_R}{\partial F^i{}_J} F^i{}_I - B_I{}^J\right] \dot{P}^I_\alpha P^{-\alpha}_J + \left[\frac{\partial \psi_R}{\partial F^i{}_I} - T_i{}^I\right] \dot{F}^i{}_I \leq 0, \quad (6.110)$$

with

$$B_I{}^J \equiv -B_I^\alpha P^J_\alpha. \quad (6.111)$$

Focusing attention on the evolutionary part only, namely:

$$-\left[\psi_R \delta_I^J - \frac{\partial \psi_R}{\partial F^i{}_J} F^i{}_I - B_I{}^J\right] (L_P)^I{}_J \leq 0, \quad (6.112)$$

and comparing with the standard expression (6.88), we conclude that, from the point of view of the standard formulation (without the extra balance law), what we have here is an entropy sink $(B_I{}^J (L_P)^I{}_J)$ regulated from outside the system. In models of growth, this is not surprising since these models deliberately leave out from what is essentially a chemically reacting mixture a significant number of components (in the simplest case, all but one). In conclusion, whether one opts for the interpretation of these extraneous agents as external material body forces or as corrections to the usual dissipation inequality (as illustrated by the term $\bar{H}$ in Equation (6.88)), there doesn't seem to be (at least in the situations just considered) a great deal of practical difference between these two points of view.

This concludes our brief presentation of some of the ideas underlying a large body of literature based on the independent balancing of the configurational forces.

## 6.6 Example: Visco-elasto-plastic theories

### 6.6.1 A simple non-trivial model

The mass consistency condition is usually imposed in material modelling, unless there are physical grounds to disregard it (such as in the case of the non-rigorous treatment of thermal stresses). Thus, the classical theories of visco-elasto-plasticity, whereby no mass creation or annihilation is envisaged, will be recovered only if the evolution function is traceless. This conclusion follows clearly from Equation (6.76), in which we set $\Pi = 0$ to obtain:

$$tr \bar{\mathbf{L}}_P = 0. \qquad (6.113)$$

Since $\bar{\mathbf{L}}_P$ is prescribed by the evolution function $\bar{\mathbf{f}}$, in accordance with Equation (6.14), we must have that the evolution function $\bar{\mathbf{f}}$ is traceless for all values of its tensor argument $\mathbf{z}$. Perhaps the simplest non-trivial example of such a function is given by:

$$\bar{\mathbf{f}}(\mathbf{z}) = k \mathbf{z}_D, \qquad (6.114)$$

where $k$ is a material constant and $\mathbf{z}_D$ is the *deviatoric part* of $\mathbf{z}$, defined as:

$$\mathbf{z}_D \equiv \mathbf{z} - \frac{1}{3} tr(\mathbf{z}) \mathbf{I}. \qquad (6.115)$$

We remark that the form (6.114) is not preserved under arbitrary changes of archetype. In other words, we are fixing a particular archetype.

Assuming that the argument of the evolution law consists exclusively of the Eshelby stress, it follows from Equations (6.12) and (6.114) that:

$$\dot{\mathbf{P}} = k J_P \, \mathbf{P} \, \mathbf{P}^T \, \mathbf{b}_D \, \mathbf{P}^{-T}. \qquad (6.116)$$

Setting the extra (non-compliant) entropy sink $\bar{H}$ to zero, this evolution law satisfies the thermodynamic dissipation inequality (6.88), if, and only if, the constant $k$ is non-negative. Indeed:

$$\begin{aligned} 0 \geq tr(\mathbf{L}_P \boldsymbol{\beta}^T) &= -tr(\mathbf{b}^T \, \mathbf{L}_P) \\ &= -k J_P \, tr(\mathbf{b}^T \, \mathbf{P} \, \mathbf{P}^T \, \mathbf{b}_D \, \mathbf{P}^{-T} \, \mathbf{P}^{-1}) \\ &= -k J_P \, tr(\mathbf{P}^{-1} \, \mathbf{b}^T \, \mathbf{P} \, \mathbf{P}^T \, \mathbf{b}_D \, \mathbf{P}^{-T}) \\ &= -k J_P \, tr(\mathbf{P}^{-1} \, \mathbf{b}_D^T \, \mathbf{P} \, \mathbf{P}^T \, \mathbf{b}_D \, \mathbf{P}^{-T}) \\ &= -k J_P \, tr([\mathbf{P}^T \, \mathbf{b}_D \, \mathbf{P}^{-T}]^T \, [\mathbf{P}^T \, \mathbf{b}_D \, \mathbf{P}^{-T}]), \end{aligned} \qquad (6.117)$$

which implies that $k \geq 0$. To further specialize this evolution law, consider the case in which the archetype is fully isotropic. Assuming it to be in a natural (stress-free) state, its symmetry group is, therefore, the orthogonal group, whose Lie algebra consists of all skew-symmetric matrices of order 3. We need to check that our evolution law satisfies the principle of actual evolution. The pull-back of the inhomogeneity velocity gradient to the archetype can be calculated from Equations (6.13) and (6.116) as:

$$\bar{\mathbf{L}}_P = \mathbf{P}^{-1} \dot{\mathbf{P}} = kJ_P \, \mathbf{P}^T \, \mathbf{b}_D \, \mathbf{P}^{-T}, \qquad (6.118)$$

whose transpose is:

$$\bar{\mathbf{L}}_P^T = kJ_P \, \mathbf{P}^{-1} \, \mathbf{b}_D^T \, \mathbf{P}. \qquad (6.119)$$

But, according to the comments following Equation (5.39), for the case of full isotropy the product $\mathbf{bD}$ is symmetric, namely:

$$\mathbf{bD} = \mathbf{D}^T \mathbf{b}^T, \qquad (6.120)$$

an equation that applies equally well to the deviatoric part of $\mathbf{b}$. Recalling that the metric $\mathbf{D}$ is given by:

$$\mathbf{D} = (\mathbf{PP}^T)^{-1}, \qquad (6.121)$$

the combination of the last four equations handily yields the result:

$$\bar{\mathbf{L}}_P^T = \bar{\mathbf{L}}_P. \qquad (6.122)$$

Thus, the proposed evolution function takes values outside the Lie algebra of the symmetry group, implying that the principle of actual evolution is satisfied.

It remains to verify that the proposed evolution law satisfies the principle of material symmetry consistency. To this end, we will find the collection of the archetype changes, as per Equation (6.21), that preserve the form of the evolution law (6.116). We therefore demand that:

$$\dot{\mathbf{P}}' = kJ_{P'} \, \mathbf{P}' \, \mathbf{P}'^T \, \mathbf{b}_D \, \mathbf{P}'^{-T}, \qquad (6.123)$$

with $\mathbf{P}' = \mathbf{PA}$. That is, we require that the equation:

$$\dot{\mathbf{P}} = kJ_P J_A \, \mathbf{PAA}^T \, \mathbf{P}^T \, \mathbf{b}_D \, \mathbf{P}^{-T} \mathbf{A}^{-T} \mathbf{A}^{-1} = kJ_P \, \mathbf{P} \, \mathbf{P}^T \, \mathbf{b}_D \, \mathbf{P}^{-T}, \quad (6.124)$$

be satisfied identically for all $\mathbf{P}$ and $\mathbf{b}_D$. This will certainly be the case if $\mathbf{A}$ is orthogonal. Since we have assumed that our archetype is a fully isotropic solid, condition (6.26) is clearly satisifed.

### 6.6.2 Some computational considerations

We have constructed an example of an evolution law that satisfies the principles of reduction to the archetype, actual evolution and symmetry consistency. Moreover, it is consistent with the dissipation inequality. Finally, by being traceless, it rules out processes of bulk growth (or resorption). To fully implement such a model for the solution of applied problems one would have to couple the evolution law with the field equations and provide initial conditions for the implant field. This coupling can be achieved, for example, by a modification of existing finite-element codes. Rather than following this route, and to focus our attention on the time-dependent part only, we will illustrate the possible application of this particular evolution law by solving an initial value problem arising from the assumption that the body is initially homogeneous and that it undergoes spatially homogeneous configurations. Inertia effects will be likewise disregarded. In this way, we will be dealing with a system of ordinary, rather than partial, differential equations. With this procedure it becomes possible to study the phenomena of creep and relaxation by alternately fixing the stress and the deformation gradient.

The system of nine (ordinary) differential equations (6.116) has, by construction, the first integral:

$$tr(\mathbf{P}^{-1}\dot{\mathbf{P}}) = 0, \tag{6.125}$$

which integrates immediately to:

$$J_P = \text{constant}, \tag{6.126}$$

a useful result from the numerical point of view.

*Remark 6.4.* In the theory of plasticity one often finds the statement that "plastic flow is incompressible". In the context of our model, this statement is a direct consequence of the conservation of mass.

Since the stored energy function must satisfy the principle of material frame indifference, it is convenient to express it in terms of the right Cauchy-Green tensor $\mathbf{C} = \mathbf{F}^T\mathbf{F}$, namely:

$$\psi_R = \psi_R(\mathbf{C}). \tag{6.127}$$

The (symmetric) second Piola-Kirchhoff stress, given by:

$$\mathbf{S} = \mathbf{F}^{-1}\,\mathbf{T}, \tag{6.128}$$

is, therefore expressible as:

$$\mathbf{S} = 2\,\frac{\partial \psi_R}{\partial \mathbf{C}}. \tag{6.129}$$

It follows that the Eshelby stress:

$$\mathbf{b} = \psi_R \mathbf{I} - \mathbf{F}^T\mathbf{T}, \tag{6.130}$$

can also be written as:

$$\mathbf{b} = \psi_R \mathbf{I} - \mathbf{C}\mathbf{S}, \tag{6.131}$$

a convenient formula for computational purposes.

### 6.6.3 Creep of a bar under uniaxial loading

Consider a prismatic member aligned with the $X^1$-axis. We assume the reference configuration to be homogeneous and in a natural state. The bar is subjected to the sudden application (at time $t=0$) of a uniaxial tension or compression which will remain constant in time for $t>0$. If we denote by $a$ the (constant) force per unit area of the cross section in the reference configuration, we obtain that the Piola stress tensor components are given by the matrix:

$$[T] = \begin{bmatrix} a & 0 & 0 \\ 0 & 0 & 0 \\ 0 & 0 & 0 \end{bmatrix}, \tag{6.132}$$

where we have assumed that the spatial and referential axes coincide. The constant $a$ is positive for tension and negative for compression. If the material is isotropic, we may assume a deformation gradient given by a diagonal matrix of the form:

$$[F] = \begin{bmatrix} m(t) & 0 & 0 \\ 0 & n(t) & 0 \\ 0 & 0 & n(t) \end{bmatrix}. \tag{6.133}$$

Considering an archetype in a natural state (so that at $t=0$ all the implants are given by the unit matrix), we write:

$$[P] = \begin{bmatrix} q(t) & 0 & 0 \\ 0 & r(t) & 0 \\ 0 & 0 & r(t) \end{bmatrix}. \tag{6.134}$$

A straightforward calculation yields the following result for the deviatoric part of the Eshelby stress:

$$[b]_D = \begin{bmatrix} -\frac{2}{3}ma & 0 & 0 \\ 0 & \frac{1}{3}ma & 0 \\ 0 & 0 & \frac{1}{3}ma \end{bmatrix}. \tag{6.135}$$

Using the evolution law given by Equation (6.116), we obtain the following two ODE's for the evolution of the implant field:

$$\dot{q} = -kqr^2(\frac{2}{3}qma), \tag{6.136}$$

$$\dot{r} = kqr^2(\frac{1}{3}rma). \tag{6.137}$$

As expected, this system has the following first integral:

$$qr^2 = 1, \tag{6.138}$$

where the initial condition $\mathbf{P}(0) = \mathbf{I}$ has been invoked. We may, therefore, rewrite the system of ODE's as:

$$\dot{q} = -k(\frac{2}{3}qma), \tag{6.139}$$

$$\dot{r} = k(\frac{1}{3}rma). \tag{6.140}$$

The quantities $q, r$, on the one hand, and $m, n$, on the other hand, are not independent, since the constitutive law must be satisfied at all times so as to yield the constant Piola stress (6.132). For the sake of the illustration, we assume the following modified neo-Hookean constitutive law for the archetype:

$$\bar{\psi}_R = \frac{\mu}{2}\left(\operatorname{tr}(\bar{\mathbf{C}}) - 2\ln(\sqrt{\det(\bar{\mathbf{C}})}\,)\right), \tag{6.141}$$

where $\mu$ is a material constant (the shear modulus). The second Piola-Kirchoff stress at the archetype is obtained as:

$$\bar{\mathbf{S}} = 2\frac{\partial \bar{\psi}_R}{\partial \bar{\mathbf{C}}} = \mu\left(\mathbf{I} - \bar{\mathbf{C}}^{-1}\right). \tag{6.142}$$

We note that this particular constitutive law reduces, in the infinitesimal limit, to that of an isotropic elastic material with vanishing Poisson's ratio.

The constitutive law at the reference configuration is given by:

$$\psi_R = J_P^{-1}\frac{\mu}{2}\left(\operatorname{tr}(\mathbf{P}^T\mathbf{C}\mathbf{P}) - 2\ln(\sqrt{\det(\mathbf{P}^T\mathbf{C}\mathbf{P})}\,)\right), \tag{6.143}$$

and the corresponding second Piola-Kirchhoff stress is:

$$\mathbf{S} = 2\frac{\partial \psi_R}{\partial \mathbf{C}} = \mu J_P^{-1}\left(\mathbf{P}\mathbf{P}^T - \mathbf{C}^{-1}\right). \tag{6.144}$$

Taking into consideration the first integral (6.138) (namely, the fact that $J_P = 1$), we obtain:

$$\mathbf{S} = 2\frac{\partial \psi_R}{\partial \mathbf{C}} = \mu\left(\mathbf{P}\mathbf{P}^T - \mathbf{C}^{-1}\right). \tag{6.145}$$

The Piola stress can now be evaluated as:

$$\mathbf{T} = \mathbf{F}\mathbf{S} = \mu\left(\mathbf{F}\mathbf{P}\mathbf{P}^T - \mathbf{F}^{-T}\right). \tag{6.146}$$

Combining this result with Equations (6.132-6.134), we obtain the relations:

$$\mu\left(mq^2 - \frac{1}{m}\right) = a, \tag{6.147}$$

$$nr^2 - \frac{1}{n} = 0. \tag{6.148}$$

Taking into account that in our setting all the quantities involved (except, possibly, $a$) are strictly positive, we obtain the following explicit restrictions imposed by the constitutive law:

$$m = \frac{\frac{a}{\mu} + \sqrt{\left(\frac{a}{\mu}\right)^2 + 4q^2}}{2q^2}, \qquad (6.149)$$

$$n = \frac{1}{r}. \qquad (6.150)$$

Substituting the expression (6.150) into (6.139) yields the following ODE for $q = q(t)$:

$$\dot{q} = -\frac{2}{3} ka \frac{\frac{a}{\mu} + \sqrt{\left(\frac{a}{\mu}\right)^2 + 4q^2}}{2q}. \qquad (6.151)$$

It is convenient to introduce the following change of variables:

$$\tau = k\mu\, t \qquad \alpha = \frac{a}{\mu}, \qquad (6.152)$$

in terms of which Equation (6.151) is written as:

$$\frac{dq}{d\tau} = -\frac{1}{3}\alpha\left(\frac{\alpha + \sqrt{\alpha^2 + 4q^2}}{q}\right), \qquad (6.153)$$

or, equivalently:

$$\frac{d(q^2)}{d\tau} = -\frac{2}{3}\alpha\left(\alpha + \sqrt{\alpha^2 + 4q^2}\right). \qquad (6.154)$$

Numerical solutions of Equation (6.153) reveal that for compressive loads ($\alpha < 0$) the bar creeps asymptotically to the zero length as time goes to infinity. As time goes on, $q(t)$ increases in an almost linear fashion while, concomitantly, $m(t)$ tends to zero in an exponential-like fashion. For tensile loads, on the other hand, $q(t)$ decreases and vanishes at a finite time, implying that the solution $m(t)$ blows up at a finite time. For small loads, i.e. for $\alpha \ll 1$, the blow-up time can be roughly estimated as $t_b \approx 3/(2ka)$ or, equivalently, $\tau_b \approx 3/(2\alpha)$. This blow-up feature is due to the chosen constitutive law. It is interesting to note that, whereas in the instantaneous elastic response at the initial time the cross sections remain undistorted ($n = 1$, no elastic Poisson effect), once the creeping process gets underway this is no longer the case. Indeed, by virtue of equations (6.138) and (6.150), we have:

$$n^2 = q(t). \qquad (6.155)$$

### 6.6.4 Evolution, rheological models and the Eshelby stress

The preceding example on creep (or, its equally simple relaxation counterpart) clearly shows that a model of the type presented resembles in qualitative behaviour a Maxwell body. We recall that this name is given to a rheological model consisting of an elastic spring and a dashpot arranged in series. The total elongation is then given by the sum of the individual elongations of the spring and the dashpot, while the total force is the same in both elements. This resemblance is not fortuitous. Indeed, the composition **FP** is a glorified version of a subtraction of deformations, with **F** representing the total and $\mathbf{P}^{-1}$ representing the anelastic (viscous) component. The elastic energy emerges from the net elastic part only, namely, from the difference between these two components[11].

Because of this similarity, the question arises as to the use of the multiplicative technique to elevate to a fully-fledged three-dimensional evolutive model other rheological paradigms such as the Kelvin solid. We recall that the Kelvin model consists of arranging a Maxwell model in parallel with an elastic spring. One way to achieve this could be as follows: Not one but rather two elastic archetypes are defined with constitutive laws given by functions $\bar{\Psi}'_R$ and $\bar{\Psi}''_R$. The energy function of the body in then reference configuration is then given by:

$$\Psi_R(\mathbf{F}, \mathbf{X}, t) = \bar{\Psi}'_R(\mathbf{FP}'(\mathbf{X}, t)) + \bar{\Psi}''_R(\mathbf{FP}''(\mathbf{X})). \tag{6.156}$$

The additive nature of this constitutive law is a reflection of the fact that in the rheological model two elements in parallel contribute additively to the total force. The two fields of implants $\mathbf{P}'$ and $\mathbf{P}''$ are deliberately chosen independently, with the first one subject to a time evolution while the second one is constant in time. As a particular case, we can assume that the latter field is everywhere equal to the unit tensor $\mathbf{I}$, namely:

$$\Psi_R(\mathbf{F}, \mathbf{X}, t) = \bar{\Psi}'_R(\mathbf{FP}'(\mathbf{X}, t)) + \bar{\Psi}''_R(\mathbf{F}). \tag{6.157}$$

If so, the "driving force" of the evolution of the inhomogeneity is clearly not the total Eshelby stress, but only that part associated with the energy function $\bar{\Psi}'_R$, namely, with the "dissipative" part of the stress. This simple observation is further reinforced by the point of view that the implant tensor $\mathbf{P}'$ acts as a collection of internal variables and that, therefore, its associated driving force is obtained by taking the partial derivative of the energy with respect to it. This simple example raises intriguing questions as to the consideration of the Eshelby stress (defined on the basis of the total energy and the total stress) as a universal concept or, rather, a useful device for the modelling and analysis of certain classes of problems arising in many applications.

## 6.7 Example: Bulk growth

To try to understand the implication of the residual inequality:

$$tr(\mathbf{L}_P \boldsymbol{\beta}^T) \leq 0, \tag{6.158}$$

---

[11] For infinitesimal deformations, it makes sense to approximate the multiplicative Lie group $GL(3; \mathbb{R})$ by its additive Lie algebra.

imagine that an isotropic and homogeneous sphere of a material susceptible to growth and resorption has been uniformly compressed into a smaller sphere so as to fit it into a rigid spherical container. Under these conditions, the Mandel stress $\boldsymbol{\beta} = \mathbf{F}^T\mathbf{T} = \mathbf{CS}$ is spherical or, more precisely, a negative multiple of the identity tensor in the reference configuration. The satisfaction of (6.158) requires, therefore, that the trace of $\mathbf{L}_P$ be positive, which implies that resorption will develop. In other words, the material will tend to partially "evaporate" so as to relax the compressive state of stress. This conclusion satisfies our intuition, although it remains to be checked against experimental results in biological tissue. On the strength of this example, a possible evolution law for a solid susceptible to growth can be suggested by choosing:

$$\bar{\mathbf{f}}(\mathbf{z}) = -k\,\mathbf{z}, \tag{6.159}$$

where $\mathbf{z}$ is now identified with the Mandel stress and $k$ is a positive constant. More explicitly, this evolution law reads

$$\dot{\mathbf{P}} = -kJ_P\,\mathbf{P}\,\mathbf{P}^T\,\boldsymbol{\beta}\,\mathbf{P}^{-T}, \tag{6.160}$$

and all the remarks made with regard to the viscoelastic counterpart (6.116) still apply. One may roughly say that the presence of the spherical part of $\boldsymbol{\beta}$ is responsible for the net growth or resorption, while the deviatoric part takes care of other, non-growth, forms of remodelling. A more realistic model, however, would apply two separate material constants, one for each of these parts, so as to control both kinds of processes independently.

### 6.7.1 Exercise stimulates growth

So as to further emphasize the potential applications of the theory, we will show that an evolution model as simple as that of Equation (6.160), in combination with an archetypal constitutive law such as (6.141), leads to net growth under harmonic loading around a zero-stress state.

Consider a spherical chunk of isotropic material subjected at each point of its boundary to a normal traction of magnitude $a = a(t)$. In particular, we will consider an oscillatory time dependence of the form:

$$a(t) = a_0\,\sin(\omega t), \tag{6.161}$$

the amplitude $a_0$ and the angular frequency $\omega$ being given. The Piola stress is spatially constant throughout the body and is given by:

$$\mathbf{T} = a(t)\,\mathbf{I}. \tag{6.162}$$

The deformation gradient and the implant field are given, respectively, by:

$$\mathbf{F} = m(t)\,\mathbf{I}, \tag{6.163}$$

and
$$\mathbf{P} = q(t)\,\mathbf{I},\qquad(6.164)$$
the time-dependent quantities $m(t)$ and $q(t)$ to be solved for. The Mandel stress is directly obtained as:
$$\boldsymbol{\beta} = m(t)a(t)\,\mathbf{I}.\qquad(6.165)$$

The evolution law (6.160) reads now:
$$\dot{q} = -kq^4 ma,\qquad(6.166)$$
and the constitutive law (6.141) yields:
$$m = \frac{\frac{aq^3}{\mu} + \sqrt{\left(\frac{aq^3}{\mu}\right)^2 + 4q^2}}{2q^2}.\qquad(6.167)$$

As a result, we obtain the following ODE for the time evolution of the implant variable $q(t)$:
$$\dot{q} = -\frac{ka}{2}q^2\left(\frac{aq^3}{\mu} + \sqrt{\left(\frac{aq^3}{\mu}\right)^2 + 4q^2}\right).\qquad(6.168)$$

Using the change of variables indicated in Equation (6.152), we may also write:
$$\frac{dq}{d\tau} = -\frac{\alpha}{2}q^2\left(\alpha q^3 + \sqrt{(\alpha q^3)^2 + 4q^2}\right).\qquad(6.169)$$

We recall that a net growth is reflected in a decrease of the value of $q$ from its initial value of $q(0) = 1$. For very small values of the ratio $\alpha_0 = a_0/\mu$ the response $q(t)$ is approximately periodic, without any gain or loss of mass. But for somewhat larger values of this ratio a net gain (that is, a net decrease in $q(t)$) can be observed from numerical evaluations. It is important to notice that a net gain is obtained regardless of the direction of the initial cycle, as can be verified by changing the sign of $\alpha_0$. Figure 6.5 shows $q$ versus $\tau$ for $\alpha_0 = \pm 0.1$ and $\omega/(k\mu) = 20$. The calculations and the graphs have been produced with the help of the Mathematica package. Note that, although for an oscillating load the general tendency is a reduction of $q$, a constant negative load would lead to an increase of $q$ (i.e., to a resorption) as expected, while a constant positive load leads to a decrease of $q$. It is also noteworthy that the general qualitative behaviour obtained is independent of the particular archetypal constitutive law used.

### 6.7.2 A challenge to Wolff's law?

One of the widely held tenets in the theory of bone remodelling and growth is the so-called Wolff's law. Formulated in as early as the latter part of the

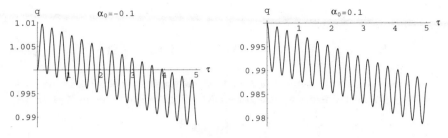

**Fig. 6.5.** Growth induced by exercise

nineteenth century [94], it is a qualitative statement to the effect that the internal architecture of a bone is determined by the stresses acting upon it. In particular, one interpretation of this statement is that the trabeculae tend to align themselves with the prevalent direction of the maximum principal stress in the bone. In fairness, few specialists in biomechanics nowadays believe this statement to be completely accurate, but rather hold the view that there must exist some connection between bone remodelling and mechanical stresses and other functional factors. Be that as it may, we propose to consider a highly idealized situation on a purely hypothetical basis and then to compare the results thus obtained with the general statement of Wolff's law.

In order to represent the inherent anisotropy of bone, we will model it as a transversely isotropic medium, the strong direction (represented by the unit vector **n**) being aligned with the local orientation of the trabeculae. For specificity in the numerical computations, we will adopt a modified (transversely isotropic) version of the archetypal isotropic constitutive law (6.141) used in previous examples, namely:

$$\bar{\psi}_R = \frac{\mu}{2} \left( \mathrm{tr}(\bar{\mathbf{C}}) - 2\ln(\sqrt{\det(\bar{\mathbf{C}})}\,) \right), \tag{6.170}$$

A convenient way to render an equation of this type orthotropic is by the introduction of a symmetric and positive-definite "fabric tensor" **M**. For the case of transverse isotropy, **M** should have a once-repeated eigenvalue. In general, given an isotropic hyperelastic constitutive equation:

$$\bar{\psi}_R = \bar{\psi}_R(\bar{\mathbf{C}}), \tag{6.171}$$

the desired modified constitutive law is then obtained by means of the replacement:

$$\bar{\psi}_R = \bar{\psi}_R \left( \mathbf{M}(\bar{\mathbf{C}} - \mathbf{I})\mathbf{M} + \mathbf{I} \right). \tag{6.172}$$

For the specific case of the constitutive law (6.170), the resulting orthotropic constitutive law is:

$$\bar{\psi}_R = \frac{\mu}{2} \left( \mathrm{tr}(\mathbf{M}(\bar{\mathbf{C}} - \mathbf{I})\mathbf{M} + \mathbf{I}) - 2\ln(\sqrt{\det(\mathbf{M}(\bar{\mathbf{C}} - \mathbf{I})\mathbf{M} + \mathbf{I})}\,) \right), \tag{6.173}$$

whence:

$$\bar{\mathbf{S}} = 2\frac{\partial \bar{\psi}_R}{\partial \bar{\mathbf{C}}} = \mu \mathbf{M}\left[\mathbf{I} - (\mathbf{M}(\bar{\mathbf{C}} - \mathbf{I})\mathbf{M} + \mathbf{I})^{-1}\right]\mathbf{M}. \qquad (6.174)$$

Adopting a basis in the archetype such that the first base-vector is aligned with the strong material direction, the tensor $\mathbf{M}$ will have the following component form:

$$[M] = \begin{bmatrix} \alpha & 0 & 0 \\ 0 & 1 & 0 \\ 0 & 0 & 1 \end{bmatrix}, \qquad (6.175)$$

where $\alpha > 1$ is a material constant representing the ratio between the strength in the trabecular direction and its transverse counterpart.

As far as the evolution law is concerned, we will consider a purely rotational remodelling, without distortion and without net growth. To achieve this end, we will assume the inhomogeneity velocity gradient to be proportional to the skew-symmetric part of the Eshelby stress, both pulled back to the archetype. In other words, denoting by $\bar{\mathbf{b}}_W$ the skew-symmetric part of the Eshelby stress in the archetype, the evolution law reads:

$$\mathbf{P}^{-1}\dot{\mathbf{P}} = k\bar{\mathbf{b}}_W, \qquad (6.176)$$

where $k$ is a material constant. It is not difficult to prove[12] that the general solution of this equation consists of a (time-dependent) orthogonal matrix multiplied to the left by a (time-independent) symmetric and positive definite matrix. If the initial condition is a (proper) orthogonal matrix, it follows that the solution remains (proper) orthogonal as time goes on. As far as the sign of the material constant $k$, if we assume that the non-compliant entropy sink $\bar{H}$ vanishes, it follows from the entropy inequality that $k$ must be non-negative. This is a crucial remark for the correct interpretation of our final result.

We will assume the initial implant to be of the form:

$$[P] = \begin{bmatrix} \cos\theta & \sin\theta & 0 \\ -\sin\theta & \cos\theta & 0 \\ 0 & 0 & 1 \end{bmatrix} \qquad (6.177)$$

where $\theta$ represents the angle of rotation of the implant, assumed to take place about the third coordinate axis. Notice that a positive $\theta$ represents a *clockwise* rotation (of the trabecular axis $\mathbf{n}$ with respect to the longitudinal axis of the bone, i.e., the first coordinate axis).

Assuming that a cylindrical specimen is subjected to a constant axial tensile force that preserves its magnitude and direction in space (a direction that we assume aligned with the first coordinate axis), the components of the Piola stress are given by:

---

[12] For example, by using the polar decomposition theorem.

$$[T] = \begin{bmatrix} a & 0 & 0 \\ 0 & 0 & 0 \\ 0 & 0 & 0 \end{bmatrix} \qquad (6.178)$$

where $a$ is a constant and where we have assumed the spatial and referential coordinate systems to coincide.

Under these circumstances, we seek a solution of the problem when the deformation gradient has the form:

$$[F] = \begin{bmatrix} b(t) & c(t) & 0 \\ 0 & d(t) & 0 \\ 0 & 0 & e(t) \end{bmatrix}. \qquad (6.179)$$

This form of the solution is chosen so that a material die originally aligned with the coordinate axes is forced to preserve the directions of the first and the third axes and to keep the deformed second axis perpendicular to the third. We remark that this form of the solution is consistent with the necessary symmetry of the Cauchy stress.

The numerical solution of this problem yields a somewhat startling result. Regardless of the initial value of $\theta$ (different from zero), the solution tends asymptotically to a final value $\theta = \pi/2$. In other words, the putative trabeculae end up aligning themselves with the *smaller* principal stress! Moreover, the zero solution (obtained by setting the initial angle to zero) is unstable[13]. We recall, however, that the entropy sink $\bar{H}$ has been assumed to vanish. If it doesn't (by virtue of some externally controlled mechanism) we can clearly change the sign of the constant $k$, thereby reversing the result and recovering Wolff's law. Alternatively, even assuming the vanishing of $\bar{H}$, one may argue that the remodelling process is initiated only after the Eshelby stress exceeds a certain threshold, in which case spots of stress concentration would play a determining role in the process. In particular, if the voids within the bone are regarded as very elongated in the axial direction, the location of the maximal principal stress may well be the apices of these voids. More importantly, the direction of these principal stresses is perpendicular to the axis of the bone, a result that would be consistent with Wolff's law, albeit in a rather indirect and unexpected way.

To dispel the impression that the numerical results obtained may be qualitatively dependent on the particular transversely isotropic constitutive law adopted, we note that the evolution equation (6.176) reduces in our case to the following differential equation:

$$\frac{d\theta}{dt} = \frac{k}{2}ac(t), \qquad (6.180)$$

as can be obtained by direct substitution of the assumed forms of $\mathbf{P}$, $\mathbf{T}$ and $\mathbf{F}$. In other words, the time evolution of the implants is dictated entirely by the

---

[13] This result was first found by DiCarlo et al. in [14].

"shear" component $c(t)$ of the deformation gradient. Although the numerical value of this component certainly depends on the particular constitutive law of the bone material, it is clear that its sign does not. If the main (strong) axis of anisotropy is rotated clockwise with respect to the bone axis (as shown in Figure 6.6), the cross section of the specimen rotates in the same direction. In other words, the sign of $c$ is the same as the sign of $\tan\theta$. By symmetry, $c$ vanishes only when $\theta = 0, \pi$ or $\theta = \pm\pi/2$. In both of these cases we obtain a constant solution, but this solution is unstable for the former case ($\theta = 0, \pi$), and stable for the latter.

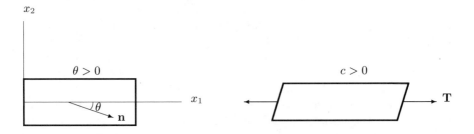

**Fig. 6.6.** Uniaxial tension of an anisotropic specimen

## 6.8 Example: Self-driven evolution

### 6.8.1 Introduction

In all the examples so far, it has been the Eshelby stress, or perhaps other quantities connected with the deformation of the body, what has constituted the driving force behind the motion of the inhomogeneities. On the other hand, it is quite possible to conceive of an evolutionary process that is *driven by the inhomogeneity pattern itself* in its natural effort to eliminate residual stresses or, even if these are absent, to achieve a defect-free structure. As we know, these processes can be enhanced by raising the temperature of the body so as to increase the probability of the atoms in overcoming potential barriers. It is also possible that an inhomogeneity (dislocation, say) pattern may lead in the opposite direction, with a dislocation pile-up arising naturally out of an initially smooth distribution of defects, a typically nonlinear phenomenon.

It behooves us to sound a word of thermodynamical warning. By not including the Eshelby stress (or, for that matter, any other deformation-induced quantity) in the evolution law, we must abandon all hope of satisfying the residual dissipation inequality in the form of Equation (6.88) with $\bar{H} = 0$. The resolution of this thermodynamic impasse (short of discarding the possibility of self-driven evolution altogether) may be suggested from different directions. Firstly, it is quite possible to reasonably claim that the Clausius-Duhem inequality in its classical form is inadequate to handle the phenomenon at hand, and that it needs to be augmented so as to include extra terms associated with the new geometrical variables, or to be interpreted in a less conventional way [24]. Secondly, one may set $\bar{H} > 0$, that is, one may imagine the existence of some extraneous entropy sink that takes care of matters, such as a distributed array of computer controlled micro-actuators with unlimited access to external power to effect the desired evolution, just as a distributed array of refrigerators might push the heat flux in the wrong direction. Finally, one may attempt to fit this kind of evolution within the framework of the alternative approach discussed in Section 6.5[14]. Be that as it may, we will consider two examples of self-driven evolution, the first of which is not much affected by the thermodynamic issue. Both examples are highly idealized, but they do show how relatively simple models of self-driven evolution can account for rather complicated nonlinear phenomena. This fact in itself is an encouraging sign of the descriptive and unifying power of the theory of continuous distributions of inhomogeneities.

### 6.8.2 A solid crystal body

**General description**

We consider a uniform body whose archetype is an elastic solid crystal, namely, an elastic material point whose symmetry group is (conjugate to) a *discrete* subgroup of the orthogonal group. In this case, as we have learned in Section 1.2.5, the material connection $\Gamma$ (at each instant of time) is unique. Moreover, as we have determined in Section 1.4.2, the presence of inhomogeneities is completely characterized in this case by the torsion tensor $\tau$ of the unique material connection. To specialize matters even more, we may assume that the archetype is a triclinic crystal, namely, that its symmetry group is trivial (effectively, this material has no symmetries, except the identity). Finally, we will confine attention to a body for which a reference configuration exists in which all material isomorphisms at all times are proper rotations. In other words, at each instant the body is in a uniform state of *contorted aelotropy* [75], also called a state of *constant strain* [27]. It follows that if one of the body points is in a natural (stress-free) state, so will all the other points of the body. If the archetype is in a natural state, the implants too are pure rotations, which we will assume to be the case. The particular evolution we have in mind, then, can be summarized as follows: the body points evolve systematically from one natural state to another in a smooth time-dependent manner. In so doing, the material implants are time-dependent rotations, smoothly distributed both

---

[14] We are grateful to Antonio DiCarlo for having suggested this possibility.

in time and material space. The reference configuration, although inhomogeneous, remains stress-free throughout the whole evolutionary process. What we have in mind, therefore, is a stress-free inhomogeneous body that is left to evolve of its own accord, without the agency of any external forces. But even if such forces were applied, we will assume that the evolution of the body will ignore their presence.

## The material connection

Let $\mathbf{E}_\alpha$ ($\alpha = 1, 2, 3$) be an orthonormal basis in the archetype, and $\mathbf{e}_I$ ($I = 1, 2, 3$) the natural basis of a Cartesian coordinate system in the reference configuration. The uniformity field $\mathbf{P}(\mathbf{X})$ maps the archetype basis into a field of bases $\mathbf{g}_\alpha(\mathbf{X})$ ($\alpha = 1, 2, 3$) in the body as follows:

$$\mathbf{g}_\alpha = P^I_\alpha \, \mathbf{e}_I. \tag{6.181}$$

In our example, the matrices $\{P^I_\alpha\}$ are, by construction, proper orthogonal matrices. The Christoffel symbols of the unique material connection are generally given by Equation (1.32), namely:

$$\Gamma^K_{IJ} = -P^{-\alpha}_I \, P^K_{\alpha,J}, \tag{6.182}$$

but, by virtue of the assumed orthogonality, they are obtained in our case as:

$$\Gamma^K_{IJ} = -P_I{}^\alpha \, P^K{}_{\alpha,J}. \tag{6.183}$$

Indices can be lowered and raised freely, due to the assumed orthonormality of all the bases at play. In particular, we note that the component representation $\{\Gamma^K_{IJ}\}$ is skew-symmetric in the indices $K$ and $I$.

## The evolution law

The inhomogeneity velocity gradient pulled back to the archetype is:

$$(\bar{L}_P)^\alpha_\beta = P^{-\alpha}_I \, \dot{P}^I_\beta = P_I{}^\alpha \, \dot{P}^I_\beta, \tag{6.184}$$

where we have used Equation (6.13) and the assumed orthogonality of the implants. Again we remark that, as a consequence of this orthogonality, the matrix $\{(\bar{L}_P)^\alpha_\beta\}$ is skew-symmetric.

As already remarked, the torsion tensor with components:

$$\tau^I_{JK} = \Gamma^I_{JK} - \Gamma^I_{KJ}, \tag{6.185}$$

is a uniquely defined instantaneous property of the body, because so is the material connection itself. Moreover, since the vanishing of this torsion is a necessary and sufficient condition for (in this case, global) homogeneity, we can certainly consider it as a true measure of the density of the continuous distribution of inhomogeneities within the body. We postulate that (regardless of the state of stress provided by the spatial configurations of the body) it is the distribution of inhomogeneities itself that drives the material evolution. Accordingly, the evolution law is assumed to be of the form:

$$\bar{\mathbf{L}}_P = \bar{\mathbf{f}}(\tau_{P^{-1}}). \tag{6.186}$$

Here we are using the concepts and notation developed earlier in Section 6.3.2 and, in particular, Equation (6.15). The pull-back $\tau_{P^{-1}}$ of the torsion tensor from the reference configuration to the archetype is given by:

$$(\tau_{P^{-1}})^{\alpha}_{\beta\gamma} = P^{-\alpha}{}_I T^I{}_{JK} P^J{}_\beta P^K{}_\gamma = P_I{}^\alpha T^I{}_{JK} P^J{}_\beta P^K{}_\gamma. \tag{6.187}$$

It represents the density of the distribution of inhomogeneities seen from the perspective of the archetype. The simplest non-trivial evolution function will establish a linear relation between the inhomogeneity velocity gradient $\bar{\mathbf{L}}_P$ and the torsion via a fifth-order tensor $\mathbf{C}$ of material constants. In components, this evolution law reads as follows:

$$(\bar{L}_P)^\rho{}_\sigma = C^\rho{}_{\sigma\alpha}{}^{\beta\gamma} \tau^\alpha{}_{\beta\gamma}, \tag{6.188}$$

or, using Equations (6.13) and (6.187):

$$\dot{P}^M{}_\sigma = C^\rho{}_{\sigma\alpha}{}^{\beta\gamma} P^M{}_\rho P_I{}^\alpha T^I{}_{JK} P^J{}_\beta P^K{}_\gamma. \tag{6.189}$$

**A two-dimensional setting**

To illustrate the range of phenomena that can fall within the scope of this model, we consider now the subclass of problems whereby the implants are, at all times, independent of, say, the third Cartesian coordinate in the reference configuration, which acts also as the axis of rotation for the implants. We thus render the problem two-dimensional and gain the added computational simplicity afforded by the explicit representation of the rotation by means of a single angular parameter. Resorting to the common notation $x, y, z$ for the Cartesian system in the reference configuration, we obtain the following explicit expression for the implant field:

$$[\mathbf{P}] = \begin{bmatrix} \cos\theta & \sin\theta & 0 \\ -\sin\theta & \cos\theta & 0 \\ 0 & 0 & 1 \end{bmatrix}, \tag{6.190}$$

where $t$ denotes time and $\theta = \theta(x,y,t)$ measures the (clockwise) rotation from the $x$-axis to the vector $\mathbf{g}_1$. Under these conditions, all the Christoffel symbols of the material connection vanish identically except the following:

$$\Gamma^1_{21} = -\Gamma^2_{11} = \theta_{,x}, \tag{6.191}$$

and

$$\Gamma^1_{22} = -\Gamma^2_{12} = \theta_{,y}. \tag{6.192}$$

The non-vanishing component of the torsion tensor are, accordingly:

$$\tau^1_{12} = -\tau^1_{21} = -\theta_{,x}, \tag{6.193}$$

and

$$\tau^2_{12} = -\tau^2_{21} = -\theta_{,y}. \tag{6.194}$$

Finally, the non-vanishing components of $\bar{\mathbf{L}}_P$ are:

$$\left(\bar{L}_P\right)^1_2 = -\left(\bar{L}_P\right)^2_1 = \theta_{,t}. \tag{6.195}$$

Introducing all these values in the proposed evolution equation (6.189), we obtain the following homogenenous quasi-linear first-order partial differential equation:

$$\theta_{,t} + (a\cos\theta - b\sin\theta)\,\theta_{,x} + (a\sin\theta + b\cos\theta)\,\theta_{,y} = 0, \tag{6.196}$$

where $a$ and $b$ are, respectively, the material constants $2C^1_{21}{}^{12}$ and $2C^1_{22}{}^{12}$. Introducing the two new constants:

$$\theta_0 = \arctan(b/a), \quad c = 1/\sqrt{a^2 + b^2}, \tag{6.197}$$

and effecting the change of variable:

$$\phi = \theta + \theta_0, \tag{6.198}$$

we can write Equation (6.196) more concisely as:

$$\phi_{,t} + c(\sin\phi)_{,x} - c(\cos\phi)_{,y} = 0. \tag{6.199}$$

The characteristic strips [15] of this equation project onto straight characteristics in the coordinate space $x, y, t$, with $\phi$ remaining constant along each characteristic. The values of the material constants, together with a given (smooth) initial condition, determine whether the characteristics tend to converge or diverge as time goes on. In the former case, the intersection of characteristics can be interpreted as the creation of dislocation pile-ups, while in the latter, the dislocations tend to dissipate with the passage of time. A more detailed discussion of these issues can be found in [35].

### 6.8.3 An isotropic solid

#### Evolution law

For the case of full isotropy, the continuity of the symmetry group (the orthogonal group) renders the material connection non-unique, thereby depriving the torsion of any of the possible material connections of its physical meaning as a measure of the dislocation density. Instead, this meaning is now conveyed by the Riemann curvature tensor $\mathbf{R}$ of the symmetric metric connection generated by the metric tensor:

$$\mathbf{G} \equiv \left(\mathbf{P}\mathbf{P}^T\right)^{-1}. \tag{6.200}$$

Accordingly, we propose (in place of Equation (6.186)) the following self-driven evolution law:

$$\bar{\mathbf{L}}_P = \bar{\mathbf{f}}(\mathbf{R}_{P^{-1}}, \nabla \mathbf{R}_{P^{-1}}), \tag{6.201}$$

in which, for extra modelling flexibility, we have included also the referential gradient $\nabla \mathbf{R}$ of the curvature. The archetype pull-backs $\mathbf{R}_{P^{-1}}$ and $(\nabla \mathbf{R})_{P^{-1}}$ are given in components respectively by:

$$R^{\alpha}{}_{\beta\gamma\rho} = R^I{}_{JKL}\, P^{-\alpha}_I\, P^J_\beta\, P^K_\gamma\, P^L_\rho, \tag{6.202}$$

and

$$R^{\alpha}{}_{\beta\gamma\rho\sigma} = R^I{}_{JKL,M}\, P^{-\alpha}_I\, P^J_\beta\, P^K_\gamma\, P^L_\rho\, P^M_\sigma. \tag{6.203}$$

We are no longer demanding that the material isomorphisms be pure rotations (this would be tantamount, in this case, to demanding homogeneity!). What this implies is that we can no longer afford the luxury of supposing that, if the body is free of external forces, there will be states of identically vanishing stress. To the contrary, residual stresses are the norm in an inhomogeneous fully isotropic body. Our proposed evolution law (6.201), however, faithful to its self-driven nature, is oblivious to the existence of residual stresses. A more realistic model might incorporate, in addition to the curvature and its gradient, the Eshelby stress and/or other deformation related quantities.

Recalling that the symmetry of the evolution law (according to the principle of symmetry consistency) must be equal to, or larger than, the material symmetry group of the archetype, the proposed evolution law can be reduced to one involving just the orthogonal invariants of the curvature and its gradient. In particular, we may, and shall, concentrate our attention on an evolution law of the form:

$$\bar{\mathbf{L}}_P = \bar{f}(\kappa_{P^{-1}}, (\nabla \kappa)_{P^{-1}})\, \mathbf{I}, \tag{6.204}$$

where $\bar{f}$ is now a scalar-valued function and $\kappa$ is the *curvature scalar*, defined as:

$$\kappa \equiv R^{IJ}{}_{IJ}. \tag{6.205}$$

Since, by construction, the outcome of (6.201) is symmetric, the principle of actual evolution is automatically satisfied.

**Inhomogeneity fading**

To get some sense of the modelling capabilities of the proposed self-driven evolution law, we consider the very special class of problems characterized by an implant field of the form:

$$[\mathbf{P}] = \begin{bmatrix} p(x,t) & 0 & 0 \\ 0 & 1/p(x,t) & 0 \\ 0 & 0 & 1 \end{bmatrix}. \qquad (6.206)$$

Using primes to indicate $x$-derivatives, the curvature scalar reduces, in this case, to the simple expression:

$$\kappa = (p^2)'', \qquad (6.207)$$

and the evolution equation boils down to:

$$\left(\frac{\dot{p}}{p}\right)^2 = \bar{f}\left((p^2)'', \left((p^2)''' \, p\right)^2\right). \qquad (6.208)$$

Consider now the case of a linear dependence on the first argument only, namely:

$$\left(\frac{\dot{p}}{p}\right)^2 = c \, (p^2)'', \qquad (6.209)$$

where $c$ is some material constant. Denoting:

$$u = p^2, \qquad (6.210)$$

we can rewrite Equation (6.209) as:

$$\dot{u} = 2 \, c \, u \, u'', \qquad (6.211)$$

which is a nonlinear heat equation. By this we mean that linearizing for small implants (namely, for $u$ close to 1), we recover the classical heat equation. This argument shows that the homogeneity will tend to fade in time as long as $c$ is positive. Note, however, that the positivity of $c$ does not arise from the undoctored Clausius-Duhem inequality, because of the self-driven assumption.

### Hypocycloidal travelling waves

Having observed that without the inclusion of the gradient of the curvature the evolution law is purely diffusive, we investigate now the case in which this gradient is included to the exclusion of the curvature itself. We are interested in particular in the possibility of the medium sustaining travelling waves. Consider, therefore, the evolution law:

$$\left(\frac{\dot{p}}{p}\right)^2 = c^2 \left((p^2)''' \, p\right)^2, \qquad (6.212)$$

where $c$ is a positive material constant. We seek solutions of the form:

$$p(x,t) = p(x - v\,t) = p(\xi), \qquad (6.213)$$

## 6.8 Example: Self-driven evolution

where $v$ is the speed of the putative travelling wave. Retaining primes to indicate $\xi$-derivatives and introducing (6.213) into (6.212), we obtain the following ordinary differential equation for $p(\xi)$:

$$v^2 \left(\frac{p'}{p}\right)^2 = c^2 \left((p^2)''' \, p\right)^2. \tag{6.214}$$

The fact that the wave speed $v$ appears squared implies that for every solution of the form $p(x-vt)$ the function $p(x+vt)$ is also a solution. We now consider the negative square root of (6.214), namely:

$$\frac{u'}{2u^{3/2}} = -\frac{c}{v} u''', \tag{6.215}$$

where we have used the change of variables (6.210). This equation has the following first integral:

$$u^{-1/2} = \frac{c}{v} u'' + D, \tag{6.216}$$

where $D$ is a constant of integration.

If we regard this last equation as the equation of "motion" of a particle, we conclude that the "mass" is $c/v$ and the "force field" is $u^{-1/2} - D$. The "potential energy" is then $-2u^{-1/2} + Du$. This analogy permits us to see immediately [1] that, provided $d$ be positive, there will be continuous periodic (bounded) solutions. The "total energy" $E$ of any such solution is given by the first integral:

$$\frac{c}{2v} (u')^2 - 2u^{-1/2} + Du = E. \tag{6.217}$$

From the particular shape of our potential energy, we conclude that the total energy $E$ is bounded as follows:

$$-\frac{1}{D} < E < 0. \tag{6.218}$$

For any given value of $E$ between these bounds, the solution will oscillate between the two values corresponding to the vanishing of the "kinetic energy", namely:

$$\frac{1-\sqrt{1+ED}}{D} < u^{1/2} < \frac{1+\sqrt{1+ED}}{D}. \tag{6.219}$$

The integration constants $D$ and $E$ are determined by the "initial" $u_0$, $u'_0$ and $u'''_0$ (for some value $\xi_0$ of $\xi$) by the relations:

$$D = u_0^{-1/2} - \frac{c}{v} u''_0 \tag{6.220}$$

and

$$E = = u_0^{-1/2} - \frac{c}{v} u_0 u''_0 + \frac{c}{2v} (u'_0)^2. \tag{6.221}$$

Equation (6.217) can be integrated exactly to yield the following general solution:

$$\sqrt{\frac{2v}{c}}\,\xi = \frac{2}{D^{3/2}}\left[\arcsin\left(\frac{D\sqrt{u}-1}{\sqrt{1+ED}}\right) - \sqrt{1+ED-(D\sqrt{u}-1)^2}\right] + F, \tag{6.222}$$

where $F$ is yet another constant of integration fixing the "initial" value $\xi_0$. It can be shown [32] that, in the space of coordinates $p$ and $\xi$, and in appropriately chosen scales, the graph of this travelling wave is a hypocycloid. The exact shape and speed of the wave are determined by its period and by the maximum and minimum values of $p$. The larger the average value of $p$ and the shorter the period, the faster the resulting speed of propagation of the dislocation wave.

# 7

# Second-grade evolution

The general lines of the theory of material evolution, as laid down in detail in the previous chapter for the case of simple materials, apply equally well to second-grade materials and, mutatis mutandis, to more general media. Nevertheless, the details of the theory get progressively more complicated and deserving of special attention. In this chapter we will study the consequences of the principles of reduction to the archetype, actual evolution and symmetry consistency as they apply to second-grade materials.

## 7.1 Introduction

As discussed in Chapter 2, second-grade material isomorphisms consist of 2-jets of local diffeomorphisms. In terms of components in a coordinate system, they are described by a pair $\{P^I_M, Q^I_{MN}\}$ with $Q^I_{MN} = Q^I_{NM}$. For a smoothly uniform second-grade body $\mathcal{B}$, the uniformity condition can, therefore, be expressed in terms of smooth implant fields $\{\mathbf{P}, \mathbf{Q}\} = \{P^I_\alpha(\mathbf{X}), Q^I_{\alpha\beta}(\mathbf{X})\}$ from a second-grade archetype to the points $\mathbf{X} \in \mathcal{B}$, where Greek indices refer to the choice of an archetypal second-order frame. Accordingly, an evolution law for a second-grade material (whose solution delivers, as in the first-grade case, a time-dependent material automorphism at one and the same point) will have the following general form in terms of a coupled system of first-order differential equations:

$$\dot{\mathbf{P}} = \mathbf{f}(\mathbf{P}, \mathbf{Q}, \mathbf{a}, \mathbf{X}), \qquad (7.1)$$

and

$$\dot{\mathbf{Q}} = \mathbf{g}(\mathbf{P}, \mathbf{Q}, \mathbf{a}, \mathbf{X}), \qquad (7.2)$$

where, as in the first-grade case, we denote by $\mathbf{a}$ a list of arguments. In many cases we can identify the list $\mathbf{a}$ with the Eshelby stress and hyperstress, as defined in Equations (5.100) and (5.101), and possibly some additional variables, such as the first and second gradients of the deformation. The evolution functions **f** and **g** cannot be completely arbitrary, but must abide by a number of criteria with which we are already familiar from our experience with the first-grade case. The implementation of these criteria is, expectedly, more involved in the case of second-grade materials, mainly because of the peculiar composition law of 2-jets, as described in Equations (2.10) and (2.11).

## 7.2 Reduction to the archetype

Under a change of reference configuration $\lambda$, keeping the archetype unaltered, the evolution equations (7.1) and (7.2) will be transformed into the equivalent expressions:

$$\dot{\hat{\mathbf{P}}} = \hat{\mathbf{f}}(\hat{\mathbf{P}}, \hat{\mathbf{Q}}, \hat{\mathbf{a}}, \lambda(\mathbf{X})), \tag{7.3}$$

and

$$\dot{\hat{\mathbf{Q}}} = \hat{\mathbf{g}}(\hat{\mathbf{P}}, \hat{\mathbf{Q}}, \hat{\mathbf{a}}, \lambda(\mathbf{X})). \tag{7.4}$$

In terms of components, if we express the change $\lambda$ of reference configuration as:

$$Y^M = Y^M(X^I), \tag{7.5}$$

the implants $\{\mathbf{P}, \mathbf{Q}\}$ and $\{\hat{\mathbf{P}}, \hat{\mathbf{Q}}\}$ are related by the following formulas:

$$\hat{P}^M_\alpha = \frac{\partial Y^M}{\partial X^I} P^I_\alpha, \tag{7.6}$$

$$\hat{Q}^M_{\alpha\beta} = \frac{\partial^2 Y^M}{\partial X^I \partial X^J} P^I_\alpha P^J_\beta + \frac{\partial Y^M}{\partial X^I} Q^I_{\alpha\beta}. \tag{7.7}$$

As far as the list of arguments represented generically by **a** is concerned, its law of transformation depends, naturally, on the elements of this list. For instance, if the list consists of the Eshelby stress and hyperstress, the following transformation equations are obtained:

$$\hat{b}^N_M = J^{-1}_{\nabla\lambda} \left[ \frac{\partial Y^N}{\partial X^I} \frac{\partial X^K}{\partial Y^M} b^I_K + \left( \frac{\partial^2 Y^N}{\partial X^I \partial X^J} \frac{\partial X^K}{\partial Y^M} + 2 \frac{\partial Y^S}{\partial X^I} \frac{\partial Y^N}{\partial X^J} \frac{\partial^2 X^K}{\partial Y^M \partial Y^S} \right) b^{IJ}_K \right]. \tag{7.8}$$

$$\hat{b}^{NS}_M = J^{-1}_{\nabla\lambda} \frac{\partial Y^S}{\partial X^J} \frac{\partial Y^N}{\partial X^I} \frac{\partial X^K}{\partial Y^M} b^{IJ}_K. \tag{7.9}$$

We note, for the record, that the Piola stress and hyperstress transform according to the following formulas:

$$\hat{T}_i^M = J^{-1}_{\nabla\lambda} \left[ \frac{\partial Y^M}{\partial X^I} T_i^I + \frac{\partial^2 Y^M}{\partial X^J \partial X^I} T_i^{JI} \right], \tag{7.10}$$

$$\hat{T}_i^{MN} = J^{-1}_{\nabla\lambda} \frac{\partial Y^M}{\partial X^I} \frac{\partial Y^N}{\partial X^J} T_i^{IJ}. \tag{7.11}$$

## 7.2 Reduction to the archetype

Choosing at a given point and instant of time a change of reference configuration whose local value coincides with the inverse of the implant, we obtain the following values for the pull-back of the Eshelby stress and hyperstress to the archetype:

$$\bar{b}^\beta_\alpha = J_P \left[ P^{-\beta}_I P^K_\alpha b^I_K + \left( -Q^L_{\rho\sigma} P^{-\rho}_I P^{-\sigma}_J P^{-\beta}_L P^K_\alpha + 2 P^{-\rho}_I P^{-\beta}_J Q^K_{\alpha\rho} \right) b^{IJ}_K \right], \tag{7.12}$$

$$\bar{b}^{\beta\gamma}_\alpha = J_P \, P^{-\beta}_I P^{-\gamma}_J P^K_\alpha b^{IJ}_K. \tag{7.13}$$

At the same time, for this particular choice, the time rates of the (now trivial) implants become:

$$\dot{\bar{P}}^\beta_\alpha = P^{-\beta}_I \dot{P}^I_\alpha, \tag{7.14}$$

and

$$\dot{\bar{Q}}^\gamma_{\alpha\beta} = P^{-\gamma}_K \left( \dot{Q}^K_{\alpha\beta} - Q^K_{\rho\beta} P^{-\rho}_I \dot{P}^I_\alpha - Q^K_{\alpha\sigma} P^{-\sigma}_J \dot{P}^J_\beta \right)$$
$$= P^{-\gamma}_K \left( \dot{Q}^K_{\alpha\beta} - Q^K_{\rho\beta} \dot{\bar{P}}^\rho_\alpha - Q^K_{\alpha\sigma} \dot{\bar{P}}^\sigma_\beta \right) \tag{7.15}$$

These quantities should be regarded as the expression of the "inhomogeneity velocity gradient" for second-grade materials at the archetype level.

In terms of all these variables, the evolution laws read:

$$\dot{\bar{P}}^\beta_\alpha = \bar{f}^\beta_\alpha(\bar{b}^\sigma_\rho, \bar{b}^{\sigma\tau}_\rho), \tag{7.16}$$

and

$$\dot{\bar{Q}}^\alpha_{\beta\gamma} = \bar{g}^\alpha_{\beta\gamma}(\bar{b}^\sigma_\rho, \bar{b}^{\sigma\tau}_\rho). \tag{7.17}$$

By virtue of Equations (7.15) and (7.16), these evolution laws can also be written as:

$$\dot{P}^I_\alpha = P^I_\beta \bar{f}^\beta_\alpha(\bar{b}^\sigma_\rho, \bar{b}^{\sigma\tau}_\rho), \tag{7.18}$$

and

$$\dot{Q}^I_{\beta\gamma} = P^I_\alpha \bar{g}^\alpha_{\beta\gamma}(\bar{b}^\sigma_\rho, \bar{b}^{\sigma\tau}_\rho) + Q^I_{\rho\gamma} \bar{f}^\rho_\beta(\bar{b}^\sigma_\rho, \bar{b}^{\sigma\tau}_\rho) + Q^I_{\beta\rho} \bar{f}^\rho_\gamma(\bar{b}^\sigma_\rho, \bar{b}^{\sigma\tau}_\rho). \tag{7.19}$$

where, for specificity, we have assumed that the list of variables **a** consists only of the Eshelby stress and hyperstress. Equations (7.15) and (7.16) represent the evolutions equations *reduced to the archetype*. In the general case, one has to calculate carefully the pullback $\bar{\mathbf{a}}$ of the list of variables to the archetype. Note that the functions $\bar{f}^\alpha_\beta$ and $\bar{g}^\alpha_{\beta\gamma}$ represent how the archetype would respond to the application of an arbitrary value of the variables **a** measured therein. This is clearly a constitutive property which, once known, permits us to calculate how any point in the reference configuration responds to the same list of arguments. We have thus reduced the evolution laws to the determination of the functions $\bar{f}^\alpha_\beta$ and $\bar{g}^\alpha_{\beta\gamma}$.

## 7.3 Actual evolution

Just as in the first-grade case, two evolutions which, at each point, differ by a constant or time-dependent element of the material symmetry group must be considered as equivalent. The immediate implication of this observation is that an evolution which consists of time-dependent elements of the symmetry group is equivalent to the zero evolution. Therefore, if we want to avoid the possibility of an evolution law ever resulting in a zero evolution, we must ensure that it instantaneously prescribes something other than an infinitesimal generator of the symmetry group. In other words, a legitimate evolution law must consists of functions $\bar{f}_\alpha^\beta$ and $\bar{g}_{\beta\gamma}^\alpha$ whose evaluation over any possible values taken by their arguments never results in an element of the Lie algebra of the material symmetry group of the archetype. This is the expression of the *principle of actual evolution*. For example, if the induced first-grade symmetry is trivial and the second-grade symmetries consists of all traceless symmetric **S** (as discussed in the example presented in Section 2.3.4), the corresponding Lie algebra consists of all pairs of the form $\{\mathbf{0}, \mathbf{S}\}$. In this case, the principle of actual evolution imposes no restrictions (other than non-vanishing) on the functions $\bar{f}_\alpha^\beta$, but the functions $\bar{g}_{\beta\gamma}^\alpha$ must be such that:

$$\bar{g}_{\alpha\gamma}^\alpha \neq 0. \qquad (7.20)$$

## 7.4 Material symmetry consistency

Under a change of archetype $\{\mathbf{A}, \mathbf{B}\}$ such as the one illustrated in Figure 2.7 we obtain the following laws of transformation for the implant maps:

$$\hat{P}_\alpha^I = P_\rho^I A_\alpha^\rho, \qquad (7.21)$$

$$\hat{Q}_{\beta\gamma}^I = Q_{\rho\sigma}^I A_\beta^\rho A_\gamma^\sigma + P_\rho^I B_{\beta\gamma}^\rho, \qquad (7.22)$$

where we have used "hats" to indicate quantities evaluated with respect to the new archetype. At the same time, the pull-backs of the Eshelby stress and hyperstress to the new archetype are given in terms of the old pull-backs by[1]:

$$\hat{b}_\alpha^\beta = J_A \left( A_\rho^{-\beta} A_\alpha^\sigma \bar{b}_\sigma^\rho - B_{\mu\sigma}^\rho A_\lambda^{-\mu} A_\tau^{-\sigma} A_\rho^{-\beta} A_\alpha^\nu \bar{b}_\nu^{\lambda\tau} + 2 B_{\alpha\rho}^\mu A_\lambda^{-\rho} A_\tau^{-\beta} \bar{b}_\mu^{\lambda\tau} \right), \qquad (7.23)$$

$$\hat{b}_\alpha^{\beta\gamma} = J_A A_\mu^{-\beta} A_\nu^{-\gamma} A_\alpha^\sigma \bar{b}_\sigma^{\mu\nu}. \qquad (7.24)$$

The laws of evolution expressed in terms of the new archetype must read:

$$\dot{\hat{P}}^I{}_\alpha = \hat{P}_\beta^I \hat{f}_\alpha^\beta(\hat{b}_\rho^\sigma, \hat{b}_\rho^{\sigma\tau}), \qquad (7.25)$$

$$\dot{\hat{Q}}^I{}_{\beta\gamma} = \hat{P}_\alpha^I \hat{g}_{\beta\gamma}^\alpha(\hat{b}_\rho^\sigma, \hat{b}_\rho^{\sigma\tau}) + \hat{Q}_{\mu\gamma}^I \hat{f}_\beta^\mu(\hat{b}_\rho^\sigma, \hat{b}_\rho^{\sigma\tau}) + \hat{Q}_{\beta\mu}^I \hat{f}_\gamma^\mu(\hat{b}_\rho^\sigma, \hat{b}_\rho^{\sigma\tau}), \qquad (7.26)$$

---

[1] Hopefully, correcting several misprints and mistakes in [29].

where $\hat{f}^\alpha_\beta$ and $\hat{g}^\alpha_{\beta\gamma}$ are the evolution functions with respect to the new archetype. Clearly, these new functions, if they are to represent the same material, must be related to the old ones, $\bar{f}^\alpha_\beta$ and $\bar{g}^\alpha_{\beta\gamma}$.

To find these relations, we start by observing that, since the change of archetype is assumed to be independent of time, the time rates of the new implants in terms of their old counterparts are obtained from Equations (7.21) and (7.22) as:

$$\dot{\hat{P}}^I{}_\alpha = \dot{P}^I_\rho A^\rho_\alpha, \tag{7.27}$$

$$\dot{\hat{Q}}^I{}_{\beta\gamma} = \dot{Q}^I_{\rho\sigma} A^\rho_\beta A^\sigma_\gamma + \dot{P}^I_\rho B^\rho_{\beta\gamma}. \tag{7.28}$$

We have now at our disposal Equations (7.18), (7.19) and (7.21-7.28), which when combined allow to read off the desired quantities as:

$$\hat{f}^\beta_\alpha = A^\rho_\alpha \, A^{-\beta}_\sigma \, \bar{f}^\sigma_\rho, \tag{7.29}$$

$$\hat{g}^\alpha_{\beta\gamma} = A^{-\alpha}_\rho \left[ A^\mu_\beta \, A^\sigma_\gamma \, \bar{g}^\rho_{\mu\sigma} + B^\sigma_{\beta\gamma} \, \bar{f}^\rho_\sigma - B^\rho_{\mu\gamma} \, A^\lambda_\beta \, A^{-\mu}_\sigma \, \bar{f}^\sigma_\lambda - B^\rho_{\beta\mu} \, A^\lambda_\gamma \, A^{-\mu}_\sigma \, \bar{f}^\sigma_\lambda \right]. \tag{7.30}$$

To enforce the *principle of material symmetry consistency* we now require that, if the transformation $\{\mathbf{A}, \mathbf{B}\}$ happens to belong to the symmetry group of the original archetype, then the functions $\hat{f}^\alpha_\beta$ and $\hat{f}^\alpha_{\beta\gamma}$ must be identical to their respective counterparts $\bar{f}^\alpha_\beta$ and $\bar{f}^\alpha_{\beta\gamma}$. Clearly, however, each of these functions must be evaluated with the corresponding arguments, which are related by Equations (7.23) and (7.24). Notice that for a material symmetry, the determinant $J_A$ is equal to 1.

## 7.5 An example

Resorting once again to the non-trivial second-grade material of practical interest discussed in Section 2.3.4, we observe that, the induced first-grade symmetry group being trivial, Equation (7.24) implies that the Eshelby hyperstress $\bar{b}^\alpha_{\beta\gamma}$ remains unaltered by the change of archetype (within the symmetry group), while the Eshelby stress $\bar{b}^\beta_\alpha$ is transformed, according to Equation (7.23) as follows:

$$\hat{b}^\beta_\alpha = \bar{b}^\beta_\alpha - S^\beta_{\rho\sigma} \, \bar{b}^{\rho\sigma}_\alpha + 2 S^\rho_{\alpha\sigma} \, \bar{b}^{\sigma\beta}_\rho, \tag{7.31}$$

where the zero-trace condition ($S^\alpha_{\alpha\beta} = 0$) is assumed to be satisfied. Moreover, the conditions (7.29) and (7.30) imposed by symmetry consistency reduce in this case, respectively, to:

$$\hat{f}^\beta_\alpha = \bar{f}^\beta_\alpha, \tag{7.32}$$

$$\hat{g}^\alpha_{\beta\gamma} = \bar{g}^\alpha_{\beta\gamma} + S^\sigma_{\beta\gamma} \, \bar{f}^\alpha_\sigma - S^\alpha_{\sigma\gamma} \, \bar{f}^\sigma_\beta - S^\alpha_{\beta\sigma} \, \bar{f}^\sigma_\gamma. \tag{7.33}$$

Therefore, according to the principle of symmetry consistency, the following conditions for the evolution functions $\bar{f}^\mu_\nu$ and $\bar{g}^\mu_{\nu\tau}$:

$$\bar{f}^\mu_\nu(\bar{b}^\alpha_\beta - S^\alpha_{\rho\sigma}\,\bar{b}^{\rho\sigma}_\beta + 2S^\rho_{\beta\sigma}\,\bar{b}^{\sigma\alpha}_\rho,\,\bar{b}^\alpha_{\beta\gamma}) = \bar{f}^\mu_\nu(\bar{b}^\alpha_\beta,\,\bar{b}^\alpha_{\beta\gamma}), \tag{7.34}$$

$$\begin{aligned}\bar{g}^\mu_{\nu\tau}(\bar{b}^\alpha_\beta - S^\alpha_{\rho\sigma}\,\bar{b}^{\rho\sigma}_\beta + 2S^\rho_{\beta\sigma}\,\bar{b}^{\sigma\alpha}_\rho,\,\bar{b}^\alpha_{\beta\gamma}) &= \bar{g}^\mu_{\nu\tau}(\bar{b}^\alpha_\beta,\,\bar{b}^\alpha_{\beta\gamma}) + \\ S^\sigma_{\nu\tau}\,\bar{f}^\mu_\sigma(\bar{b}^\alpha_\beta,\,\bar{b}^\alpha_{\beta\gamma}) - S^\mu_{\sigma\tau}\,\bar{f}^\sigma_\nu(\bar{b}^\alpha_\beta,\,\bar{b}^\alpha_{\beta\gamma}) &- S^\mu_{\nu\sigma}\,\bar{f}^\sigma_\tau(\bar{b}^\alpha_\beta,\,\bar{b}^\alpha_{\beta\gamma}).\end{aligned} \tag{7.35}$$

must be satisfied identically for all traceless **S**, and for all values of the remaining variables.

The rigorous general consequences of identity (7.34) are not available, although it appears that the 15 independent components of **S** might give enough freedom to claim that the functions $\bar{f}^\mu_\nu$ are independent of the first argument $\bar{b}^\alpha_\beta$. Be that as it may, we can certainly assume that for a given archetype this is the case, namely, the functions $\bar{f}^\mu_\nu$ are assumed to depend on the variables $\bar{b}^\alpha_{\beta\gamma}$. Under such assumption, the condition (7.34) is satisfied identically. If we also assume that, for the given archetype, the functions $\bar{g}^\mu_{\nu\tau}$ are independent of the first argument $\bar{b}^\alpha_\beta$, we obtain that the relation (7.35) is violated unless $\bar{f}^\mu_\nu = 0$. In other words, we can obtain a consistent second-grade evolution equation by assuming that the first-grade structure does not evolve at all and that the second-grade structure evolves as an arbitrary function of the Eshelby hyperstress alone.

*Remark 7.1.* It is interesting to note that the $(\alpha,\beta)$-trace of the combined last three terms of Equation (7.33) vanishes. Therefore, if we consider the extended version of the principle of symmetry consistency (which combines it with the principle of actual evolution, as we have seen in Remark 6.3), it is not necessary to assume that the functions $\bar{f}^\alpha_\beta$ vanish identically. In other words, as long as for the given archetype the functions $\bar{f}$ and $\bar{g}$ depend on the Eshelby hyperstress alone (and not on the Eshelby stress), we obtain a consistent evolution law for a material of the type under consideration.

The example just presented, which may be of relevance in modelling phenomena of mass diffusion, represents possibly the closest one can get to a purely second-grade evolution. At the other end of the spectrum, we have the Toupin materials, whose symmetry group is of the type $\{\mathcal{G}, \mathbf{0}\}$, or a conjugate thereof, as we have seen in Chapter 2. A good example is provided by a second-grade material whose induced first-grade structure is that of an isotropic solid, but which has no second-grade symmetries whatsoever. In this case, the evolution functions $\bar{f}$ and $\bar{g}$ must be isotropic matrix-valued functions with respect to each of their matrix arguments.

## 7.6 Concluding remarks

In this chapter we have presented the outline of a consistent theory of second-grade evolution. The more general case of Cosserat media can be handled in a similar way, but will not be pursued here. Another interesting generalization of the theory of material evolution is the one provided by a general relativistic framework. One of the appealing features of this context is that uniformity and evolution can be regarded, respectively, as the space-like and time-like aspects of the same phenomenon. A detailed development of this point of view has recently appeared in print [34].

# Part III

# Mathematical Foundations

# 8
# Basic geometric concepts

In this introductory chapter we summarize, in a rather selective way, those basic differential geometric concepts, notions and facts which we feel are relevant and useful in our exposition of the theory of material inhomogeneities and their evolution. Many definitions and statements introduced in this part can be easily generalized to encompass situations broader than those considered here. Yet, it is not our intention to present a comprehensive introduction to Differential Geometry but rather only the material relevant to enhance the understanding of the mechanical theory discussed in this monograph. A thorough treatment of Differential Geometry, which in spirit and in some cases to the letter, we follow in this chapter, can be found in the classical exposition of Kobayashi & Nomizu [57].

We assume that the reader is familiar with the fundamentals of real analysis, abstract algebra and linear algebra, and understands the concept of a topological space.

## 8.1 Differentiable manifolds

As much as a material body is the fundamental object of Continuum Mechanics, a (differentiable) manifold is the fundamental object of Differential Geometry. These are the entities on which "everything happens", both in geometry and in mechanics. Although material bodies, as shown in the first part of our exposition, are usually defined as particularly simple instances of differentiable manifolds, we present here the general definition of this concept. The reason for this general approach is that the concept of a differentiable manifold is needed not only to deal with a material body as such but is also required to understand other mathematical and mechanical objects appearing in the theory of inhomogeneities such as frame bundles, G-structures and connections.

## 8 Basic geometric concepts

There are several different ways of defining a differentiable manifold, an object whose main feature is that it looks locally like the Euclidean space $\mathbb{R}^n$. Rather than following this line of direct approach we shall define a manifold by introducing first the concept of a pseudo-group of transformations of the topological space a manifold will be modeled after.

A *pseudo-group of transformations* on a topological space[1] $S$ is a set $\Gamma(S)$ of mappings of $S$ such that:

1. Each $f \in \Gamma(S)$ is a homeomorphism[2] of an open set of $S$ onto another open set of $S$.
2. If $f \in \Gamma(S)$ then any restriction of $f$ to an open subset of $S$ is in $\Gamma(S)$.
3. Let $U$ be an open set in $S$ and assume that $U$ is a union of open sets, say $U_j$. A homeomorphism $f : U \to S$ is in $\Gamma(S)$ provided its restriction to every subset $U_j$ is in $\Gamma(S)$.
4. For any open set $U \subset S$ the identity transformation of $U$, $id_U$, is in $\Gamma(S)$.
5. If $f \in \Gamma(S)$ then its inverse $f^{-1} \in \Gamma(S)$.
6. If $f, h \in \Gamma(S)$ then the composition $h \circ f$ is in $\Gamma(S)$, i.e., given $f : U \to V$ and $h : U' \to V'$ such that $V \cap U' \neq \emptyset$, where $U, U', V, V'$ are open subsets of $S$, $h \circ f : f^{-1}(V \cap U') \to S$ is in $\Gamma(S)$.

Let us assume that $M$ is a topological Hausdorff space[3]. An *atlas* on $M$ compatible with the given pseudo-group of transformations $\Gamma(S)$ is a family of pairs $\{(U_i, \varphi_i)\}$, called *charts*, such that:

1. Each $U_i$ is an open subset of $M$ and the union $\bigcup U_i = M$.
2. Each mapping $\varphi_i$ is a homeomorphism of $U_i$ onto an open subset of $S$.
3. If $U_i \cap U_j \neq \emptyset$ then $\varphi_i \circ \varphi_j^{-1}$ restricted to $\varphi_j(U_i \cap U_j)$ is an element of $\Gamma(S)$.

A *complete atlas* of $M$ compatible with the pseudo-group of transformations $\Gamma(S)$ is an atlas of $M$ which is not contained in any other atlas of $M$ compatible with $\Gamma(S)$. Note that given an atlas $\{(U_i, \varphi_i)\}$ compatible with $\Gamma(S)$ there is a unique complete atlas compatible with $\Gamma(S)$ containing $\{(U_i, \varphi_i)\}$. Indeed, consider a family of pairs $\{(U_I, \varphi_I)\}$ such that each $\varphi_I$ is a homeomorphism of an open set $U_I \subset M$ onto an open subset of $S$ and $\varphi_i \circ \varphi_I^{-1} : \varphi_I(U_I \cap U_i) \to \varphi_i(U_I \cap U_i)$ is an element of $\Gamma(S)$ whenever $U_I \cap U_i \neq \emptyset$. Then, the atlas $\{(U_i, \varphi_i)\}$ is contained in $\{(U_I, \varphi_I)\}$ and the latter is a complete atlas of $M$. Note also, that if the pseudo-group of transformations of S, say $\tilde{\Gamma}(S)$, is contained in $\Gamma(S)$ then an atlas of $M$ compatible with $\tilde{\Gamma}(S)$ is also compatible with $\Gamma(S)$.

---

[1] We are particularly interested in the Euclidean space $\mathbb{R}^n$.
[2] A *homeomorphism* of a topological space is a one-to-one and onto continuous mapping such that its inverse is also continuous.
[3] A topological space $M$ is called Hausdorff if for any two points, say $p \neq q$, there exist disjoint open sets $U_p$ and $U_q$ such that $p \in U_p$ and $q \in U_q$.

## 8.1 Differentiable manifolds

We are ready now to introduce the concept of a (*differentiable*) *manifold*. Namely, a *manifold* of class $C^r$ and dimension $n$ is a topological Hausdorff space $M$ with a fixed complete atlas compatible with $\Gamma^r(\mathbb{R}^n)$, where $\Gamma^r(\mathbb{R}^n)$ denotes the pseudo-group of r-differentiable transformations on $\mathbb{R}^n$ and where $r$ is a positive integer or infinity. A manifold will be called *smooth* when it is of class $C^\infty$. Henceforth, we will only be dealing with smooth manifolds.

Given a (complete) atlas $\{(U_i, \varphi_i)\}$ on an $n$-dimensional manifold $M$ the chart $(U_i, \varphi_i)$ introduces on $U_i$ a *local coordinate system*

$$u^l(p) = (\varphi_i(p))^l, \quad l = 1, \ldots, n \tag{8.1}$$

by assigning to a point $p \in U_i$ the n-tuple of numbers $\varphi_i(p) \in \mathbb{R}^n$. Having two coordinate systems available on two overlapping charts, the *transition functions* $\varphi_i \circ \varphi_j^{-1} : \varphi_j(U_i \cap U_j) \to \mathbb{R}^n$ manifest themselves via $n$ invertible differentiable functions relating coordinates of points in the overlap of charts.

Consider now two differentiable manifolds, say $M$ and $N$, and a mapping $f : M \to N$. We say that the mapping $f$ is *differentiable* if for every chart $(U_i, \varphi_i)$ of $M$ and every chart $(V_k, \psi_k)$ of $N$ such that $f(U_i) \subset V_k$ the mapping

$$\psi_k \circ f \circ \varphi_i^{-1} : \varphi_i(U_i) \to \psi_k(V_k) \tag{8.2}$$

is differentiable. In particular, viewing $\mathbb{R}$ as a trivial differentiable manifold, a *differentiable curve* on a manifold $M$ is a differentiable mapping (*parametrization*) $x : [a, b] \subset \mathbb{R} \to M$. The image $x([a,b]) \subset M$ of the curve $x$ will be called its *graph*. Note that the same graph can be parameterized in a variety of ways.

In order to develop differential calculus on a manifold and, in particular, to be able to calculate derivatives in a specific direction, we introduce the concept of a tangent vector to a differentiable manifold. Our definition is intrinsic in nature as it does not refer to a concept of an *ambient space* the manifold may sit in, e.g., a sphere sitting in $\mathbb{R}^3$. We say that, given a point $p \in M$, the *tangent vector* to a curve $x(t)$ on the manifold $M$ at the point $p = x(t_0)$ is the linear functional

$$X_p : \mathcal{F}(p) \to \mathbb{R} \tag{8.3}$$

on the *algebra of all differentiable functions* at $p$ (i.e., functions differentiable in some open neighbourhood of $p$) defined by

$$X_p(f) \equiv \frac{df(x(t))}{dt}\Big|_{t_0}. \tag{8.4}$$

Note that, in addition to being linear, the mapping $X_p$ satisfies the following product rule (Leibniz rule)

$$X_p(fg) = g(p)X_p(f) + f(p)X_p(g) \tag{8.5}$$

for every $f, g \in \mathcal{F}(p)$. One can show that the set $T_pM$ of all tangent vectors to $M$ at $p$, called the *tangent space* at $p$, is a real vector space.[4]

Given a curve $x(t)$ through the point $p = x(t_0)$ and a coordinate system $u^1, \ldots, u^n$ in an open neighbourhood of $p$, let

$$u^j(t) = u^j(x(t)) \tag{8.6}$$

and let $\frac{\partial}{\partial u^j}\big|_p : \mathcal{F}(p) \to \mathbb{R}$ be such that

$$\frac{\partial}{\partial u^j}\big|_p(f) \equiv \frac{\partial f}{\partial u^j}\big|_p. \tag{8.7}$$

The set of vectors

$$\frac{\partial}{\partial u^1}\big|_p, \ldots, \frac{\partial}{\partial u^n}\big|_p \tag{8.8}$$

not only spans the space $T_pM$, as

$$X_p(f) = \frac{df(x(t))}{dt}\big|_{t_0} = \sum_{j=1}^n \frac{\partial f}{\partial u^j}\big|_p \frac{du^j}{dt}\big|_{t_0}, \tag{8.9}$$

but it is, in fact, a basis thereof.

A vector field on a manifold is a result of selecting at each point of a manifold a tangent vector. Formalizing this process we say that a *vector field* on the manifold $M$ is a mapping

$$X : M \to \bigcup_{p \in M} T_pM \tag{8.10}$$

such that $X(p) \in T_pM$ for every $p \in M$. $X$ is called a *differentiable* vector field if, for every differentiable function $f$, $X(f)$, viewed as a real-valued function on $M$, is differentiable in a neighbourhood of every point. Namely, a vector field $X$ is differentiable if, for every point $p \in M$, there exists an open neighbourhood, say $U_p$, such that $X_q(f)$ is differentiable at every $q \in U_p$, and for every $f \in \mathcal{F}(q)$. Representing $X_p$ in the given coordinate induced basis $\frac{\partial}{\partial u^1}\big|_p, \ldots, \frac{\partial}{\partial u^n}\big|_p$ as

$$X_p = \sum_{j=1}^n \xi_j(p) \frac{\partial}{\partial u^j}\big|_p \tag{8.11}$$

one obtains the $\xi_j$ as the *components of the vector field* $X$ in the coordinate system $u^1, \ldots, u^n$. These components are real-valued functions on the manifold $M$. In fact, it can be shown [57] that:

---

[4] An equivalent definition of a tangent vector is given by identifying differentiable curves through a point, say $p$, having the same derivatives at $p$ in one, and thus every, coordinate system.

## 8.1 Differentiable manifolds

**Proposition 8.1.** *A vector field $X$ on a manifold $M$ is differentiable if, and only if, its components in one, and therefore in every, coordinate system are differentiable functions on $M$.*

The set $\mathcal{X}(M)$ of all differentiable vector fields on $M$ is a real vector space with point-wise addition and multiplication by scalars. In fact, it is an algebra with *bracket operation* defined by

$$[X,Y](f) \equiv X(Y(f)) - Y(X(f)) \tag{8.12}$$

for any differentiable function $f : M \to \mathbb{R}$, and any pair of vector fields $X, Y \in \mathcal{X}(M)$. Indeed, it is possible to show that:

**Proposition 8.2.** *$[X,Y]$ is a vector field on a manifold $M$ provided $X$ and $Y$ are vector fields on $M$. Moreover, as the bracket operation $[\cdot,\cdot]$ is skew-symmetric and it satisfies the* Jacoby identity

$$[[X,Y],Z] + [[Y,Z],X] + [[Z,X],Y] = 0, \tag{8.13}$$

*the space $\mathcal{X}(M)$ is, in fact, a Lie algebra over the set of real numbers.*

As we plan to deal later with manifolds which are stratified (fibred) in one or more ways, we digress now briefly to consider a *product manifold*. To this end, assume that $M$ and $N$ are two manifolds of dimension $m$ and $n$, respectively. Given an atlas $\{(U_i, \varphi_i)\}$ on $M$ and an atlas $\{(V_j, \psi_j)\}$ on $N$, it is only natural to consider on the topological space $M \times N$ the differentiable structure defined by the atlas $\varphi_i \times \psi_j : U_i \times V_j \to \mathbb{R}^{m+n}$ where $(\varphi_i \times \psi_j)(p,q) \equiv (\varphi_i(p), \psi_j(q))$ for every $(p,q) \in M \times N$. Note that this atlas is not complete even if the atlases on $M$ and $N$ are complete but, as indicated earlier, it can be completed to its unique maximal atlas. The tangent space to $M \times N$ at a point $(p,q)$, namely $T_{(p,q)}M \times N$, can be identified easily with the direct sum of the individual tangent spaces, that is, $T_p M \oplus T_q N$. Indeed, let $Z_{(p,q)} \in T_{(p,q)}M \times N$. Select a curve $z(t) = (x(t), y(t))$ on $M \times N$ through $(p,q) = (x(t_0), y(t_0))$ such that $Z_{(p,q)}$ is tangent to it at $(p,q)$. Let $\widetilde{X}_{(p,q)} \in T_{(p,q)}M \times N$ be the vector tangent to the curve $(x(t), q)$ at $(p,q)$ and let $\widetilde{Y}_{(p,q)} \in T_{(p,q)}M \times N$ be, respectively, the vector tangent to the curve $(p, y(t))$ at $(p,q)$. Then $Z_{(p,q)} = \widetilde{X}_{(p,q)} + \widetilde{Y}_{(p,q)}$. Moreover, vectors $\widetilde{X}_{(p,q)}$ and $\widetilde{Y}_{(p,q)}$ can be identified easily with the corresponding vectors on $M$ and $N$ through simple projections of the product $M \times N$ onto the first and second factor, respectively. Given a function $f : M \times N \to \mathbb{R}$,

$$Z_{(p,q)}(f) = \frac{df(x(t), y(t))}{dt}\bigg|_{t_0} = \frac{df(x(t), y(t_0))}{dt}\bigg|_{t_0}$$
$$+ \frac{df(x(t_0), y(t))}{dt}\bigg|_{t_0} = \widetilde{X}_{(p,q)}(f) + \widetilde{Y}_{(p,q)}(f). \tag{8.14}$$

As we mentioned earlier, the tangent space of the manifold $M$ at the point $p \in M$ is a real vector space. Its dual space $T_p^*M$ is called the space of *co-vectors* at $p$. A smooth field of co-vectors

$$\omega : M \to \bigcup_{p \in M} T_p^*M \tag{8.15}$$

such that $\omega_p \equiv \omega(p) \in T_p^*M$ is called a 1-*form* on $M$. In other words, a 1-form $\omega$ on $M$ is a linear mapping from the space $\mathcal{X}(M)$ of all vector fields on $M$ into the algebra of all differentiable functions $\mathcal{F}(M)$ on $M$ such that

$$\omega(X)(p) = \omega_p(X_p), \quad X_p \in \mathcal{X}(M), \quad \omega_p \in T_p^*M, \quad p \in M. \tag{8.16}$$

In particular, given a differentiable function $f : M \to \mathbb{R}$, its *total differential* is the 1-form $df$ defined at each $p \in M$ by

$$df_p(X_p) \equiv X_p(f) \tag{8.17}$$

for every $X_p \in T_pM$. If $u^1, \ldots, u^n$ is a local coordinate system in a neighbourhood of $p$, the total differentials $du_p^1, \ldots, du_p^n$ form a basis of $T_p^*M$. Moreover, according to the definition of a differential

$$du_p^j(X_p) = du_p^j(\xi^k(p)\frac{\partial}{\partial u^k}\big|_p) = \xi^j(p) \tag{8.18}$$

for any

$$X_p = \sum_{k=1}^n \xi^k(p)\frac{\partial}{\partial u^k}\big|_p \tag{8.19}$$

and any $j = 1, \ldots, n$. Thus, given a coordinate system $u^1, \ldots, u^n$ in an neighbourhood of a point $p$, any 1-form $\omega$ can be represented locally as

$$\omega = \sum_{k=1}^n f_k du^k, \tag{8.20}$$

where the functions $f_k$, called *components* of $\omega$ in the coordinates $u^1, \ldots, u^n$, are differentiable (in the neighbourhood of $p$) real-valued functions.

Generalizing the concept of a 1-form we say that a (differentiable) $r$-*form* on a $n$-dimensional manifold $M$ is a skew-symmetric r-linear mapping of the Cartesian product $\times^r \mathcal{X}(M) \equiv \mathcal{X}(M) \times \cdots \times \mathcal{X}(M)$ (r-times) into $\mathcal{F}(M)$. We denote by $D^r(M)$ the set of all differentiable r-forms on $M$, where $r = 0, 1, \ldots, n$, and where by convention $D^0(M) \equiv \mathcal{F}(M)$. Each set $D^r(M)$ is a real vector space as well as an $\mathcal{F}(M)$-module. Namely, if $f \in \mathcal{F}(M)$ and $\omega \in D^r(M)$ then $f\omega \in D^r(M)$ is viewed as an r-form such that $(f\omega)_p = f(p)\omega_p$, for any $p \in M$. An alternative way of defining an r-form is to consider the exterior algebra $\bigwedge T_p^*M$ with an alternating product $\wedge$ defined as follows[5]. If $\omega_1, \ldots, \omega_r$ are 1-forms on $M$ and if $X_p^1, \ldots, X_p^r$ are vectors at $p \in M$, then

$$(\omega_1 \wedge, \cdots, \wedge \omega_r)_p(X_p^1, \ldots, X_p^r) \equiv \det\{\omega_j(X_p^k)\}, \quad j,k = 1, \ldots, r. \tag{8.21}$$

An $r$-form $\omega$ evaluated at $p \in M$ is an element of degree $r$ in $\bigwedge T_p^*M$. In a local coordinate system $u^1, \ldots, u^n$ the form $\omega$ can therefore be expressed uniquely as

$$\omega = \sum_{i_1 < i_2 < \cdots < i_r} f_{i_1 \ldots i_r} du^{i_1} \wedge \cdots \wedge du^{i_r}. \tag{8.22}$$

Let $D(M)$ denote the totality of differential forms on $M$. The *exterior differential* $d : D(M) \to D(M)$ is a linear mapping such that

1. $d(D^r(M)) \subset D^{r+1}(M)$.
2. If $f \in \mathcal{F}(M)$ then $df$ is the total differential of $f$.
3. If $\omega \in D^r(M)$ and $\lambda \in D^s(M)$ then

$$d(\omega \wedge \lambda) = d\omega \wedge \lambda + (-1)^r \omega \wedge d\lambda. \tag{8.23}$$

4. $d^2 \equiv d \circ d = 0$.

The concept of a differential form can be generalized further to include differential forms with values in a vector space. That is, let $V$ be an $m$-dimensional real vector space. A $V$-valued $r$-form at $p \in M$ is a skew-symmetric $r$-linear mapping $\omega$ of the product $\times^r T_pM$ into $V$. Given a basis $v^1, \ldots, v^m$ in $V$ one can write

$$\omega_p = \sum_{j=1}^m \omega_j(p) v^j \tag{8.24}$$

where $\omega_j$ are usual $r$-forms on $M$. Indeed,

$$\omega_p(X_p^1, \cdots, X_p^m) = \sum_{j=1}^m \omega_j(p)(X_p^1, \cdots, X_p^m) v^j \tag{8.25}$$

for any $X_p^1, \ldots, X_p^m \in T_pM$. The exterior derivative of $\omega$ is simply

$$d\omega \equiv \sum_{j=1}^m d\omega_j(p) v^j. \tag{8.26}$$

By definition, the form $\omega$ is differentiable if each form $\omega_j$ is differentiable. In what follows we will only consider differentiable forms, both real and vector-valued.

---

[5] The definition of an exterior algebra over a vector space can be found in any multilinear algebra text. The reader may also consult [84], [92] and [61]

Given an object on a manifold, such as a vector or a form, and a mapping between manifolds, the important question to ask is how such an object gets "moved" by this mapping from one manifold to the other. In particular, given a differentiable mapping $f : N \to M$ from a manifold $N$ into another manifold $M$, we define the *tangent map* induced by $f$ at $p \in N$ as the linear mapping $f_* : T_pN \to T_{f(p)}M$ acting on tangent vectors in the following way. For each $X_p \in T_pN$ take a curve $x(t)$ on $N$ through $p$ such that $X_p$ is tangent to it at $p = x(t_0)$. The image $f_*(X_p)$ is then selected as the vector tangent to the image $f(x(t))$ at $f(p)$. As we shall see later, this way of moving geometric objects between manifolds is not universal, it works for vectors but it does not work, for example, for 1-forms.

A mapping $f : N \to M$ from a manifold $N$ into another manifold $M$ is said to be of *rank* $r$ at $p \in N$ if the dimension of $f_*(T_pN)$ is $r$. If the rank of $f$ at $p$ is equal to the dimension of $N$, $f_*$ is *injective* (or "into") at $p$. If the rank of $f$ at $p$ equals the dimension of $M$, $f_*$ is *surjective* (or "onto") at $p$. The mapping $f$ is called an *immersion* if $f_*$ is injective for every point $p \in N$. The implicit function theorem [92] implies that:

**Proposition 8.3.** *Let $f$ be a mapping from a manifold $N$ into a manifold $M$ and let $p \in N$.*

- *If $f_*$ is injective at $p$, then there exists a neighbourhood of $p$, say $U_p$, such that $f$ is a homeomorphism of $U_p$ onto $f(U_p)$.*
- *If $f_*$ is surjective at $p$ then $f : U_p \to M$ is an open mapping for some open neighbourhood $U_p$ of $p$.*
- *If $f_*$ is a linear isomorphism of $T_pN$ onto $T_{f(p)}M$ then $f$ defines a homeomorphism of an open neighbourhood of $p$ onto an open neighbourhood of $f(p)$ such that the inverse $f^{-1}$ is differentiable.*

An immersion $f$ is called an *embedding* of $N$ into $M$ if it is a homeomorphism onto its image (in the topology induced by $M$)[6]. We say then that the image $f(N)$ is an *embedded submanifold* of $M$. On the other hand, an open subset $N$ of a manifold $M$, with the induced topology from $M$, is called an *open submanifold* of $M$. A *diffeomorphism* of a manifold $N$ onto a manifold $M$ is a homeomorphism $f$ such that both $f$ and its inverse $f^{-1}$ are differentiable mappings.

The dual map to the tangent map $f_*$ induced by $f : N \to M$ at the point $p$ is the linear mapping $f^* : T^*_{f(p)}M \to T^*_pN$, called the *pull-back* of $f$ at $p$, such that, given a 1-form $\omega$ on $M$, $f^*\omega$ is a 1-form on $N$ defined by

$$f^*\omega(X_p) \equiv \omega(f_*X_p) \tag{8.27}$$

---

[6] Note that not every one-to-one immersion is an embedding. A nice illustration of this point can be found in [84].

for every $X_p \in T_pN$ and $p \in N$. Note that in contrast to the tangent map (and its push-forward action), the pull-back always defines a 1-from on $M$, even if $f$ is just injective.

A *distribution* $\mathcal{D}$ of dimension $r$ on a manifold $M$ of dimension $n \geq r$ is an assignment to each point $p \in M$ of an $r$-dimensional vector subspace $\mathcal{D}_p$ of $T_pM$. The distribution $\mathcal{D}$ is called *differentiable* if given a point $p \in M$ there exists an open neighbourhood $U_p$ of $p$ and $r$ differentiable vector fields $X^1, \ldots, X^r$ on $U_p$ such that $X_q^1, \ldots, X_q^r$ form a basis of $\mathcal{D}_q$ for every $q \in U_p$. A vector field $X$ on $M$ belongs to the distribution $\mathcal{D}$ if $X_p \in \mathcal{D}_p$ for every $p \in M$, . Moreover, the distribution $\mathcal{D}$ is called *involutive* if the bracket $[X,Y]$ (Equation (8.12)) belongs to $\mathcal{D}$ whenever $X$ and $Y$ belong to $\mathcal{D}$.

A connected submanifold $K \subset M$ is called an *integral manifold of the distribution* $\mathcal{D}$ if there exists an embedding $f : K \to M$ such that $f_*(T_pK) = \mathcal{D}_p$ for every $p \in M$. The integral manifold $K$ is called *maximal* if there is no other integral manifold of $\mathcal{D}$ containing $K$. The classical *Frobenius theorem* states:

**Proposition 8.4.** *Let $\mathcal{D}$ be an involutive distribution on a manifold $M$. Given a point $p \in M$ there exists a unique maximal integral manifold $K$ of $\mathcal{D}$ through $p$. Any other integral manifold of $\mathcal{D}$ through $p$ is an open submanifold of $K$.*

Let us consider now as a special case of a distribution a single vector field $X$ on $M$. A curve $x : (a,b) \to M$ is said to be an *integral curve of* $X$ if for every value of the parameter $t \in (a,b)$ the vector $X_{x(t)}$ is tangent to the curve $x$ at $x(t)$. The classical existence and uniqueness theorem for ordinary differential equations [61] guarantees that given a vector field $X$ and a point $p \in M$, there exists $\epsilon > 0$ and a unique curve $x(t)$ such that it is an integral curve of $X$ for $|t| < \epsilon$ and $x(0) = p$. A more "kinematic" way of looking at the integrability of a vector field is by using the concept of a *local one-parameter group of diffeomorphisms* of a manifold $M$, that is, a family of diffeomorphisms of $M$ with the property that for each point $p \in M$ there exists an open neighbourhood, say $U_p$, of $p$ and an open interval $I_\epsilon \equiv (-\epsilon, \epsilon) \subset \mathbb{R}$ such that:

1. For each $t \in I_\epsilon$, $\varphi_t : U_p \to M$ is a diffeomorphism of $U_p$ onto the open set $\varphi_t(U_p) \subset M$.
2. If $t, s, t+s \in I_\epsilon$ and if $\varphi_s(p) \in U_p$, then $\varphi_{t+s}(p) = \varphi_t(\varphi_s(p))$.

A (local) 1-parameter group of transformations $\varphi_t$ on $M$ induces a vector field $X$ by selecting at every point $p \in M$ the vector tangent to the curve $x(t) = \varphi_t(p)$ at $p = \varphi_0(p)$. The relation between the integrability of a vector field and the existence of a local 1-parameter group of diffeomorphisms is summarized in the following statement:

**Proposition 8.5.** *Let $X$ be a vector field on a manifold $M$. Then, for each point $p \in M$ there exists an open neighbourhood $U_p$ of $p$, a positive number*

$\epsilon$, and a local 1-parameter group of diffeomorphisms $\varphi : I_\epsilon \times M \to M$ which induces the given vector field $X$.

Indeed, note that given the initial point $p$, the diffeomorphism $\varphi : I_\epsilon \times \{p\} \to M$ traces an integral curve in $M$. Note also that given a diffeomorphism $f : M \to N$ from one manifold to another and a vector field $X$ on $M$ generating a local 1 parameter group of diffeomorphisms $\varphi_t$, the vector field $f_* X$ on $N$ generates a local 1-parameter group of diffeomorphisms of $N$, namely, $f \circ \varphi_t \circ f^{-1}$. In particular, if $f : M \to M$ is a diffeomerphism of $M$ and if the vector field $X$ on $M$ is $f$-invariant, that is, $f_*(X_p) = X_{f(p)}$ for every $p \in M$, the mapping $f$ commutes with every local 1-parameter group of diffeomorphisms generated by $X$. The converse is also true.

## 8.2 Lie groups

A group $\mathcal{G}$ is called a *Lie group* if it is also a differentiable manifold such that the group operation
$$(a, b) \in \mathcal{G} \times \mathcal{G} \mapsto ab^{-1} \in \mathcal{G} \tag{8.28}$$
is a differentiable mapping for any $a, b \in \mathcal{G}$. Let $L_a : \mathcal{G} \to \mathcal{G}$ (respectively $R_a : \mathcal{G} \to \mathcal{G}$) be the *left* (respectively *right*) *translation* of $\mathcal{G}$ by an element $a \in \mathcal{G}$. We shall write,
$$L_a(g) \equiv ag, \quad R_a(g) \equiv ga \tag{8.29}$$
where $a, g \in \mathcal{G}$. Also, let $\mathrm{ad}_a : \mathcal{G} \to \mathcal{G}$ denote the inner automorphism of $\mathcal{G}$ defined by
$$\mathrm{ad}_a(g) \equiv aga^{-1}, \quad g \in \mathcal{G}. \tag{8.30}$$

A *left* (respectively *right*) *invariant vector field* on $\mathcal{G}$ (viewed as a manifold) is a vector field $X$ invariant under the left (right) translations $L_a$ (respectively $R_a$). Thus, a vector field $X \in \mathcal{X}(\mathcal{G})$ is left invariant if
$$L_{a*}(X_g) = X_{ag} \tag{8.31}$$
for every $a, g \in \mathcal{G}$. It is easy to see that any left (right) invariant vector field on a Lie group is differentiable.

Consider now $T_e \mathcal{G}$, that is, the tangent space to $\mathcal{G}$ at the identity $e$. Let $X_e \in T_e \mathcal{G}$ and let $X_a \equiv L_{a*}(X_e)$ for some $a \in \mathcal{G}$. Hence, varying the element $a$ through the entire group $\mathcal{G}$ we obtain a left invariant vector field on $\mathcal{G}$. In fact, the left translation $L_a$ establishes, through its tangent map, an isomorphism between $T_e \mathcal{G}$ and the algebra of all left invariant vector fields on $\mathcal{G}$, say $\mathfrak{g}$, called the *Lie algebra*[7] of $\mathcal{G}$. Thus, the Lie algebra $\mathfrak{g}$ is an $n$-dimensional Lie subalgebra of the algebra of all vector fields $\mathcal{X}(\mathcal{G})$ on $\mathcal{G}$.

---

[7] Note that the bracket $[\cdot, \cdot]$ of two left-invariant vector fields on $\mathcal{G}$ is a left-invariant vector field on $\mathcal{G}$.

Every $A \in \mathfrak{g}$, as a vector field on $\mathcal{G}$, generates a (global) 1-parameter group of diffeomorphisms of $\mathcal{G}$. Indeed, let $\varphi_t$ be a local 1-parameter group of diffeomorphisms of $\mathcal{G}$ generated by $A \in \mathfrak{g}$. Then, $\varphi_t e$ is defined locally. Moreover, as the vector field $A$ is left-invariant the diffeomorphism $\varphi_t$ commutes with the left transaltions, i.e., $L_a(\varphi_t e) = \varphi_t(L_a e) = \varphi_t a$ and it is defined locally for every $a \in \mathcal{G}$. As the measure of locality of the action of $\varphi_t a$ is independent of $a$, $a_t \equiv \varphi_t a$ is defined for all $t \in \mathbb{R}$ and every $a \in \mathcal{G}$. In fact, $a_t$ is the unique solution to the differential equation $\dot{a}_t = a_t A_e$ with $a_0 = e$. If we denote $\exp A \equiv a_1 = \varphi_1 e$ then $a_t = \exp t A$ for all $t \in \mathbb{R}$. Thus defined mapping $A \mapsto \exp A$ of the Lie algebra $\mathfrak{g}$ into the group $\mathcal{G}$ is called the *exponential mapping*.

A subgroup $\mathcal{H}$ of the Lie group $\mathcal{G}$ is a *Lie subgroup* of $\mathcal{G}$ if $\mathcal{H}$ is a submanifold of $\mathcal{G}$ and a Lie group with respect to the submanifold's differentiable structure. The Lie algebra $\mathfrak{h}$ of $\mathcal{H}$ can be identified with a subalgebra of $\mathfrak{g}$. Conversely, any subalgebra $\mathfrak{h}$ of $\mathfrak{g}$ is the Lie algebra of the connected Lie subgroup of $\mathcal{G}$ [57]:

**Proposition 8.6.** *There exists a one-to-one correspondence between connected Lie subgroups of a Lie group $\mathcal{G}$ and the Lie subalgebras of the corresponding Lie algebra $\mathfrak{g}$.*

Let $\phi : \mathcal{G} \to \mathcal{G}$ be an arbitrary automorphism of a Lie group $\mathcal{G}$. If $A \in \mathfrak{g}$ then $\phi_*(A)$ is again a left invariant vector field on $\mathcal{G}$ and $\phi_*([A, B]) = [\phi_*(A), \phi_*(B)]$ for every $A, B \in \mathfrak{g}$. That is, the tangent map $\phi_*$ of an automorphism of a Lie group is an automorphism of its Lie algebra $\mathfrak{g}$. In particular, the automorphism $\mathrm{ad}_{a*}$ generated by the inner automorphism $\mathrm{ad}_a : \mathcal{G} \to \mathcal{G}$, and denoted again by $\mathrm{ad}_a$, is called the *adjoint representation* of $\mathcal{G}$ (in the Lie algebra $\mathfrak{g}$), $a \mapsto \mathrm{ad}_a$. Specifically, given $a \in \mathcal{G}$ and $A \in \mathfrak{g}$ we have that

$$\mathrm{ad}_a(A) = L_{a*}(R_{a^{-1}*}(A)) = R_{a^{-1}*}(L_{a*}(A)) = R_{a^{-1}*}(A) \tag{8.32}$$

as $A$ is left invariant and right and left translations commute with each other.

A differential form $\omega$ on a Lie group $\mathcal{G}$ is said to be *left invariant* if its pull-back $L_a^*(\omega)$ is identical to $\omega$ for every $a \in \mathcal{G}$. The vector space $\mathfrak{g}^*$ of all left invariant 1-forms on $\mathcal{G}$ is the dual space of the Lie algebra $\mathfrak{g}$. In particular, if $A \in \mathfrak{g}$ and $\omega \in \mathfrak{g}^*$ then $\omega(A)$ is constant on $\mathcal{G}$. Note that if the 1-form $\omega$ is left invariant then its differential $d\omega$ is also left invariant as the exterior differentiation commutes, by definition, with the pull-back. One can therefore show that, given a left-invariant 1-form $\omega$ on $\mathcal{G}$ and taking into account the fact that the bracket of two left-invariant vector fields is left-invariant, we obtain the *Maurer-Cartan equation*

$$d\omega(A, B) = -\frac{1}{2}\omega([A, B]) \tag{8.33}$$

for any $A, B \in \mathfrak{g}$.

A Lie group $\mathcal{G}$ is said to *act differentially* on the right on a (differentiable) manifold $M$ if:

1. Every element $a \in \mathcal{G}$ induces a transformation

$$R_a : M \to M, \quad R_a(x) \equiv xa, \tag{8.34}$$

where $x \in M$.
2. The mapping $R : \mathcal{G} \times M \to M$ defined by $R(a,x) \equiv R_a(x)$ is a differentiable mapping.
3. $R_{ab}(x) = R_b(R_a(x))$ for all $a, b \in \mathcal{G}$ and $x \in M$.

In a similar manner we can introduce a left action of $\mathcal{G}$ on $M$, that is $L_a(x) \equiv ax$, by requiring that $L_{ab}(x) = L_a(L_b(x))$ for all $a, b \in \mathcal{G}$ and $x \in M$. Henceforth, by the action of a Lie group $\mathcal{G}$ on a manifold $M$ we shall mean its right action. Note also that, as long as it invites no confusion, we shall use the same symbol to indicate the action of a group on a manifold as well as the translation within a group.

We say that $\mathcal{G}$ acts *effectively* on a manifold $M$ if $R_a(x) = x$ for all $x \in M$ implies that $a = e$. The action $R$ is said to be *free* if $R_a(x) = x$ for some $x \in M$ implies that $a = e$. If the group $\mathcal{G}$ acts on a manifold $M$ on the right it assigns to each element $A$ of its Lie algebra $\mathfrak{g}$ a vector field on $M$. That is, given $A \in \mathfrak{g}$, the action of the 1-parameter subgroup $a_t = \exp tA$ on $M$ induces a vector field on $M$ which we will denote by $A^*$. Hence, given the right action of the group $\mathcal{G}$ on a manifold we have a mapping $\sigma : \mathfrak{g} \to \mathcal{X}(M)$ such that $\sigma(A) = A^*$. Another way to introduce the mapping $\sigma$ is to consider for every $x \in M$ the right action $\sigma_x \equiv R(\cdot, x) : \mathcal{G} \to M$. The mapping $\sigma$ can then be defined by requiring that $\sigma_{x*}(A_e) = (\sigma(A))_x$, where obviously $A_e$ is identified with the field $A$. It follows immediately from the definition that $\sigma$ is a linear mapping. In fact, we have [57]:

**Proposition 8.7.** *Assume that the Lie group $\mathcal{G}$ acts on a manifold $M$ on the right. Then, the mapping $\sigma : \mathfrak{g} \to \mathcal{X}(M)$ such that $\sigma(A) = A^*$ is a Lie algebra homomorphism. Moreover, if the action of $\mathcal{G}$ on $M$ is effective then $\sigma$ is an isomorphism of $\mathfrak{g}$ into $\mathcal{X}(M)$. If the action of $\mathcal{G}$ on $M$ is free then $\sigma(A)$ does not vanish on $M$ as long as $A \in \mathfrak{g}$ is non-zero.*

## 8.3 Fibre bundles

Suppose that $M$ and $F$ are smooth manifolds of arbitrary finite dimensions. A (differentiable) *fibre bundle* on $M$ with the *standard fibre* $F$ is a manifold $P$ satisfying the following two conditions:

1. There exists a smooth surjective map $\pi : P \to M$, called the *projection* .

2. The manifold $P$ is locally trivial in the sense that every point $p \in M$ has an open neighbourhood $U$ and a diffeomorphism $\psi \equiv \pi \times \varphi : \pi^{-1}(U) \to U \times F$ such that $\psi(u) = (\pi(u), \varphi(u)) \in U \times F$ for all $u \in \pi^{-1}(U)$, where $\pi^{-1}(U)$ denotes the inverse image of $U$ by the projection $\pi$.

A fibre bundle will be denoted as $P(M, F)$, or simply $\pi : P \to M$, where $P$ is called the *total space* with the *base manifold* $M$ and the projection $\pi$. The pair $(U, \psi)$ is called a *bundle chart* (or a local *trivialization*). In fact, the fibre bundle $P$ is called *trivializable*[8] if it is diffeomorphic to the the product manifold $M \times F$. The following properties of the projection $\pi$ are of great importance:

**Proposition 8.8.** *The projection $\pi : P \to M$ of a smooth fibre bundle $P(M, F)$ is a submersion, that is, for each $u \in P$ the induced tangent map $\pi_* : T_u P \to T_{\pi(u)} M$ is surjective. Furthermore, for each $p \in M$ the fibre $\pi^{-1}(p)$ is a closed submanifold of $P$ diffeomorphic to the standard fibre $F$.*

Assume that $\{U_i\}$ is an open covering of the base manifold $M$ and that $(U_\alpha, \pi \times \varphi_\alpha)$ and $(U_\beta, \pi \times \varphi_\beta)$ are two bundle charts on $P$ such that $U_\alpha \cap U_\beta$ is non-empty. The restrictions of $\pi \times \varphi_\alpha$ and $\pi \times \varphi_\beta$ to $\pi^{-1}(U_\alpha \cap U_\beta)$ are also, in general different, bundle charts. Consider the map

$$\varphi_{\beta\alpha} : U_\alpha \cap U_\beta \to \mathrm{Diff}(F) \tag{8.35}$$

where $\mathrm{Diff}(F)$ denotes the set of all diffeomorphisms of $F$ and where

$$\varphi_{\beta\alpha}(p) \equiv \varphi_\beta|_{\pi^{-1}(p)} \circ (\varphi_\alpha|_{\pi^{-1}(p)})^{-1} \tag{8.36}$$

is, for each $p \in U_\alpha \cap U_\beta$, a diffeomorphism of the standard fibre $F$ into itself. The maps $\varphi_{\beta\alpha}$ have the property that

$$\varphi_{\gamma\alpha}(p) = \varphi_{\gamma\beta}(p) \circ \varphi_{\beta\alpha}(p) \tag{8.37}$$

for each $p \in U_\alpha \cap U_\beta \cap U_\gamma$. The maps $\varphi_{\beta\alpha}$ corresponding to the covering $\{U_\alpha\}$ are called the *transition functions* of the bundle $P(M, F)$. Note that the transition functions corresponding to a particular covering of $M$ form a subgroup of $\mathrm{Diff}(F)$ called the *structure group of the fibre bundle*.

We finish this section by introducing the concept of a fibre bundle induced by a mapping, also called the pull-back of a fibre bundle. Let $\pi_M : P \to M$ be a fibre bundle over $M$ with the standard fibre $F$, and let $f : N \to M$ be a smooth mapping from a manifold $N$ into a manifold $M$. Then, there exists a unique fibre bundle $\pi_N : Q(N, F) \to N$ and a *bundle homomorphism*[9] $\tilde{f} :$

---

[8] We differentiate here between a trivializable and a trivial fibre bundle; the latter being simply a Cartesian product of two manifolds, say $P = M \times F$.
[9] A homomorphism of a fibre bundle into another fibre bundle is a smooth mapping between fibre bundles which preserves fibres.

$Q(N, F) \to P(M, F)$ such that $\pi_M(\widetilde{f}(u)) = f(\pi_N(u))$ for every $u \in Q(N, F)$. The bundle $Q(N, F)$ defined this way is called the *fibre bundle induced by $f$ from $P(M, F)$* and it is often denoted as $\widetilde{f}P$. The map $\sigma : N \to P(M, F)$ such that $\pi_M \circ \sigma = f$ is called the *lift of $f$ to $P$*.

### 8.3.1 Principal fibre bundles

Replace in the definition of a fibre bundle the manifold $F$ by a Lie group $\mathcal{G}$, and assume that $\mathcal{G}$ acts freely on the total space $P$ on the right, that is, there exists a differentiable mapping $R : \mathcal{G} \times P \to P$ such that $R(ab, u) = R(b, R(a, u))$ for every $a, b \in \mathcal{G}$ and $u \in P$. Moreover, let any local trivialization $\pi \times \varphi : \pi^{-1}(U) \to U \times \mathcal{G}$ be such that it commutes with the right action of $\mathcal{G}$ on $P$, i.e.,

$$\varphi(R_a(u)) = R_a(\varphi(u)) \tag{8.38}$$

for all $u \in \pi^{-1}(U)$ and every $a \in \mathcal{G}$. The fibre bundle $P(M, \mathcal{G})$ obtained this way is called a *principal fibre bundle* over $M$ with the group $\mathcal{G}$ as its *structure group*, or simply a *principal $\mathcal{G}$-bundle*. Note that all fibres $\pi^{-1}(p)$ of $P(M, \mathcal{G})$, $p \in M$, are isomorphic to $\mathcal{G}$ and that the base manifold $M$ is isomorphic to the quotient space $P/\mathcal{G}$. Indeed [79], let $u, \widetilde{u} \in \pi^{-1}(p)$, $p \in M$, and let $\pi \times \varphi$ be a bundle chart on $P$ over some open neighbourhood of $p$. Set $g \equiv \varphi(u)$ and $h \equiv \varphi(\widetilde{u})$. Then, in the light of (8.38), we have that

$$(\pi \times \varphi)(ug^{-1}h) = (p, \varphi(u)g^{-1}h) = (p, h) = (\pi(\widetilde{u}), h) = (\pi \times \varphi)(\widetilde{u}). \tag{8.39}$$

Since $\pi \times \varphi$ is injective, $\widetilde{u} = ug^{-1}h$. Note also that the transition functions $\varphi_{\beta\alpha}$ corresponding to a covering $\{U_\alpha\}$ of $M$ take value in the group $\mathcal{G}$ if we define

$$\varphi_{\beta\alpha}(p) = \varphi_\beta(u)(\varphi_\alpha(u))^{-1} \tag{8.40}$$

for any $p \in M$ and any $u \in \pi^{-1}(p)$. Moreover, the composition property (8.37) becomes

$$\varphi_{\gamma\alpha}(p) = \varphi_{\gamma\beta}(p)\varphi_{\beta\alpha}(p). \tag{8.41}$$

In other words, a fibre bundle is a principal $\mathcal{G}$-bundle with the structure group $\mathcal{G}$ when the transition functions (8.35) take value in the group $\mathcal{G}$, as summarized in the following existence statement [57]:

**Proposition 8.9.** *Let $M$ be a smooth manifold with an open covering $\{U_\alpha\}$ and let $\mathcal{G}$ be a Lie group. Given, for every non-empty set $U_\alpha \cap U_\beta$, a mapping $\varphi_{\beta\alpha} : U_\alpha \cap U_\beta \to \mathcal{G}$ satisfying the composition property (8.41), there exists a principal $\mathcal{G}$-bundle $P(M, \mathcal{G})$ with transition functions $\varphi_{\beta\alpha}$.*

Given a principal $\mathcal{G}$-bundle $P(M, \mathcal{G})$, the action of the group $\mathcal{G}$ on the manifold $P$ induces, according to Proposition 8.7, a homomorphism $\sigma$ of the Lie algebra $\mathfrak{g}$ of $\mathcal{G}$ into the Lie algebra $\mathcal{X}(P)$ of all vector fields on $P$. For each $A \in \mathfrak{g}$, $A^* \equiv \sigma(A)$ is called the *fundamental vector field* corresponding to $A$. To see how this is works, select $u \in P$. Then $A_u^*$ is not only tangent to $P(M, \mathcal{G})$ but it is also tangent to the fibre $\pi^{-1}(\pi(u)) \subset P$, as the right action of $\mathcal{G}$ on $P$ preserves fibres. Moreover, as $\mathcal{G}$ acts freely on $P$, $A^*$ does not vanish as long as $A \neq 0$. Finally, as each fibre is isomorphic to $\mathcal{G}$, the dimension of its tangent space (at any point) is equal to the dimension of $\mathfrak{g}$. Therefore, the mapping $A \mapsto A_u^*$ of $\mathfrak{g}$ into $T_u P$ is a linear isomorphism at each $u \in P$. Note also that:

**Proposition 8.10.** *If $A^*$ is the fundamental vector field corresponding to $A \in \mathfrak{g}$ then, for each $a \in \mathcal{G}$, $R_{a*}(A^*)$ is the fundamental vector field corresponding to $\mathrm{ad}_{a^{-1}}(A) \in \mathfrak{g}$.*

Next consider a smooth map $f : M \to N$ from a manifold $M$ into another, not necessarily different, manifold $N$. Given a principal $\mathcal{G}$-bundle $\pi_N : P(N, \mathcal{G}) \to N$, the map $f$ induces a principal $\mathcal{G}$-bundle $\pi_M : P(M, \mathcal{G}) \to M$ and a mapping $f^* : P(N, \mathcal{G}) \to P(M, \mathcal{G})$ (by translating fibres to points on $M$) such that

$$\pi_N(f^*(u)) = f(\pi_M(u)) \tag{8.42}$$

for any $u \in P(N, \mathcal{G})$. Note that $P(M, \mathcal{G})$ is nothing else but the $\mathcal{G}$-bundle on $M$ induced by $f$. It is immediately clear that the mapping $f^*$ commutes with the action of $\mathcal{G}$ on the bundles. Reversing the action of the induced map, we say that a *homomorphism* of an $\mathcal{H}$-bundle $P(M, \mathcal{H})$ into another $\mathcal{G}$-bundle $Q(N, \mathcal{G})$ is a mapping $f : P \to Q$ and a homomorphism $f^\natural : \mathcal{H} \to \mathcal{G}$ such that $f(ua) = f(u) f^\natural(a)$ for all $u \in P$ and $a \in \mathcal{H}$. Every homomorphism $f : P(M, \mathcal{H}) \to Q(N, \mathcal{G})$ maps a fibre into a fibre inducing therefore a mapping $\widehat{f} : M \to N$ satisfying (8.42). We say that the bundles $P(M, \mathcal{H})$ and $Q(N, \mathcal{G})$ are homomorphic along $\widehat{f} : M \to N$. If the homomorphism $f : P \to Q$ is an embedding and if $f^\natural : \mathcal{H} \to \mathcal{G}$ is a group monomorphism then we say that $P(M, \mathcal{H})$ is a *subbundle* of $Q(N, \mathcal{G})$ where, in fact, we have identified $f(P)$ with $P$ and $f^\natural(\mathcal{H})$ with $\mathcal{H}$. If, in addition, $M = N$ and the induced (by the homomorphism $f$) mapping $\widehat{f} : M \to M$ is the identity, then the mapping $f : P(M, \mathcal{H}) \to Q(M, \mathcal{G})$ is called a *reduction* of the structure group $\mathcal{G}$ of $Q(M, \mathcal{G})$ to $\mathcal{H}$, and the bundle $P(M, \mathcal{H})$ is called the *reduced bundle*. In particular, given a $\mathcal{G}$-bundle $P(M, \mathcal{G})$ and a Lie subgroup $\mathcal{H} \subset \mathcal{G}$ we say that the structure group $\mathcal{G}$ is reducible to its subgroup $\mathcal{H}$ if there exists a reduced bundle $Q(M, \mathcal{H})$. Thus, as the logical consequence of Proposition 8.9, we have that:

**Proposition 8.11.** *The structure group $\mathcal{G}$ of the $\mathcal{G}$-bundle $P(M, \mathcal{G})$ is reducible to a Lie subgroup $\mathcal{H}$ if, and only if, there exists an open covering $\{U_\alpha\}$ of $M$ with the corresponding transition functions $\varphi_{\beta\alpha}$ taking values in the subgroup $\mathcal{H}$.*

Before proceeding any further with the presentation of the theory of fibre bundles we shall now discuss briefly a couple of important examples. In all examples we assume that $M$ is a $n$-dimensional, real, smooth manifold.

208     8 Basic geometric concepts

*Example 8.12. Frame bundle of a manifold*:   A *(linear) frame* at a point $p$ of a manifold $M$ is an ordered basis $E_1,\ldots,E_n$ of the tangent space $T_pM$. Let $L(M)$ denote the collection of all such ordered bases at all points of $M$. Also, let $\pi: L(M) \to M$ be the projection assigning to the frame $u$ at $p$ the point $p$. The *general linear group* $GL(n;\mathbb{R})$, that is, the set of all non-singular $n \times n$ matrices, acts on $L(M)$ on the right by assigning to a frame, say $E_1,\ldots,E_n$ of $T_pM$, the frame $F_1,\ldots,F_n$ of $T_pM$ defined by

$$F_i = \sum_{j=1}^n a_i^j E_j, \quad i=1,\ldots,n \tag{8.43}$$

where the matrix $\{a_i^j\} \in GL(n;\mathbb{R})$. In order to introduce a differentiable structure on $L(M)$, consider a coordinate system $x^1,\ldots,x^n$ on an open neighbourhood $U$ of $p \in M$. Given a frame $E_1,\ldots,E_n$ at $q \in U$ one can represent it uniquely as

$$E_i = \sum_{j=1}^n E_i^j \frac{\partial}{\partial x^j}\Big|_q, \quad i=1,\ldots, \tag{8.44}$$

where $\{E_i^j\}$ is a non-singular matrix. Thus, we have a one-to-one correspondence between $\pi^{-1}(U)$ and $U \times GL(n;\mathbb{R})$. Furthermore, we are able to establish a differential structure on $L(M)$ by simply accepting as a coordinate system on $\pi^{-1}(U)$ the collection $\{x^i, E_i^j\}$, $i,j=1,\ldots,n$. This implies that $L(M)$ can be made into a principal $GL(n;\mathbb{R})$-bundle over $M$ called the *frame bundle of $M$* or the *bundle of linear frames of $M$*. An alternative definition of the bundle of linear frames of a manifold, using the concept of a jet of a function, will be introduced later.

*Example 8.13. $\mathcal{G}$-structure*:   Let $L(M)$ be the bundle of linear frames on a manifold $M$. Assume that $\mathcal{G} \subset GL(n;\mathbb{R})$ is a closed Lie subgroup of $GL(n;\mathbb{R})$. Let $L_\mathcal{G}(M)$ be a $\mathcal{G}$-principal subbundle of $L(M)$. Then, the inclusion $L_\mathcal{G}(M) \hookrightarrow L(M)$ is a reduction of $GL(n;\mathbb{R})$ of $L(M)$ to its subgroup $\mathcal{G}$. The corresponding reduced subbundle $L_\mathcal{G}(M)$ is called a *$\mathcal{G}$-structure*. In particular, let $O(n) \subset GL(n;\mathbb{R})$ be the orthogonal subgroup of $GL(n;\mathbb{R})$. Then, the bundle $Q(M)$ of all orthogonal frames on $M$ is a reduced subbundle of $L(M)$ (see also Example 9.36). Other $\mathcal{G}$-structures will be discussed later. Note that the process of reducing a structure group to its subgroup may not be unique.

### 8.3.2 Associated fibre bundles

Having the concept of a principal $G$-bundle available already we shall proceed now to construct fibre bundles associated with it. To this end, let $P(M,\mathcal{G})$ be a $\mathcal{G}$-bundle and let $E$ be a differentiable manifold on which $\mathcal{G}$ acts on the left. Consider the product manifold $P \times E$ and let $\mathcal{G}$ act on it on the right according to the following definition:

$$R_a(u, \zeta) \equiv (R_a(u), L_{a^{-1}}(\zeta)) \equiv (ua, a^{-1}\zeta) \tag{8.45}$$

where $(u, \zeta) \in P \times E$ and $a \in \mathcal{G}$. It is easy to check that this indeed constitutes a right action. The action of the group $\mathcal{G}$ on $P \times E$ induces the quotient space $E_\mathcal{G} \equiv (P \times E)/\mathcal{G}$. Let $\pi : P \to M$ be the standard projection of $P(M, \mathcal{G})$ and let the projection $\pi : P \times E \to M$, denoted by the same letter, be such that $\pi(u, \zeta) = \pi(u)$ for each $u \in P$ and every $\zeta \in E$. The projection $\pi : P \times E \to M$ induces in a natural way a projection $\pi_E : E_\mathcal{G} \to M$ on the quotient $(P \times E)/\mathcal{G}$. Indeed, if $u\zeta \in E_\mathcal{G}$, where $u\zeta$ denotes an equivalence class in $(P \times E)/\mathcal{G}$ generated by the element $(u, \zeta)$, then $\pi_E(u\zeta) \equiv \pi(u)$, and the set $\pi_E^{-1}(p)$ is called the fibre of $E_\mathcal{G}$ over $p \in M$. In order to introduce a differentiable structure on $E_\mathcal{G}$ we require that, given an open set $U \subset M$, the local bundle $\pi_E^{-1}(U)$, which by the right action of the group $\mathcal{G}$ is isomorphic to $U \times F$, must be an open submanifold of $E_\mathcal{G}$. $E_\mathcal{G}(M)$ is called the *fibre bundle over the base manifold $M$ associated with* the principal $\mathcal{G}$-bundle $P(M, \mathcal{G})$. A somewhat different (possibly deeper) perspective is provided by the following statement:

**Proposition 8.14.** *Let $P(M, \mathcal{G})$ be a principal $\mathcal{G}$-bundle and let $E$ be a manifold on which $\mathcal{G}$ acts on the left. Consider a fibre bundle $E_\mathcal{G}(M)$ associated with $P(M, \mathcal{G})$. For each $u \in P$ and every $\zeta \in E$, let $u\zeta$ denote the equivalence class in $E_\mathcal{G}$ generated by the pair $(u, \zeta) \in P \times E$. Then, each $u \in P$ can be viewed as a mapping of the standard fibre $E$ onto $\pi_E^{-1}(\pi(u))$ which commutes with the action of the group $\mathcal{G}$ on $P$ and $E$, i.e.,*

$$R_a(u)\zeta = uL_a(\zeta) \tag{8.46}$$

*for every $u \in P$, $a \in \mathcal{G}$ and $\zeta \in E$.*

In particular, viewing $u, v \in P$, where $\pi(u) = p$ and $\pi(v) = q$, as diffeomorphisms of $E$ onto $\pi_E^{-1}(p)$ and $\pi_E^{-1}(q)$, respectively, $v \circ u^{-1}$ represents a diffeomorphism of $\pi_E^{-1}(p)$ onto $\pi_E^{-1}(q)$. Therefore, if $u, \widetilde{u} \in \pi^{-1}(p)$ then $u \circ \widetilde{u}^{-1}$ is an automorphism of the fibre $\pi_E^{-1}(p)$. As $\widetilde{u} = ua$ for some $a \in \mathcal{G}$, the automorphism $u \circ \widetilde{u}^{-1}$ can be represented as $u \circ (a^{-1}u^{-1})$ showing that the group of automorphisms of any fibre $\pi_E^{-1}(p)$ is isomorphic with the structure group $\mathcal{G}$ [57].

To illustrate the concept of an associated bundle we present a couple of typical examples.

*Example 8.15. Tangent bundle of a manifold* : Define by $TM$ the union of all tangent spaces $T_pM$ over all $p \in M$, and consider it as the total space of a fibre bundle with the natural projection $\pi_{TM} : TM \to M$. The general linear group $GL(n; \mathbb{R})$ acts on the bundle of linear frames $L(M)$ on the right by (8.43). On the other hand, let $GL(n; \mathbb{R})$ act on $\mathbb{R}^n$ on the left by:

$$\sum_{j=1}^n r^j a_j^i \tag{8.47}$$

where $(r^1,\ldots,r^n) \in \mathbb{R}^n$. Consider a vector $X_p \in T_pM$ and a linear frame $(F_1,\ldots,F_n) \in L(M)$ at $p \in M$. We can represent $X_p$ uniquely as $X_p = \sum_j X^j F_j$, where $(X^1,\cdots,X^n) \in \mathbb{R}^n$. On the other hand, given $\{a_j^i\} \in GL(n;\mathbb{R})$ there exists a linear frame $E_1,\ldots,E_n$ at $p$ such that $F_j = \sum_i a_j^i F_i$, $j = 1,\cdots,n$. Also,

$$X_p = \sum_{j=1}^{n} X^j F_j = \sum_{j=1}^{n} X^j (\sum_{i=1}^{n} a_j^i E_i) = \sum_{i=1}^{n} (\sum_{j=1}^{n} X^j a_j^i) E_i, \tag{8.48}$$

showing, in fact, that we can identify the *tangent bundle* $TM$ with the quotient space $(L(M) \times \mathbb{R}^n)/GL(n;\mathbb{R})$, where the group action on the product is specified by the equation (8.45). Hence, $TM$ can be viewed as a fibre bundle over $M$ associated with $L(M)$, with $\mathbb{R}^n$ as its standard fibre. Indeed, given a pair $(u,r) \in L(M) \times \mathbb{R}^n$, where $u = (E_1,\ldots,E_n)$ and $r = (r^1,\ldots,r^n)$, the equivalence class $ur$ is a vector $\sum_i r^i E_i$ in $TM$. Viewing, in the spirit of Proposition 8.14, a frame $u = (E_1,\ldots,E_n)$ as a mapping $u : \mathbb{R}^n \to TM$ such that

$$u(r^1,\ldots,r^n) = \sum_{i=1}^{n} r^i E_i, \tag{8.49}$$

Equation (8.48) shows that $u$ commutes with the actions of $GL(n;\mathbb{R})$ on $L(M)$ and $\mathbb{R}^n$.

*Example 8.16. Cotangent bundle*: We have just shown that the tangent bundle of a manifold is an associated bundle of the bundle of linear frames. Similarly, one can show that the *cotangent bundle*

$$T^*M \equiv \bigcup_{p \in M} T_p^*M \tag{8.50}$$

is also an associated bundle of the bundle of linear frames, with the same standard fiber $\mathbb{R}^n$, but the covariant left action of $GL(n;\mathbb{R})$ on it as dictated by the requirement that for every $p \in M$, given $X_p \in T_pM$ and $\omega_p \in T_p^*M$, the value $\omega_p(X_p)$ is a scalar and therefore is invariant under the action of the general linear group.

*Example 8.17. Geometric object on a manifold*: The tangent space of a manifold viewed, as presented in the previous example, as an associated bundle of the bundle of linear frames is an illustration of what classically has been called a *geometric object* on a manifold [85]. Namely, selecting $\mathbb{R}^m$ as the standard fibre $E$ and imposing the action of $GL(n;\mathbb{R})$ on $E$ on the left we create the associated bundle $(L(M) \times E)/GL(n;\mathbb{R})$. Given $(u,\xi) \in L(M) \times \mathbb{R}^m$, the element $u\xi$ has an absolute meaning but the numbers $v^{-1}(u\xi)$ by which we represent it depend on the particular frame (system of coordinates) $v \in L(M)$.

### 8.3.3 Sections of fibre bundles

A *cross section* (or, simply, a *section*) of a fibre bundle $P(M, F)$, with projection $\pi : P \to M$, is a mapping $s : M \to P$ such that $\pi \circ s$ is the identity on $M$. A cross section is called smooth if $s$ is a smooth mapping. The set of all cross sections of $P$ will be denoted by $\Gamma(P)$. A *local section* of $P$ is a cross section of the subbundle $\pi^{-1}(U)$ of $P$ where $U$ is an open subset of $M$. The existence of local sections is trivially guaranteed by the definition of a fibre bundle. In particular, if $P(M, \mathcal{G})$ is a principal $\mathcal{G}$-bundle over $M$, the local smooth sections are defined canonically. Indeed, let $\pi \times \varphi : U \to U \times \mathcal{G}$ be a local chart on $U \subset M$, then there exists a (local) smooth section $s : U \to P$ defined as $s(p) \equiv (p, \varphi^{-1}(e))$ where $p \in U$ and $e \in \mathcal{G}$ is the identity element of the group. Any other section on $U$ can be obtained from $s$ by the differentiable action of the group $\mathcal{G}$ on the points of the image of the section $s$. On the other hand, the existence of a global section has far reaching implications [57]:

**Proposition 8.18.** *Let $P(M, \mathcal{G})$ be a principal fibre bundle over $M$. $P$ admits a (global) smooth cross section $s : M \to P$ if, and only if, $P$ is isomorphic to the trivial bundle $M \times \mathcal{G}$.*

Knowing that only the existence of a local smooth section of a principal frame bundle is guaranteed, we shall now turn our attention to the question of extending a cross section of a fibre bundle defined on a subset of the base manifold. To this end, let us consider a mapping $f : W \to N$ from a subset $W$ of $M$ into another manifold $N$. Recall that $f$ is said to be differentiable at $p \in W$ if there exists an open neighbourhood $U$ of $p$ and a differentiable map $\widehat{f} : U \to N$ such that $f = \widehat{f}$ on $U \cap W$. Given a fibre bundle $P$ and a subset $W$ of $M$, by a cross section of $P$ on $W$ we mean a differentiable mapping $\sigma : W \to P$ such that $\sigma \circ \pi = \mathrm{id}_W$.

**Proposition 8.19.** *Let $E_\mathcal{G}(M)$ be an associated fibre bundle on $M$. Assume that the base manifold $M$ is paracompact[10] and that the standard fibre $E$ is diffeomorphic to $\mathbb{R}^m$, for some integer $m$. Let $W$ be a closed (possibly empty) subset of $M$. Then, any cross section $\sigma : W \to E_\mathcal{G}(M)$ can be extended to a cross section defined on the whole of $M$. In other words, $E_\mathcal{G}(M)$ admits always a global cross section* [57].

The key assumption of this proposition is that the manifold $M$ is paracompact. Indeed, the most important property of a paracompact space is that it admits partitions of unity (as summarized below in Theorem 8.20). To be able to sketch what this means and how it comes about we need first a few definitions. The *support* of a differentiable function $f : M \to \mathbb{R}$ is the closure of the set $\{q \in M : f(q) \neq 0\}$. Let $\{U_i\}_{i \in I}$ be a *locally finite covering* of a manifold, i.e., a covering such that every point $p \in M$ has an open neighbourhood that

---

[10] A topological space $M$ is *paracompact* if it is Hausdorff and every open covering of $M$ has an open, locally finite, refinement.

intersects with only finitely many sets $U_i$. A family of differentiable functions $\{f_i\}_{i\in I}$ on $M$ is called a *partition of unity subordinate to the locally finite covering* $\{U_i\}_{i\in I}$ if:

1. $f_i(q) \in [0,1]$ for every $q \in M$ and any $i \in I$.
2. The support of each function $f_i$ is contained in the corresponding $U_i$ set.
3. $\sum_{i\in I} f_i(q) = 1$ at every $q \in M$.

Note that as the local covering $\{U_i\}_{i\in I}$, to which the partition of unity $\{f_i\}_{i\in I}$ is subordinate, is locally finite, at each point $q$ the sum in (3) extends over a finite number of functions only.

**Theorem 8.20.** *Let $\{U_i\}_{i\in I}$ be a locally finite open covering of a paracompact manifold $M$. Then there exists a partition of unity $\{f_i\}_{i\in I}$ subordinate to $\{U_i\}_{i\in I}$.*

Therefore, if the base manifold $M$ is paracompact and $\{U_i\}_{i\in I}$ is a locally finite open covering of $M$ such that the sections $s_i : U_i \to \pi_E^{-1}(U_i) \subset E_{\mathcal{G}}(M)$ exist, then the mapping

$$s \equiv \sum_{i\in I} g_i s_i \qquad (8.51)$$

defines the desired (Proposition 8.19) global section of $E_{\mathcal{G}}(M)$, where $g_i \equiv f_i \circ \pi_E$.

*Example 8.21. Natural associated bundle of a $\mathcal{G}$-bundle :* Let $L(M)$ be the bundle of linear frames on $M$ and let $\mathcal{G}$ be a closed Lie subgroup of $GL(n;\mathbb{R})$. Consider the homogeneous space $GL(n;\mathbb{R})/\mathcal{G}$ the elements of which are, by definition, the cosets $\{a\mathcal{G}\}$, $a \in GL(n;\mathbb{R})$. The group $GL(n;\mathbb{R})$ acts on $GL(n;\mathbb{R})/\mathcal{G}$ in the natural way on the left. As $GL(n;\mathbb{R})$ acts on the principal bundle $L(M)$ on the right the fibre bundle $L(M)/\mathcal{G}$ can be identified[11] with the associated bundle of $L(M)$ over $M$ with standard fibre $GL(n;\mathbb{R})/\mathcal{G}$. Thus, we have the following alternative way of checking the availability of a reduction of the structure group of $L(M)$ to a subgroup [57]:

**Proposition 8.22.** *The structure group $GL(n;\mathbb{R})$ of $L(M)$ is reducible to its closed subgroup $\mathcal{G}$ if, and only if, the associated bundle $L(M)/\mathcal{G}$ admits a (global) cross section.*

Combining Propositions 8.19 and 8.22 we conclude that the structure group of the bundle of linear frames of a manifold $M$ can be reduced to a closed subgroup $\mathcal{G}$ provided the base manifold $M$ is paracompact.

---

[11] This construction can be carried out for any principal $\mathcal{G}$-bundle and any closed subgroup of its structure group.

# 9
# Theory of connections

In this chapter we present some aspects of the theory of connection on principal $G$-bundles and their associated bundles. Our main objective is to familiarize the reader with the concept a linear connection, a mechanism which enables one to compare what "happens" at different points of a manifold.

## 9.1 Connections on principal $G$-bundles

Let $P(M,\mathcal{G})$ be a principal fibre bundle over a manifold $M$ with structure group $\mathcal{G}$ and $\pi : P \to M$ as its standard projection. Given $u \in P(M,\mathcal{G})$ let us denote by $\mathcal{V}_u$ the *vertical vector subspace* of $T_u P$ consisting, by definition, of all vectors tangent to the fibre $\pi^{-1}(\pi(u)) \subset P(M,\mathcal{G})$ viewed as a closed submanifold of $P$. By a *connection* on $P(M,\mathcal{G})$ we mean a differentiable distribution $\mathcal{H}$ on $P$ such that:

1. At each $u \in P$, the tangent space $T_u P$ is a direct sum of $\mathcal{V}_u$ and $\mathcal{H}_u$, that is, $T_u P = \mathcal{V}_u \oplus \mathcal{H}_u$.
2. The distribution $\mathcal{H}$ is invariant under the action of the group $\mathcal{G}$ on $P$, i.e., $\mathcal{H}_{ua} = R_{a*}(\mathcal{H}_u)$ for every $u \in P$ and $a \in \mathcal{G}$.

We call the distribution $\mathcal{H}$ *horizontal* and the space $\mathcal{H}_u$ the *horizontal subspace* of $T_u P$. A vector $X_u \in T_u P$ is called *horizontal* if $X_u \in \mathcal{H}_u$ and *vertical* if $X_u \in \mathcal{V}_u$. According to the definition of a connection any vector $X_u \in T_u P$ can be uniquely represented as $X_u = vX_u + hX_u$, where $vX_u \in \mathcal{V}_u$ and $hX_u \in \mathcal{H}_u$ are called the *vertical* and *horizontal components* of $X_u$, respectively. Incidently, at each $u \in P$, the horizontal space $\mathcal{H}_u$ is isomorphic to the tangent space $T_{\pi(u)} M$ via the tangent map $\pi_* : T_u P \to T_{\pi(u)} M$.

Suppose now that $X$ is a vector field on the manifold $M$ and a connection $\mathcal{H}$ on $P(M,\mathcal{G})$ is available. A *horizontal lift* of $X$ is a vector field $X^*$ on $P$ such that $X^*$ is horizontal and $\pi_*(X_u^*) = X_{\pi(u)}$ at every $u \in P$. It can be shown [57] that the horizontal lift of a vector field on $M$ is unique and that it is invariant under the right action of the structure group of $P(M,\mathcal{G})$. The converse is also true. That is, any horizontal vector field on a principal $\mathcal{G}$-bundle, which is left $\mathcal{G}$-invariant, is a lift of some vector field on the base manifold $M$. Moreover, the set of horizontal lifts of all vector fields on $M$ is closed with respect to the addition, scalar multiplication and the Lie bracket operation [79]:

**Proposition 9.1.** *Let $\mathcal{H}$ be a connection on a principal bundle $P(M,\mathcal{G})$. Given a pair of vector fields $X, Y \in \mathcal{X}(M)$ and a function $f \in \mathcal{F}(M)$,*

1. $X^* + Y^* = (X+Y)^*$.
2. $(fX)^* = (f \circ \pi)X^*$.
3. *The horizontal component of the bracket $[X^*, Y^*]$ is the horizontal lift of $[X, Y]$, i.e., $h[X^*, Y^*] = [X, Y]^*$.*

Having a connection $\mathcal{H}$ on $P(M,\mathcal{G})$ available we can proceed now to define a connection form on $P$ as an alternative way of characterizing the horizontal distribution $\mathcal{H}$. To this end, recall that, as summarized in Proposition 8.7, the action of the group $\mathcal{G}$ on $P$ introduces a homomorphism assigning to every $A \in \mathfrak{g}$ a fundamental vector field $A^*$ (see Section 8.3.1). This, in turn, establishes at any $u \in P$ a linear isomorphism of the Lie algebra $\mathfrak{g}$ with the vertical subspace $\mathcal{V}_u$. We define the *connection form* $\omega$ as a $\mathfrak{g}$-valued 1-form on $P(M,\mathcal{G})$ such that

$$(\omega(X_u))^* = vX_u \tag{9.1}$$

for every $X_u \in T_u P$ and $u \in P$. Putting this in words, we say that the connection form $\omega$ assigns to a tangent vector $X_u$ an element $\omega(X_u)$ of the Lie algebra $\mathfrak{g}$ such that its corresponding fundamental vector at $u$ is the vertical component of $X_u$. As evident from the definition, the connection form characterizes uniquely the corresponding horizontal distribution $\mathcal{H}$ as it vanishes on a vector field $X$ if, and only if, $X \in \mathcal{H}$. In addition, the connection 1-form $\omega$ satisfies the following condition:

$$\omega(A^*) = A \tag{9.2}$$

for any $A \in \mathfrak{g}$. Moreover, $\omega$ is an *equivariant* 1-form on $P$, i.e., for any vector field $X$ on $P$

$$R_a^*\omega(X) \equiv \omega(R_{a*}(X)) = \mathrm{ad}_{a^{-1}}(\omega(X)) \tag{9.3}$$

as implied by Proposition 8.10. Conversely [57]:

**Proposition 9.2.** *Given a $\mathfrak{g}$-valued 1-form $\omega$ on $P(M,\mathcal{G})$ satisfying conditions (9.2) and (9.3) there exists a unique connection $\mathcal{H}$ on $P(M,\mathcal{G})$ whose connection form is $\omega$.*

Suppose now that a connection is given locally only and that the base manifold $M$ is paracompact, thus admitting a partition of unity. Then, as in the case a cross section of a fibre bundle (Proposition 8.19), it is possible to extend the connection to the entire manifold. Indeed:

## 9.1 Connections on principal $G$-bundles

**Theorem 9.3.** *Let $P(M,\mathcal{G})$ be a principal $\mathcal{G}$-bundle and let $W$ be a closed, possibly empty, subset of $M$. If $M$ is paracompact then any connection defined on $\pi^{-1}(W)$ can be extended to a connection on the whole bundle. In particular, $P(M,\mathcal{G})$ admits a connection provided the base manifold $M$ is paracompact.*

If $\omega_i$ is a connection form on $\pi^{-1}(U_i)$ then the extension

$$\omega = \sum_{i \in I} (f_i \circ \pi) \omega_i \tag{9.4}$$

defines a connection on the whole bundle, where $\{f_i\}_{i \in I}$ denotes a partition of unity subordinate to the locally finite covering $\{U_i\}_{i \in I}$ of $M$.

We have shown so far that there exists a one-to-one correspondence between a connection on a principal fibre bundle $P(M,\mathcal{G})$ and a properly defined $\mathfrak{g}$-valued 1-form $\omega$ on $P$. Our next step is to show that the connection form $\omega$ can actually be represented by forms defined locally on the base manifold $M$. Assume that $\{(U_\alpha, \psi_\alpha)\}$, where $\psi_\alpha = \pi \times \varphi_\alpha : \pi^{-1}(U_\alpha) \to U_\alpha \times \mathcal{G}$, is a trivialization of $P(M,\mathcal{G})$. Let $\varphi_{\beta\alpha}$ are the corresponding transition functions (8.37) and let $s_\alpha : U_\alpha \to P$ be the trivialization-induced local section defined for any $p \in U_\alpha$ by $s_\alpha(p) = \psi_\alpha^{-1}(p, e)$, where $e$ denotes the identity of the structure group $\mathcal{G}$. Also, let $\theta$ be the left-invariant $\mathfrak{g}$-valued *canonical 1-form* defined uniquely on the group $\mathcal{G}$ by the requirement that

$$\theta(A) = A \tag{9.5}$$

for any $A \in \mathfrak{g}$. Assume that $U_\alpha \cap U_\beta$ is non-empty and denote by $\theta_{\beta\alpha}$ the pull-back $\varphi_{\beta\alpha}^* \theta$ of the form $\theta$ to the manifold $M$. Correspondingly, let

$$\omega_\alpha \equiv s_\alpha^* \omega \tag{9.6}$$

denote the $\mathfrak{g}$-valued 1-form on $U_\alpha$ obtained by pulling back the connection $\omega$ by the trivialization-induced local section. Note also that $s_\beta(p) = s_\alpha(p)\varphi_{\alpha\beta}(p)$ for any $p \in U_\alpha \cap U_\beta$ and that given $X \in T_p(U_\alpha \cap U_\beta)$, $s_{\beta*}(X) \in T_{s_\beta(p)}P$. Moreover, using Leibniz formula for a derivative and the fact that $s_{\beta*}(X) = (s_\alpha \varphi_{\alpha\beta})_*(X)$ we obtain that

$$s_{\beta*}(X) = s_{\alpha*}(X)\varphi_{\alpha\beta}(p) + s_\alpha(p)\varphi_{\alpha\beta*}(X) \tag{9.7}$$

for any $X \in T_p(U_\alpha \cap U_\beta)$. Evaluating the connection form $\omega$ on a vector field $X$ and utilizing the definition (9.6) as well as the Leibniz formula (9.7) we obtain that:

**Proposition 9.4.** *The forms $\theta_{\beta\alpha}$ and $\omega_\alpha$ are such that*

$$\omega_\beta(p) = \mathrm{ad}(\varphi_{\beta\alpha}(p))^{-1} \omega_\alpha(p) + \theta_{\beta\alpha}(p) \tag{9.8}$$

*for any $p \in U_\alpha \cap U_\beta$. Conversely, given a family of $\mathfrak{g}$-valued 1-forms $\omega_\alpha$ on the covering $\{U_\alpha\}$ of $M$ satisfying the relation (9.8), there exists a unique connection $\mathcal{H}$ on $P(M,\mathcal{G})$ with the connection form $\omega$ such that the relation (9.6) holds.*

## 9.1.1 Parallelism in a principal $G$-bundle

A connection $\mathcal{H}$ on a principal $\mathcal{G}$-bundle $P(M,\mathcal{G})$ induces the notion of a parallel transport along curves in the base manifold $M$. Let $x : [a,b] \subset \mathbb{R} \to M$ be a piecewise smooth curve on $M$. A curve in a principal bundle $u : [a,b] \to P$ is called *horizontal* if the vectors tangent to it are all horizontal. A *horizontal lift of a curve* $x(t)$ is a horizontal curve $x^*(t)$ in the principle bundle $P(M,\mathcal{G})$ such that $\pi(x^*(t)) = x(t)$ for every $t \in [a,b]$. The notion of the horizontal lift of a curve corresponds to that of the horizontal lift of a vector, and the uniqueness of one implies the uniqueness of the other. Namely:

**Proposition 9.5.** *Let $x : [0,1] \to M$ be a continuously differentiable curve in $M$. Given $u_0 \in P$ such that $\pi(u_0) = x(0)$ there exists a unique horizontal lift $x^*(t)$ of $x(t)$ such that $\pi(x^*(0)) = x(0)$. In particular, if $X^*$ is the horizontal lift of a vector field $X$ on $M$ then the integral curve of $X^*$ through the point $u_0$ is the lift of the integral curve of $X$ through $x(0)$.*[1]

The *parallel displacement* (or *parallel transport*) on $P(M,\mathcal{G})$ can now be defined as a family of isomorphisms between fibres. Consider a smooth curve $x : [0,1] \to M$ and let $u_0$ be an arbitrary element of $P(M,\mathcal{G})$ such that $\pi(u_0) = x(0)$. If $x^* : [0,1] \to P$ is the unique horizontal lift of the curve $x(t)$ then it sends the element $x^*(0) \in \pi^{-1}(x(0))$ onto $x^*(1)$ of $\pi^{-1}(x(1))$. Varying $u_0$ through the fibre $\pi^{-1}(\pi(u_0)) = \pi^{-1}(x(0))$ we obtain a mapping $\rho$ of the fibre $\pi^{-1}(x(0))$ into the fibre $\pi^{-1}(x(1))$. Recall that, as per the definition of a connection, the parallel transport along any curve commutes with the right action of the structure group $\mathcal{G}$ on $P$. Thus, the parallel displacement $\rho$ is an isomorphism of fibres. One can also show [57] that:

**Proposition 9.6.** *Let $x(t)$ be a piecewise smooth curve on a manifold $M$ and let $\rho$ denote the corresponding parallel transport along $x(t)$. Then,*

1. *The parallel transport $\rho$ is independent of any specific parametrization of the curve $x(t)$.*
2. *The parallel transport $\rho^{-1}$ along the reversed curve $y(t) = x(1-t)$ is the inverse of the parallel transport along $x(t)$.*
3. *The parallel transport along the composition of two curves $x(t)$ and $y(t)$ such that $x(1) = y(0)$ is the composition of the corresponding parallel transports along $x(t)$ and $y(t)$.*

---

[1] The proof of this proposition draws from the fact that, by the local triviality of the principal bundle, for every curve $x(t)$ on $M$ there exists a curve $w(t)$ on $P(M,\mathcal{G})$ such that $\pi(w(t)) = x(t)$ for $t \in [0,1]$. Any other curve in $P(M,\mathcal{G})$ projecting onto $x(t)$ can be obtained by a differentiable action of the structure group. That is, to find the horizontal lift $x^*(t)$ of $x(t)$ one must determine a curve $a(t)$ in the structure group $\mathcal{G}$ such that $x^*(t) = R_{a(t)}w(t)$ at every $t \in [0,1]$, where $a(0)$ is the identity of $\mathcal{G}$ [57].

We would like to point out that the concept of the parallel transport, that is the system of lifts of curves from the base manifold $M$ to the bundle space $P$, can be introduced axiomatically without any reference to the notion of a horizontal distribution [79]. Furthermore, one can show that there is a one-to-one correspondence between a connection on a principal $\mathcal{G}$-bundle $P(M,\mathcal{G})$ and such a system of lifts of curves from $M$ to $P$.

### 9.1.2 Reduction of a connection

In Section 8.3.1 we investigated mappings between principal fibre bundles and, in particular, the process of reducing a structure group of a principal fibre bundle to a subgroup. Here, we will look at the effect of these processes on connections on principal bundles. We are especially interested in learning how a reduction to a subgroup, resulting in a principal subbundle over the same base manifold, affects the existing principal bundle connection.

**Proposition 9.7.** *Let* $f : Q(N,\mathcal{K}) \to P(M,\mathcal{G})$ *be a homomorphism of principal fibre bundles such that* $f^\natural : \mathcal{K} \to \mathcal{G}$ *is the corresponding homomorphism of structure groups mapping* $\mathcal{K}$ *isomorphically into* $\mathcal{G}$. *Let* $\widehat{f} : N \to M$ *denote the underlying diffeomorphism of the base manifolds* $N$ *and* $M$. *Given a connection* $\mathcal{H}$ *on* $P(M,\mathcal{G})$ *with* $\omega$ *as its connection form, we have that:*

1. *There is a unique connection in* $Q(N,\mathcal{K})$ *such that the horizontal subspace* $\widehat{\mathcal{H}}$ *of* $TQ$ *is mapped into the horizontal subspace* $\mathcal{H}$ *of* $TP$.
2. *The connection form* $\widehat{\omega}$ *of* $\widehat{\mathcal{H}}$ *is such that*

$$\omega(f_*(X)) = f^\natural(\widehat{\omega}(X))$$

*for every* $X \in TQ$.

The connection $\widehat{\mathcal{H}}$ on $Q(N,\mathcal{K})$ is called the *f-induced connection* from $\mathcal{H}$. In particular, if $\mathcal{K} = \mathcal{G}$ and $f^\natural$ is the identity, $\widehat{\omega} = f^*\omega$. Furthermore, if $\mathcal{K}$ is a Lie subgroup of $\mathcal{G}$ and $Q(M,\mathcal{K})$ is a reduced subbundle of $P(M,\mathcal{G})$, where $\widehat{f} : M \to M$ is the identity transformation, we say that the connection $\mathcal{H}$ on $P(M,\mathcal{G})$ is *reducible* to a connection $\widehat{\mathcal{H}}$ on $Q(M,\mathcal{K})$. In particular [57]:

**Proposition 9.8.** *Let* $Q(M,\mathcal{K})$ *be a subbundle of* $P(M,\mathcal{G})$ *where* $\mathcal{K}$ *is a Lie subgroup of* $\mathcal{G}$. *Assume that* $\mathfrak{k}$ *is the Lie algebra of* $\mathcal{K}$ *and that* $\mathfrak{g} = \mathfrak{k} \oplus \mathfrak{m}$ *for some subspace* $\mathfrak{m}$ *of* $\mathfrak{g}$. *Moreover, let* $\mathfrak{m}$ *be such that* $\mathrm{ad}_k(\mathfrak{m}) = \mathfrak{m}$ *for every* $k \in \mathcal{K}$. *If* $\widehat{\omega}$ *denotes a restriction of a connection form* $\omega$ *on* $P(M,\mathcal{G})$ *to the subbundle* $Q(M,\mathcal{K})$ *then the* $\mathfrak{k}$-*component of* $\widehat{\omega}$ *is a connection form on* $Q(M,\mathcal{K})$. *In particular, the connection* $\omega$ *on* $P(M,\mathcal{G})$ *is reducible to a connection* $\widehat{\omega}$ *on the reduced bundle* $Q(M,\mathcal{K})$ *if, and only if, the restriction* $\widehat{\omega}$ *of* $\omega$ *to* $Q(M,\mathcal{K})$ *is* $\mathfrak{k}$-*valued. In other words,* $\widehat{\omega}$ *on* $Q(M,\mathcal{K})$ *is a reduction of* $\omega$ *on* $P(M,\mathcal{G})$ *if* $\widehat{\omega} = \omega$ *on* $Q(M,\mathcal{K})$.

Consider now as an example of a principal fibre bundle on a manifold $M$ its bundle of linear frames $L(M)$. The structure group of $L(M)$ is reducible to a closed subgroup if its natural associated bundle admits a global section (see Example 8.21). One wonders how can the reduction of a connection on $L(M)$ be characterized in terms of such a section. To be able to investigate possible answers to this question we need first to introduce the concept of a parallel transport on a bundle associated with a principal bundle. Thus, let $\mathcal{H}$ be a connection on a principal bundle $P(M, \mathcal{G})$ and let $E_{\mathcal{G}}(M)$ be an arbitrary bundle associated with $P(M, \mathcal{G})$. At every point $q$ of the associated bundle $E_{\mathcal{G}}(M)$ we must introduce a decomposition of its tangent space $T_q E_{\mathcal{G}}$ into a direct sum of a vertical and a horizontal subspaces. To this end, let the *vertical subspace* at $q \in E_{\mathcal{G}}$ be simply the space of all vectors tangent at $q$ to the fibre $\pi_E^{-1}(p)$, where $p = \pi_E(q) \in M$. To define the *horizontal subspace* $\mathcal{H}_q$ select a pair $(u, \zeta) \in P \times E$ representing $q$ via the identification $(P \times F)/\mathcal{G} = E_{\mathcal{G}}(M)$. Fix $\zeta \in E$ and consider a mapping sending the elements $v$ of $P$ into the elements $v\zeta \in E_{\mathcal{G}}(M)$. Take as the horizontal subspace $\mathcal{H}_q \in T_q E_{\mathcal{G}}$ the image of the horizontal subspace $\mathcal{H}_u \subset T_u P$ under the corresponding tangent map. It should be easy to see that the choice of $\mathcal{H}_q$ is independent of the choice of a representative $(u, \zeta)$ of $q$. It is also true that $T_q E_{\mathcal{G}} = T_q \pi_E^{-1}(p) \oplus \mathcal{H}_q$ for any $q \in E_{\mathcal{G}}$ and $p = \pi_E(q)$. Given $q_0 \in E_{\mathcal{G}}$ such that $q_0 = u_0 \zeta$ for some $(u_0, \zeta) \in P \times E$, the *horizontal lift* $q^*(t)$ of a curve $q(t)$ in $E_{\mathcal{G}}$ is $q^*(t) = x^*(t)\zeta$ where $x^*(t)$ is the horizontal lift of the curve $x(t) = \pi_E(q(t))$. A local cross section $s : U \subset M \to E_{\mathcal{G}}(M)$ is called *parallel* if $s_*(T_p M) = \mathcal{H}_{s(p)}$ for every $p \in U$. In other words, a section $s$ of an associated bundle $E_{\mathcal{G}}(M)$ is parallel if the parallel transport of $s(x(0))$ along any curve $x : [0, 1] \to U$, $x(0) = p$, gives $s(x(1))$. What follows is a natural consequence of Proposition 8.22 (see also Example 8.21):

**Proposition 9.9.** *Consider the bundle of linear frames of the manifold $M$, $L(M)$, and let $\mathcal{G}$ be a closed Lie subgroup of $GL(n; \mathbb{R})$. Denote by $\mathcal{K}$ the homogenous space $GL(n; \mathbb{R})/\mathcal{G}$ and let $s : M \to (L(M) \times \mathcal{K})/GL(n; \mathbb{R})$ be the global cross section defining the reduction of $GL(n; \mathbb{R})$ to the subgroup $\mathcal{G}$ (Proposition 8.22). The connection $\mathcal{H}$ on $L(M)$ is reducible to a connection $\widehat{\mathcal{H}}$ on the reduced bundle $L(M)/\mathcal{K}$ (Proposition 9.8) if, and only if, $s$ is parallel with respect to the connection $\mathcal{H}$.*

### 9.1.3 Structure equation, curvature and holonomy

Consider again a principal $\mathcal{G}$-bundle $P(M, \mathcal{G})$. Let $\mathcal{H}$ be a connection on $P(M, \mathcal{G})$ and let $\omega : TP \to \mathfrak{g}$ be the corresponding connection form. Also, let $h : TP \to \mathcal{H}$ be the horizontal projection assigning to a tangent vector of $P$ its unique horizontal component. Assume that $\varrho$ is a representation[2] of

---

[2] A *representation* of a group $\mathcal{G}$ in a vector space $\mathbf{V}$ is a homomorphism $\varrho$ of the group $\mathcal{G}$ with the group $GL(\mathbf{V})$ of all linear automorphisms of $\mathbf{V}$. Hence, $\varrho(a) \in GL(\mathbf{V})$ for any $a \in \mathcal{G}$.

the group $\mathcal{G}$ in a finite dimensional vector space $\mathbf{V}$. A *pseudotensorial form* of degree $r$ on $P$ of type $(\varrho, \mathbf{V})$ is a $\mathbf{V}$-valued $r$-form $\phi$ such that

$$R_a^*\phi(X_1, \cdots, X_r) = \varrho(a^{-1})\phi(X_1, \cdots, X_r) \tag{9.9}$$

for any $a \in \mathcal{G}$, any $u \in P$, and any selection of vector fields $X_1, \cdots, X_r \in T_u P$. If the representation $\varrho$ is the adjoint representation of the group $\mathcal{G}$ (see Equation (8.32)), then the pseudotensorial form $\phi$ is said to be of type ad $\mathcal{G}$. The form $\phi$ is called *horizontal* if $\phi(X_1, \cdots, X_r)$ vanishes when at least one of its $r$ arguments is vertical. In other words, $\phi$ is horizontal if does not vanish on horizontal vectors only.

Given a pseudotensorial $r$-form $\phi$ on $P$ of type $(\varrho, \mathbf{V})$ let us define a new $(r+1)$-form $D\phi$ by requiring that at any $u \in P$ and for any selection of vector $X_1, \ldots, X_{r+1} \in T_u P$

$$D\phi(X_1, \cdots, X_{r+1}) \equiv d\phi(h(X_1), \cdots, h(X_{r+1})). \tag{9.10}$$

The form $D\phi \equiv d\phi \circ h$ is called the *exterior covariant derivative* of the form $\phi$. It is an elementary consequence of the above definition that:

**Proposition 9.10.** *If $\phi$ is a pseudotensorial $r$-form of type $(\varrho, \mathbf{V})$ on $P$ then its exterior covariant derivative $D\phi$ is horizontal.*

One can now show [57, 79] that:

**Proposition 9.11.** *Let $\omega$ be a connection form on $P(M, \mathcal{G})$ and let $\phi$ be a horizontal 1-form of type ad $\mathcal{G}$ on $P$. Then, for any $u \in P$ and any $X, Y \in T_u P$*

$$D\phi(X, Y) = d\phi(X, Y) + \frac{1}{2}[\phi(X), \omega(Y)] + \frac{1}{2}[\omega(X), \phi(Y)]. \tag{9.11}$$

The proof of this proposition is rather straightforward and uses techniques which become standard in the analysis of connections. It is first and foremost based on the fact that every tangent vector $X \in TP$ can be uniquely decomposed into its vertical and horizontal parts. Therefore, to verify that the identity holds it is sufficient to consider the following three cases: i) $X$ and $Y$ are both horizontal, ii) $X$ is horizontal and $Y$ is vertical, and iii) $X$ and $Y$ are both vertical. The following facts are also utilized:

- If $X, Y$ are vectors on a manifold and if $X$ generates a local 1-parameter group of transformations $\varphi_t$ then

$$[X, Y]_p = \lim_{t \to 0} \frac{1}{t}[Y_p - (\varphi_{t*}(Y))_p] \tag{9.12}$$

  for any point $p$.
- If $A^*$ is the fundamental vector field corresponding to $A \in \mathfrak{g}$ and $X$ is a horizontal vector field then $[X, A^*]$ is horizontal.

Consider the connection 1-form $\omega$ on $P$. Its exterior covariant derivative $\Omega \equiv D\omega$ is a 2-form of type ad $\mathcal{G}$. We shall call it the *curvature form* of $\omega$ (or of the corresponding connection $\mathcal{H}$). If we replace the form $\phi$ in the Equation (9.11) by the connection form $\omega$ we obtain a relation known as the *structure equation* of the connection $\mathcal{H}$. Namely:

**Theorem 9.12.** *Let $\omega$ be a connection 1-form on $P(M,\mathcal{G})$ and let $\Omega$ be its curvature form. Then, at any $u \in P$ and for any pair of vectors $X, Y \in T_u P$ the following holds:*

$$d\omega(X,Y) = -\frac{1}{2}[\omega(X),\omega(Y)] + \Omega(X,Y).^3 \qquad (9.14)$$

*Moreover, the curvature form $\Omega$ is covariantly constant, i.e.,*

$$D\Omega \equiv 0.^4 \qquad (9.15)$$

In particular, as the identity (9.14) gives a decomposition of $d\omega$ into its horizontal and vertical parts, we have that:

**Corollary 9.13.** *If $X$ and $Y$ are horizontal vector fields on $P$ then*

$$\omega([X,Y]) = -2\Omega(X,Y). \qquad (9.16)$$

Using the definition of the curvature and the corresponding structure equation we can also show [79] that:

**Corollary 9.14.** *The curvature form $\Omega$ is an equivariant 2-form, that is,*

$$R_a^* \Omega = \mathrm{ad}_{a^{-1}} \Omega \qquad (9.17)$$

*for every $a \in \mathcal{G}$.*

We complete our investigation of the curvature form by looking at the coordinate representation of the structure equation. Let $E_1, \ldots, E_m$ be a basis of the Lie algebra $\mathfrak{g}$. Knowing that the Lie algebra is closed with respect to the bracket operation, let the constants $c^i_{jk}$, called the *Lie algebra constants* (or *structure constants*), be defined as

$$[E_j, E_k] = \sum_{i=1}^{m} c^i_{jk} E_i, \quad j,k = 1, \ldots m. \qquad (9.18)$$

---

[3] This equation is often simbolically written as

$$d\omega = -\frac{1}{2}[\omega,\omega] + \Omega. \qquad (9.14)$$

It is instructive to compare this structure equation with the Maurer-Cartan equation (8.33), the latter being just a restiction of the structure equation to a fibre.

[4] This is known as the *Bianchi identity*.

9.1 Connections on principal $\mathcal{G}$-bundles

Represent
$$\omega = \sum_{i=1}^{m} \omega^i E_i \quad \text{and} \quad \Omega = \sum_{i=1}^{m} = \Omega^i E_i \tag{9.19}$$

where the forms $\omega^i$ and $\Omega^i$ are now $\mathbb{R}$-valued. Then, the structure equation (9.14) can be written as

$$d\omega^i = -\frac{1}{2} \sum_{j,k=1}^{m} c^i_{jk} \omega^j \wedge \omega^k + \Omega^i, \quad i = 1, \ldots, m. \tag{9.20}$$

We finish this section by introducing the concept of a holonomy group thus providing the reader with a somewhat different perspective of the curvature form of a connection. To this end, let $\mathcal{H}$ be a connection on a principal $\mathcal{G}$-bundle $P(M, \mathcal{G})$. The *holonomy group* of the connection $\mathcal{H}$ at $p \in M$ is the group of parallel displacements $\rho$ (see Proposition 9.6) along all piecewise smooth curves in $M$ starting and ending at $p$. We will denote the holonomy group at $p \in M$ by $\Phi(p)$. Note that by the definition of a principal $\mathcal{G}$-bundle the holonomy group $\Phi(p)$ based at $p \in M$ is isomorphic to a subgroup of the structure group $\mathcal{G}$. Indeed, given $p \in M$, and selecting $u \in \pi^{-1}(p)$ establishes such an isomorphism via parallel translation. We shall denote this isomorphic subgroup of $\mathcal{G}$ by $\Phi(u)$ and call it again the *holonomy group* of the connection $\mathcal{H}$ at $u$. If $u, \widehat{u} \in \pi^{-1}(p)$ then $\Phi(u)$ and $\Phi(\widehat{u})$ are conjugate subgroups of $\mathcal{G}$ where the conjugation is established by the unique $a \in \mathcal{G}$ such that $\widehat{u} = ua$. Furthermore, $\Phi(u) = \Phi(\widehat{u})$ if, and only if, $u$ and $\widehat{u}$ can be joined in $P$ by a horizontal curve. In fact, the following Reduction Theorem is true [79]:

**Proposition 9.15.** *Let $\mathcal{H}$ be a connection on a principal $\mathcal{G}$-bundle $P(M, \mathcal{G})$. Assume that the base manifold $M$ is connected and paracompact. Select $u_0 \in P$ and let $P(u_0)$ be the set of all points in $P$ which can be joined from $u_0$ by a horizontal curve in $P$. Then, $P(u_0)$ is a principal subbundle of $P$ with $\Phi(u_0)$ as its structure group. Moreover, the connection $\mathcal{H}$ on $P$ is reducible to a connection on $P(u_0)$.*

The subbundle $P(u_0)$ is called the *holonomy bundle* of the connection $\mathcal{H}$ at $u_0 \in P$.

The relation between the holonomy group of a connection and its curvature form is established by the Ambrose-Singer Theorem [57]:

**Theorem 9.16.** *Assume that $M$ is a connected, paracompact smooth manifold. Let $P(M, \mathcal{G})$ be a principal $\mathcal{G}$-bundle with a connection $\mathcal{H}$ and the corresponding curvature 2-form $\Omega$. If $\Phi(u)$ is the holonomy group of $\mathcal{H}$ at $u \in P$ and $P(u)$ is the corresponding holonomy bundle, then the Lie algebra of $\Phi(u)$ is a subalgebra of $\mathfrak{g}$ spanned by all elements of the form $\Omega_v(X,Y)$, where $v \in P(u)$ and $X$ and $Y$ are arbitrary horizontal vectors at $T_v P$.*

### 9.1.4 Flat connections

Consider the trivial principal $\mathcal{G}$-bundle $P(M,\mathcal{G}) = M \times \mathcal{G}$. Let us define on $P$ a connection $\mathcal{H}$ by requiring that at any $u \in P$, where $u = (p, a)$, $\mathcal{H}_u = T_p M$. This is obviously a well defined connection on $P$. It is called the *canonical flat connection* on the trivial bundle $\pi : M \times \mathcal{G} \to M$. Let $\theta$ be the canonical left-invariant 1-form on the group $\mathcal{G}$, as defined in (9.5). Then, the connection form of the canonical flat connection on $M \times \mathcal{G}$ is given by the pull-back

$$\omega \equiv f^*\theta, \tag{9.21}$$

where $f : M \times \mathcal{G} \to \mathcal{G}$ is the natural projection onto the second factor. The Maurer-Cartan equation (8.33) and the fact that the exterior differentiation commutes with the pull-back operation imply immediately that the canonical flat connection has zero curvature.

Given an arbitrary $\mathcal{G}$-bundle $P(M,\mathcal{G})$, a connection $\mathcal{H}$ on $P$ is called a *flat connection* if at every point $p \in M$ one can select an open neighbourhood $U \ni p$ and an isomorphism $\psi : \pi^{-1}(U) \to U \times \mathcal{G}$ such that $\psi_*(\mathcal{H}_u) = T_{\pi(u)}M$ at each $u \in \pi^{-1}(U)$. In other words, the connection $\mathcal{H}$ on $P(M,\mathcal{G})$ is flat if it is locally isomorphic to the canonical flat connection on $M \times \mathcal{G}$. The Ambrose-Singer Theorem 9.16 and the Reduction Theorem 9.15 imply that:

**Theorem 9.17.** *A connection on a principal $\mathcal{G}$-bundle $P(M,\mathcal{G})$ is flat if, and only if, the curvature form vanishes identically.*

Moreover:

**Corollary 9.18.** *Let $\mathcal{H}$ be a connection on $P(M,\mathcal{G})$ such that its curvature form $\Omega$ vanishes. If the base manifold $M$ is paracompact and simply connected then $P$ is isomorphic with the trivial bundle $M \times \mathcal{G}$ (is trivializable) and the connection $\mathcal{H}$ is isomorphic with the canonical flat connection on $M \times \mathcal{G}$.*

Finally, one also deduces [79] from Theorem 9.17 (see also Corollary 9.13) that:

**Corollary 9.19.** *A connection on a principal $\mathcal{G}$-bundle $P(M,\mathcal{G})$ is flat if the corresponding horizontal distribution $\mathcal{H}$ is involutive.*

## 9.2 Linear connections

Consider the bundle of linear frames $L(M)$ (cf. Example 8.12) of an $n$-dimensional, real, differentiable manifold $M$. Let $\pi : L(M) \to M$ denote its standard projection. A *canonical form* on $L(M)$ is an $\mathbb{R}^n$-valued 1-form $\vartheta$ on $L(M)$ defined at any linear frame $u \in L(M)$ and any vector $X \in T_u L(M)$ as

$$\vartheta(X) \equiv u^{-1}(\pi_*(X)), \tag{9.22}$$

where $u$ is viewed as a linear mapping from $\mathbb{R}^n$ to $T_{\pi(u)}M$ assigning to an $n$-tuple of numbers $\xi$ a vector tangent to $M$ having $\xi$ as its coordinates in the frame $u$ (cf. Example 8.15). The canonical form $\vartheta$ on the frame bundle $L(M)$ is a pseudotensorial 1-form of type $(\mathrm{id}, \mathbb{R}^n)$. That is

$$(R_a^* \vartheta)(X) = a^{-1}(\vartheta(X)) \tag{9.23}$$

where $a^{-1}(\vartheta(X))$ is the standard action of the general linear group $GL(n; \mathbb{R})$ on $\mathbb{R}^n$ and where

$$a^{-1}(\vartheta(X)) = a^{-1}(u^{-1}(\pi_*(X))) = (ua)^{-1}(\pi_*(X)) \tag{9.24}$$

for every $X \in T_{\pi(u)}L(M)$.

A connection on the bundle of linear frames $L(M)$ is called a *linear connection* on $M$. Given such a connection $\mathcal{H}$ on $L(M)$ and using the identification map $u : \mathbb{R}^n \to TM$, where $u$ is any linear frame, we can associate with every horizontal vector $X_u \in T_u L(M)$ an element $\xi \in \mathbb{R}^n$ by requiring that $\pi_*(X_u) = u(\xi)$. Conversely, given $\xi \in \mathbb{R}^n$, there exists a unique horizontal vector field $X(\xi)$, called the *standard horizontal vector field* of $\xi$, such that the above relation is satisfied at every $u \in L(M)$. Note that if $X(\xi)$ is the standard horizontal vector field of $\xi \in \mathbb{R}^n$ then

$$\vartheta(X(\xi)) = u^{-1}(\pi_*(X(\xi))) = u^{-1}(u(\xi)) = \xi. \tag{9.25}$$

Moreover:

**Proposition 9.20.** *Let $X(\xi)$ be the standard horizontal vector field of $\xi \in \mathbb{R}^n$ on $L(M)$. Then, for any $a \in GL(n; \mathbb{R})$ and $\xi \in \mathbb{R}^n$*

1. $R_{a*}(X(\xi)) = X(a^{-1}\xi)$.
2. *If $\xi \neq 0$ then $X(\xi)$ does not vanish.*

*Furthermore, given the connection form $\omega$ of the linear connection $\mathcal{H}$ on $L(M)$ and selecting $\xi \in \mathbb{R}^n$, the conditions $\vartheta(X(\xi)) = \xi$ and $\omega(X(\xi)) = 0$ completely determine the standard horizontal vector field $X(\xi)$.*

Let $\vartheta$ be the canonical form on the bundle of linear frames $L(M)$. Its exterior covariant derivative $\Theta \equiv D\vartheta$ is a horizontal 2-form on $L(M)$ called the *torsion form* of the linear connection $\omega$. The form $\Theta$ satisfies the following *second structure equation*:

**Theorem 9.21.** *Let $\omega$ be the connection form of a linear connection on $M$ and let $\Theta$ be its torsion 2-form. Then,*

$$d\vartheta(X,Y) = -\frac{1}{2}(\omega(X)\vartheta(Y) - \omega(Y)\vartheta(X)) + \Theta(X,Y) \tag{9.26}$$

224   9 Theory of connections

for every $u \in L(M)$ and $X, Y \in T_u L(M)$.[5]

One can also show the validity of the following *second Bianchi identity*, relating the torsion 2-form $\Theta$ and the curvature form $\Omega$.

**Theorem 9.22.** *Given a linear connection $\omega$ on $M$ where $\Theta$ and $\Omega$ are the corresponding torsion and curvature forms,*

$$D\Theta(X, Y, Z) = (\Omega \wedge \vartheta)(X, Y, Z) \tag{9.28}$$

*at any $u \in L(M)$ and for any selection of vectors $X, Y, Z \in T_u L(M)$.*

Let us try now to express a linear connection on a manifold $M$ in a local coordinate system. Consider a coordinate neighbourhood $U \subset M$ with a local coordinate system $x^1, \ldots, x^n$. A linear frame $u \in L(M)$ such that $\pi(u) = p \in U$ can be expressed uniquely as

$$(\sum_{i=1}^n u_1^i \frac{\partial}{\partial x^i}\big|_p, \ldots, \sum_{i=1}^n u_n^i \frac{\partial}{\partial x^i}\big|_p) \tag{9.29}$$

where $\det\{u_j^i\} \neq 0$ due to the linear independence of the coordinate-induced base vectors $\frac{\partial}{\partial x^1}\big|_p, \ldots, \frac{\partial}{\partial x^n}\big|_p$. Identifying $\pi^{-1}(U) \subset L(M)$ with $U \times GL(n; \mathbb{R})$ we take $\{x^i, u_k^j\}$, $i, j, k = 1, \ldots, n$, as a local coordinate system on $\pi^{-1}(U)$. Let $e_1, \ldots, e_n$ be the natural basis of $\mathbb{R}^n$. Then, the canonical form $\vartheta$ (cf. Equation (9.22)) can be expressed at any $p \in M$ as

$$\vartheta(X_p) = \sum_{i=1}^n \vartheta^i(X_p) e_i \tag{9.30}$$

where $X_p \in T_p M$ and $\vartheta^i$, $i = 1, \ldots, n$, are real-valued 1-forms on $M$. Moreover, as the linear frame $u \in \pi^{-1}(p)$ maps each $n$-tuple $e_i$ into a vector

$$\sum_{j=1}^n e_i^j \frac{\partial}{\partial x^j}\big|_p \tag{9.31}$$

the forms $\vartheta^i$, $i = 1, \ldots, n$, can be expressed as

---

[5] The proof of this identity is based, among others facts, on the realization that if $A^*$ is a fundamental vector field corresponding to the element $A$ of the Lie algebra $gl(n; \mathbb{R})$ of $GL(n; \mathbb{R})$ then given $\xi \in \mathbb{R}^n$ and the corresponding standard horizontal vector field $X(\xi)$

$$[A^*, X(\xi)] = X(A\xi). \tag{9.27}$$

The product $A\xi$ is the image of the standard action of the algebra $gl(n; \mathbb{R})$ of $n \times n$ matrices on $\mathbb{R}^n$. Note that the product $\omega(\cdot)\vartheta(\cdot)$ in the Equation (9.26) should be understood the same way.

$$\vartheta^i = \sum_{j=1}^{n} \vartheta^i_j dx^j. \tag{9.32}$$

Suppose that $\{E^i_j\}$, $i,j = 1,\ldots,n$, is a basis of the Lie algebra $gl(n;\mathbb{R})$. Then, the $gl(n;\mathbb{R})$-valued connection form $\omega$ can be represented as

$$\omega = \sum_{i,j=1}^{n} \omega^i_j E^j_i \tag{9.33}$$

where $\omega^i_j$, $i,j = 1,\ldots,n$, are real-valued 1-forms on $L(M)$. Given the coordinate system $\{x^i, u^j_k\}$ on $\pi^{-1}(U)$, let $s: U \to L(U)$ be a local cross section assigning to each point $p \in U$ the frame $(\frac{\partial}{\partial x^1}|_p, \cdots, \frac{\partial}{\partial x^n}|_p)$. Consider the pullback $s^*\omega$. The form $s^*\omega$ is a $gl(n;\mathbb{R})$-valued 1-form on $U \subset M$ and as such can now be represented in a local coordinate system on $U$ as

$$s^*\omega = \sum_{i,j=1}^{n} s^*\omega^i_j E^j_i = \sum_{i,j=1}^{n} (\sum_{k=1}^{n} \Gamma^i_{kj} dx^k) E^j_i. \tag{9.34}$$

The real-valued functions $\Gamma^i_{kj}$ defined in the neighbourhood $U \subset M$ are called the *Christoffel symbols of the linear connection* $\omega$ with respect to the local coordinate system $x^1, \ldots, x^n$. Taking this procedure one step further and utilizing the definition of the pull-back operator and the representation (9.32) of the standard form $\vartheta$, one can show that:

$$\omega^i_j = \sum_{k=1}^{n} u^i_k (du^k_j + \sum_{l,m=1}^{n} \Gamma^k_{ml} u^l_j dx^m), \tag{9.35}$$

where we benefited from the fact that the section $s$ is represented in the induced coordinate system $\{x^i, u^j_k\}$ on $\pi^{-1}(U)$, as $\{x^i, \delta^j_k\}$, $i,j,k = 1,\ldots,n$. Going now back to the transformation rule for the connection forms (9.8), we obtain the following equivalent transformation law for the Christoffel symbols:

**Proposition 9.23.** *Let $\omega$ be a linear connection on a manifold $M$. Suppose that $\Gamma^i_{jk}$ and $\widehat{\Gamma}^i_{jk}$ are the Christoffel symbols of the connection $\omega$ with respect to the coordinate systems $x^1,\ldots,x^n$ and $\widehat{x}^1,\ldots,\widehat{x}_n$ on the coordinates neighbourhoods $U$ and $\widehat{U}$, respectively. Assume that $U \cap \widehat{U} \neq 0$. Then, at any point $p \in U \cap \widehat{U}$*

$$\widehat{\Gamma}^\alpha_{\beta,\gamma} = \sum_{i,j,k=1}^{n} \Gamma^i_{jk} \frac{\partial x^j}{\partial \widehat{x}^\beta} \frac{\partial x^k}{\partial \widehat{x}^\gamma} \frac{\partial \widehat{x}^\alpha}{\partial x^i} + \sum_{i=1}^{n} \frac{\partial^2 x^i}{\partial \widehat{x}^\beta \partial \widehat{x}^\gamma} \frac{\partial \widehat{x}^\alpha}{\partial x^i}. \tag{9.36}$$

## 9.3 Connections in an associated bundle

Let $P(M, \mathcal{G})$ be a principal $\mathcal{G}$-bundle and let $E_\mathcal{G}(M)$ denote an associated bundle with $E$ as its standard fibre[6]. Given $p \in M$, each fibre $\pi_E^{-1}(p)$ of $E_\mathcal{G}(M)$ is isomorphic to $E$. Moreover, if $u \in P$ and $\pi(u) = p$ then $u$ can be viewed as an isomorphism of $E$ onto $\pi_E^{-1}(p)$ (see Example 8.15). Let $\Gamma(E_\mathcal{G})$ denote the collection of all cross sections of the bundle $E_\mathcal{G}(M)$. If the typical fibre $E = \mathbb{R}^m$ then $\Gamma(E_\mathcal{G})$ is an infinitely dimensional vector space over the set of real numbers with point addition and scalar multiplication. $\Gamma(E_\mathcal{G})$ is also an $E$-module, i.e., if $\lambda : M \to E_\mathcal{G} = \mathbb{R}^m$ is a function on $M$ then $(\lambda s)(p) = \lambda(p)s(p)$ and $\lambda s \in \Gamma(E_\mathcal{G})$ for any $s \in \Gamma(E_\mathcal{G})$ and $p \in M$.

Suppose that $\mathcal{H}$ is a connection in $P(M, \mathcal{G})$. Consider a curve $x : [a, b] \to M$ and its horizontal lift $x^* : [a, b] \to P$. For each fixed $\xi \in E$ define $x^*\xi$, as we did at the end of Section 9.1.2, as the horizontal lift of $x$ to $E_\mathcal{G}(M)$, where $x^*\xi$ is viewed as an equivalence class in $(P \times E)/\mathcal{G}$. It defines in a natural way the notion of parallel transport (cf. Proposition 9.6) as an isomorphism $\rho$ between fibres of $E_\mathcal{G}(M)$, identifying objects at different fibres having the same coordinates along a horizontal curve. This enables us to introduce the concept of a covariant derivative. But first, we say that a section $s : M \to E_\mathcal{G}(M)$ is defined on the curve $x : [a, b] \to M$ if $\pi_E(s(x(t))) = x(t)$ for all $t \in [a, b]$. The *covariant derivative of the section $s$ defined on the curve $x$* in the direction of $\dot{x}(t)$ is given as

$$\nabla_{\dot{x}(t)} s \equiv \lim_{h \to \infty} \frac{1}{h} [\rho_t^{t+h}(s(x(t+h))) - s(x(t))] \tag{9.37}$$

where $\rho_t^{t+h}$ denotes the parallel displacement of the fibre $\pi_E^{-1}(x(t+h))$ along the curve $x$ from $x(t+h)$ to $x(t)$. In general, given $X \in T_pM$ and a cross section $s$ defined in an neighbourhood of the point $p$, the covariant derivative of $s$ in the direction of $X$ is $\nabla_X s \equiv \nabla_{\dot{x}(t)} s$ where $x : [-\varepsilon, \varepsilon] \to M$ is the integral curve of $X$ such that $X = \dot{x}(0)$.

Having defined the concept of a covariant derivative of a section of an associated bundle, let us, for the remainder of this section, restrict our attention to bundles associated with the bundle of linear frames $L(M)$ and linear connections only. Subject to these restrictions, we shall look at the covariant differentiation from yet another angle. Indeed, suppose that a cross section $s$ of a vector bundle $E_\mathcal{G}(M)$ associated with $L(M)$ is defined on an open subset $U \subset M$. Associate with the section $s : U \to E_\mathcal{G}(M)$ an $\mathbb{R}^m$-valued function $f : \pi^{-1}(U) \subset L(M) \to \mathbb{R}^m$ such that

$$f(u) = u^{-1}(s(\pi(u))) \tag{9.38}$$

---

[6] If $E = \mathbb{R}^m$, $\rho : \mathcal{G} \to GL(m; \mathbb{R})$ is a representation of $\mathcal{G}$ in $GL(m; \mathbb{R})$, and $\mathcal{G}$ acts on $E$ through $\rho$, then $E_\mathcal{G}(M)$ is called a *vector bundle*.

for every $u \in \pi^{-1}(U)$, where $u^{-1}$ is viewed again as a mapping (see the identity (9.22)). Let $X \in T_pM$ and let $X^* \in T_uL(M)$ be its horizontal lift, where $\pi(u) = p$. Consider the derivation $X^*f$ to realize that it takes values in $\mathbb{R}^m$. Note also that $u(X^*f)$ belongs to the fibre $\pi_E^{-1}(p)$. Comparing this construction with the definition of the covariant derivative using the concept of the parallel transport (9.37), we obtain:

**Proposition 9.24.** *Suppose that $s$ is a local section of a vector bundle $E_\mathcal{G}(M)$ defined in a neighbourhood of a point $p \in M$. Let $\nabla_X s$ be the covariant derivative of a section $s$ in the direction of $X \in T_pM$ and let $f$ be defined by (9.38). Then,*

$$\nabla_X s = u(X^*f). \tag{9.39}$$

Incidentally, the above proposition shows that the concept of covariant derivative is independent of the choice of the underlying curve. Moreover, one can now identify easily the following properties of the covariant differentiation operator:

**Proposition 9.25.** *Let $X, Y \in T_pM$ and let $s$ and $\widehat{s}$ be two cross sections of a vector bundle $E_\mathcal{G}(M)$ defined in a neighbourhood of $p \in M$. Then,*

1. $\nabla_{X+Y} s = \nabla_X s + \nabla_Y s$.
2. $\nabla_X (s + \widehat{s}) = \nabla_X s + \nabla_X \widehat{s}$.
3. $\nabla_{\lambda X} s = \lambda \nabla_X s$ *for any $\lambda \in \mathbb{R}$.*
4. $\nabla_X (\lambda s) = \lambda(p) \nabla_X s + (X\lambda) s(p)$ *for any $\mathbb{R}^m$-valued function $\lambda$ defined in a neighbourhood of $p$.*

Given a vector bundle $E_\mathcal{G}(M)$ over the manifold $M$, with $\mathbb{R}^m$ as its standard fibre, a *field $K$ of a quantity $E$* on $M$ is a cross section of $E_\mathcal{G}(M)$. We can define a covariant derivative of $K$ at $p \in M$ in the direction of $X \in T_pM$ as $\nabla_X K$ in (9.39). In particular, if $X$ is a vector field on $M$ and $K$ is a field of a quantity $E$ on $M$ then $\nabla_X K$ is a field of the same quantity. In other words, $\nabla_X K$ is still a section of $E_\mathcal{G}(M)$. Also, if $g$ is a smooth real-valued function on $M$ then $\nabla_X g = Xg$. Thus, looking at the tangent space $TM$ as a vector bundle (rather than an associated bundle like in Example 8.15), the following is an immediate corollary to Proposition 9.25:

**Proposition 9.26.** *If $X, Y$ and $Z$ are vector fields on $M$ and $g$ is a smooth real-valued function on $M$ then*

1. $\nabla_X (Y + Z) = \nabla_X Y + \nabla_X Z$.
2. $\nabla_{X+Y} Z = \nabla_X Z + \nabla_Y Z$.
3. $\nabla_{gX} Y = f \nabla_X Y$.
4. $\nabla_X (gY) = g \nabla_X Y + (Xg) Y$.

Conversely, it is possible to show [57] that any mapping $\mathcal{X}(M) \times \mathcal{X}(M) \to \mathcal{X}(M)$ assigning to a pair of vector fields $(X, Y)$ another vector field, say $\nabla_X Y$, satisfying all four conditions of Proposition 9.26 is a covariant derivative on $TM$ with respect to some linear connection on $M$.

A field $K$ on $M$ (viewed as a cross section of the vector bundle $E_\mathcal{G}(M)$) is considered *parallel* if, and only if, $\nabla_X K \equiv 0$ for all $X \in T_p M$ and $p \in M$. That is, as shown in Proposition 9.24, $K$ is parallel if it is covariantly (horizontally) constant.

Having the torsion form $\Theta$ and the curvature form $\Omega$ of the linear connection $\omega$ defined already we proceed now to define the corresponding *torsion field* $\mathcal{T}$ and the *curvature field* $\mathcal{R}$. Namely, let $\mathcal{T} : TM \times TM \to TM$ be given point-wise as

$$\mathcal{T}(X,Y) \equiv u(2\Theta(X^*, Y^*)) \tag{9.40}$$

for any pair $X, Y \in T_p M$, where $u \in L(M)$, $\pi(u) = p$, and the horizontal vectors $X^*, Y^* \in T_u L(M)$ are such that $\pi_*(X^*) = X$ and $\pi_*(Y^*) = Y$. A straightforward calculation shows that as the form $\Theta$ is pseudotensorial and horizontal, $\mathcal{T}(X,Y)$ is independent of the choice of $u$ and the vectors $X^*, Y^*$. Moreover, at each point $p \in M$, $\mathcal{T}$ is a skew-symmetric (see the identity (9.26)) bilinear mapping $T_p M \times T_p M \to T_p M$, i.e.,

$$\mathcal{T}(X,Y) = -\mathcal{T}(Y,X) \tag{9.41}$$

for every pair of vectors $X, Y \in T_p M$.

Similarly, we define the *curvature field* $\mathcal{R}$ by requiring that

$$\mathcal{R}(X,Y)Z \equiv u(2\Omega(X^*, Y^*))u^{-1}(Z) \tag{9.42}$$

for any choice of $X, Y, Z \in T_p M$, where $u \in L(M)$, $\pi(u) = p$, and $X^*, Y^*, Z^* \in T_u L(M)$ are such that $\pi_*(X^*) = X$, $\pi_*(Y^*) = Y$ and $\pi_*(Z^*) = Z$. As in the case of the torsion, one can show that the definition of the curvature field $\mathcal{R}$ is independent of the choice of $u$ and the vectors $X^*, Y^*, Z^*$. Realize also that given $X, Y \in T_p M$, $\mathcal{R}(X,Y)$ can be viewed as an endomorphism of $T_p M$. It also follows from the definition of the curvature $\Omega$ (see Theorem 9.12) that

$$\mathcal{R}(X,Y) = -\mathcal{R}(Y,X) \tag{9.43}$$

for every pair of vector fields $X, Y \in \mathcal{X}(M)$.

Utilizing Proposition 9.24 one can represent the covariant derivative of a vector field in the following way:

**Lemma 9.27.** *Suppose that $X^*, Y^*$ are the horizontal lifts of the vector fields $X, Y \in \mathcal{X}(M)$. Then,*

$$(\nabla_X Y)_p = u(X_u^*(\vartheta(Y_u^*))) \tag{9.44}$$

*for every $p \in M$ and any frame $u \in L(M)$ such that $\pi(u) = p$.*

Subsequently, using the definitions of the torsion and the curvature, one can show [57] that:

## 9.3 Connections in an associated bundle

**Theorem 9.28.** *Given a linear connection on $M$, the corresponding fields of torsion $\mathcal{T}$ and curvature $\mathcal{R}$ satisfy the following identities:*

$$\mathcal{T}(X,Y) = \nabla_X Y - \nabla_Y X - [X,Y] \tag{9.45}$$

*and*

$$\mathcal{R}(X,Y)Z = [\nabla_X, \nabla_Y]Z - \nabla_{[X,Y]}Z \tag{9.46}$$

*where $X, Y$ and $Z$ are vector fields on $M$.*

Adopting a local coordinate system on a manifold $M$ and the corresponding coordinate system on the bundle of linear frames $L(M)$, we can express the components of a linear connection in terms of the appropriate covariant derivatives. Indeed, it follows from Theorem 9.28 and Lemma 9.27 that:

**Proposition 9.29.** *Let $x^1, \ldots, x^n$ be a local coordinate system in $M$ and let $\omega$ be a linear connection form. Then, the components $\Gamma^i_{jk}$ of the connection $\omega$ with respect to the given coordinate system are:*

$$\nabla_{e_j} e_k = \sum_{i=1}^n \Gamma^i_{jk} e_i \tag{9.47}$$

*where $e_j \equiv \frac{\partial}{\partial x^j}$, $j = 1, \ldots, n$. Moreover, given a vector fields*

$$Y = \sum_{i=1}^n y^i e_i, \tag{9.48}$$

$$\nabla_{e_i} y^j = \frac{\partial y^j}{\partial x^i} + \sum_{k=1}^n \Gamma^j_{ik} y^k. \tag{9.49}$$

If one defines the components of the torsion and curvature fields as

$$\mathcal{T}(\frac{\partial}{\partial x^j}, \frac{\partial}{\partial x^k}) = \sum_{i=1}^n T^i_{jk} \frac{\partial}{\partial x^i}, \tag{9.50}$$

$$\mathcal{R}(\frac{\partial}{\partial x^k}, \frac{\partial}{\partial x^l})\frac{\partial}{\partial x^j} = \sum_{i=1}^n R^i_{jkl} \frac{\partial}{\partial x^i}, \tag{9.51}$$

Theorem 9.28 and Proposition 9.29 imply that:

**Proposition 9.30.** *Let $x^1, \ldots, x^n$ be a local coordinate system on $M$ and let $\Gamma^i_{jk}$ denote the components of a linear connection on $M$ with respect to the given coordinate system. Then,*

$$T^i_{jk} = \Gamma^i_{jk} - \Gamma^i_{kj} \tag{9.52}$$

*and*

$$R^i_{jkl} = \left(\frac{\partial \Gamma^i_{lj}}{\partial x^k} - \frac{\partial \Gamma^i_{kj}}{\partial x^l}\right) + \sum_{m=1}^n (\Gamma^m_{lj}\Gamma^i_{km} - \Gamma^m_{kj}\Gamma^i_{lm}). \tag{9.53}$$

## 9.4 G-structures

In Section 8.3.1 we introduced the process of reducing a structure group of an arbitrary $\mathcal{G}$-principal bundle $P(M, \mathcal{G})$ to a subgroup $\mathcal{K}$ resulting in a principal $\mathcal{K}$-subbundle $P(M, \mathcal{K})$. In this chapter we shall scrutinize this process when applied to the bundle of linear frames of a manifold. We are particularly interested in such reductions as they are central in the mathematical theory of inhomogeneities. Thus, given a smooth manifold $M$, consider its bundle of linear frames $L(M)$, and a subgroup $\mathcal{G}$ of the general linear group $GL(n; \mathbb{R})$. A $\mathcal{G}$-*structure* on $M$, denoted as $B_{\mathcal{G}}(M)$, is a reduction of the principal bundle $L(M)$ to the subgroup $\mathcal{G}$. In other words, $B_{\mathcal{G}}(M)$ is a principal subbundle of $L(M)$ with the property that for any $u \in B_{\mathcal{G}}(M)$ and any $a \in GL(n; \mathbb{R})$ an element $R_a(u) \in B_{\mathcal{G}}(M)$ if, and only if, $a \in \mathcal{G}$.

What interests us first and foremost is the question as to whether a particular manifold (e.g., the given body manifold) admits a reduction of a frame bundle to a specific subgroup. The answer to this question can be formulated in at least two different ways. First, rephrasing Proposition 8.11, we can say that a manifold $M$ admits a $\mathcal{G}$-structure $B_{\mathcal{G}}(M) \subset L(M)$, for some subgroup $\mathcal{G} \subset GL(n; \mathbb{R})$, if, and only if, there exists an open covering $\{U_\alpha\}$ of $M$ and the corresponding local sections $s_\alpha : U_\alpha \to L(M)$ such that the transition functions $\varphi_{\beta\alpha}$, relating the sections $s_\alpha$ via

$$s_\beta(p) = \varphi_{\beta\alpha}(p) s_\alpha(p) \tag{9.54}$$

at every $p \in U_\alpha \cap U_\beta \neq \emptyset$, take values in the subgroup $\mathcal{G}$. On the other hand, as stated in Proposition 8.22, $B_{\mathcal{G}}(M)$ is a $\mathcal{G}$-structure on the manifold $M$, where $\mathcal{G}$ is a closed subgroup of $GL(n; \mathbb{R})$, when there exists a one-to-one correspondence between the subbundles $B_{\mathcal{G}}(M)$ and the global sections of the corresponding associated bundle $L(M)/\mathcal{G}$ (cf. Example 8.21).

Consider a diffeomorphism $f : M \to N$ of a manifold $M$ onto a manifold $N$. It induces in a natural way the linear diffeomorphism $f_* : TM \to TN$ as well as the principal-bundle isomorphism (denoted by the same symbol) $f_*$ from $L(M)$ into $L(N)$. Suppose that $B_{\mathcal{G}}(M)$ and $B_{\mathcal{G}}(N)$ are two $\mathcal{G}$-structures on $M$ and $N$, respectively. If the isomorphism $f_*$ is such that $f_*(B_{\mathcal{G}}(M)) = B_{\mathcal{G}}(N)$ we say that the given $\mathcal{G}$-structures are $f$-*isomorphic*. Conversely, two $\mathcal{G}$-structures $B_{\mathcal{G}}(M)$ and $B_{\mathcal{G}}(N)$, where the manifolds $M$ and $N$ do not necessarily need to be different, are called *isomorphic* if there exists a diffeomorphism $f : M \to N$ such that $f_*(B_{\mathcal{G}}(M)) = B_{\mathcal{G}}(N)$. Furthermore, we say that the bundles $\pi_M : B_{\mathcal{G}}(M) \to M$ and $\pi_N : B_{\mathcal{G}}(N) \to N$ are *locally equivalent*[7] at $x \in B_{\mathcal{G}}(M)$ and $y \in B_{\mathcal{G}}(N)$ if there exist open neighbourhoods

---

[7] As pointed out in [84], a somewhat different problem of equivalence is often of interest. Namely, we say that $B_{\mathcal{G}}(M)$ and $B_{\mathcal{K}}(M)$ are *conjugate structures* if there exists $a \in GL(n; \mathbb{R})$ such that $R_a(B_{\mathcal{G}}(M)) = B_{\mathcal{K}}(M)$ thus implying that $a^{-1}\mathcal{G}a = \mathcal{K}$. The conjugate equivalence problem is to determine when one $\mathcal{G}$-structure is conjugate equivalent to another structure.

$U$ of $x$ and $V$ of $y$, and an isomorphism $f_*$ of $B_{\mathcal{G}}(U)$ onto $B_{\mathcal{G}}(V)$, such that the underlying base diffeomorphism $f$, where $\pi_N \circ f_* = f \circ \pi_M$, has the property that $f(x) = y$. In particular, consider as the manifold $M$ a finite dimensional vector space, say $\mathbb{R}^n$. At any $x \in \mathbb{R}^n$ the vector space structure of $\mathbb{R}^n$ provides the identification of the tangent space $T_x \mathbb{R}^n$ with $\mathbb{R}^n$. Let $\mathcal{G}$ be a subgroup of $GL(n;\mathbb{R})$ and consider at each $x \in \mathbb{R}^n$ all frames obtained from the standard frame by the action of the group $\mathcal{G}$. The collection of frames obtained through this process is a principal $\mathcal{G}$-bundle called the *standard flat $\mathcal{G}$-structure*. Given an arbitrary $\mathcal{G}$-structure $B_{\mathcal{G}}(M)$, we say that it is *locally flat* if it is locally equivalent to the standard flat $\mathcal{G}$-structure.

Looking at $\mathcal{G}$-structures from a different perspective, we say that a $\mathcal{G}$-structure is *integrable* if every point $p \in M$ has a coordinate neighbourhood $U$ and a coordinate system $x^1, \ldots, x^n$ on $U$ such that the induced field of bases $\left(\frac{\partial}{\partial x^1}, \ldots, \frac{\partial}{\partial x^n}\right)$ is a local cross section of $B_{\mathcal{G}}(M)$. Such a local coordinate system on $M$ will be called *admissible* with respect to the given $\mathcal{G}$-structure $B_{\mathcal{G}}(M)$. In particular, if $x^1, \ldots, x^n$ is an admissible (in $B_{\mathcal{G}}(M)$) local coordinate system on $U$ and $y^1, \ldots, y^n$ is yet another admissible local coordinate system on $V$, where $U \cap V \neq \emptyset$, then at any $p \in U \cap V$ the Jacobian matrix $\left\{\frac{\partial y^i}{\partial x^j}(p)\right\}$, $i,j = 1, \ldots, n$, is an element of the group $\mathcal{G}$. It is a simple exercise, as shown in [56], to confirm that the local flatness of a $\mathcal{G}$-structure and its integrability are two different names for the same property. Indeed:

**Proposition 9.31.** *A $\mathcal{G}$-structure $B_{\mathcal{G}}(M)$ on $M$ is locally flat if, and only if, it is integrable. That is, $B_{\mathcal{G}}(M)$ is locally flat if, and only if, there exists an atlas $\{(U_\alpha, \varphi_\alpha)\}$ on $M$ and the corresponding local coordinate systems $x^1_\alpha, \ldots, x^n_\alpha$ on $U_\alpha$ such that each section*

$$\left(\frac{\partial}{\partial x^1_\alpha}, \cdots \frac{\partial}{\partial x^n_\alpha}\right) : U_\alpha \to L(U_\alpha) \tag{9.55}$$

*lies in $B_{\mathcal{G}}(M)$.*

Suppose now that $B_{\mathcal{G}}(M)$ is an integrable $\mathcal{G}$-structure on $M$. Let $U \subset M$ be a coordinate neighbourhood with an admissible coordinate system $x^1, \cdots, x^n$. Consider a linear connection $\omega$ on $B_{\mathcal{G}}(U)$ such that the coordinate-induced vector fields $\frac{\partial}{\partial x^1}, \cdots, \frac{\partial}{\partial x^n}$ are parallel. This fact implies immediately that the canonical form $\vartheta$ on $L(M)$ (cf. Equation (9.22)), restricted to $B_{\mathcal{G}}(U)$, is covariantly constant. Thus, recalling the definition of the torsion and Theorem 9.3 one is able to conclude [56] that:

**Theorem 9.32.** *If a $\mathcal{G}$-structure $B_{\mathcal{G}}(M)$ over a paracompact manifold $M$ is integrable, then it admits a torsion-free connection.*

The converse is not always true.

### 9.4.1 Examples of $G$-structures

We present here a number of important, both from the mathematical and the applications standpoint, examples of different $\mathcal{G}$-structures.

*Example 9.33. Complete parallelism*: Let $\mathcal{G} = \{e\}$, where $e$ is the identity of $GL(n;\mathbb{R})$. The corresponding $\mathcal{G}$-structure of $M$ is a smooth choice of a single frame at each point of the manifold $M$. That is, as stated in Proposition 8.22, a manifold admits an $\{e\}$-structure if the bundle of linear frames admits a global section. In fact, such a manifold $M$ is said to be *parallelizable*, a terminology justified as follows. Let $s : M \to L(M)$ be a global section. Consider the frame $u \in s(M)$ at $p \in M$. It identifies the tangent space $T_pM$ with $\mathbb{R}^n$ as $u^{-1}(X_p) \in \mathbb{R}^n$ for every $X_p \in T_pM$ (cf. Example 8.15). In turn, the unique frame $w \in s(M)$ at $q \in M$ identifies $u^{-1}(X_p)$ with the vector $w(u^{-1}(X_p)) \in T_qM$. Thus, given a global section of $L(M)$, we do obtain a completely path-independent parallel translation of tangent vectors on $M$. For this reason an $\{e\}$-structure on $M$ is often called a *complete parallelism* on $M$. One of the most important examples of a complete parallelism on a manifold is a Lie group, whereby the left-invariant vector fields give an identification of a tangent space at a point with the corresponding Lie algebra.

*Example 9.34. Differentiable distribution*: Let $\mathcal{D}$ be an $r$-dimensional differentiable distribution on $M$. Take as $B_\mathcal{G}(M)$ the collection of all the frames $u \in L(M)$ such that at any point $p \in M$ $u^{-1}(\mathcal{D}_p) = V$, where $V$ is a fixed $r$-dimensional vector subspace of $\mathbb{R}^n$. Given $a \in GL(n;\mathbb{R})$ and a frame $u \in B_\mathcal{G}(M)$, consider the frame $R_a(u) \equiv ua$. Then, $(ua)^{-1}(\mathcal{D}_p) = a^{-1}(u^{-1}(\mathcal{D}_p)) = a^{-1}(V)$ and $ua \in B_\mathcal{G}(M)$ if, and only if, $a^{-1}(V) = V$. Hence, it is easy to see that the set $B_\mathcal{G}(M)$ selected above is a reduction of the bundle of linear frames of $M$ to the group $\mathcal{G} = GL(V) \subset GL(n;\mathbb{R})$ containing all elements of $GL(n;\mathbb{R})$ which preserve the vector subspace $V \subset \mathbb{R}^n$. The Frobenius Theorem 8.4 implies that a $GL(V)$-structure is integrable if, and only if, the corresponding distribution $\mathcal{D}$ is involutive.

*Example 9.35. Volume form*: Assume that $M$ is an n-dimensional manifold. Select as the group $\mathcal{G} \subset GL(n;\mathbb{R})$ the special linear group $SL(n;\mathbb{R})$, that is, the set of all linear transformations $a \in GL(n;\mathbb{R})$ with unit determinant. Given $p \in M$, consider the fibre $\pi^{-1}(p) \subset L(M)$ to realize that the natural action of $GL(n;\mathbb{R})$ on $L(M)$ renders the quotient $\pi^{-1}(p)/SL(n;\mathbb{R})$ one-dimensional. In fact, $\pi^{-1}(p)/SL(n;\mathbb{R})$ is isomorphic to $\mathbb{R} - \{0\}$ with the induced action of $GL(n;\mathbb{R})$ such that $R_a(r) = (\det a)r$ for any $a \in GL(n;\mathbb{R})$ and $r \in \mathbb{R} - \{0\}$. It follows, that the cross sections of $L(M)/SL(n;\mathbb{R})$ are in one-to-one correspondence with the volume elements on $M$, that is, the nowhere vanishing $n$-forms on $M$. Indeed, suppose that a $n$-form $\nu \equiv \varrho dx^1 \wedge \cdots \wedge dx^n$ is given in some coordinate neighbourhood $U$ with the corresponding coordinate system $x^1, \ldots, x^n$. Then, according to (8.18),

$$\nu(\frac{\partial}{\partial x^1}, \cdots, \frac{\partial}{\partial x^n}) = \varrho \det\left\{dx^i(\frac{\partial}{\partial x^j})\right\} = \varrho. \tag{9.56}$$

establishing the required isomorphism.

Although not every manifold admits a $SL(n;\mathbb{R})$-structure, if it does, the structure is integrable [56][8]. To see how this comes about, consider once gain a locally defined volume form $\nu \equiv \varrho dx^1 \wedge \cdots \wedge dx^n$ where $\varrho$ is a smooth function on the coordinate neighbourhood $U$. Assume, without loss of generality, that there exists a function $y(x^1, \cdots, x^n)$ such that $\varrho = \frac{\partial y}{\partial x^1}$. Then, $\nu \equiv \varrho dx^1 \wedge \cdots \wedge dx^n = dy \wedge dx^2 \wedge \cdots \wedge dx^n$ proving that the coordinate system $y, x^2, \cdots, x^n$ is admissible with respect to the given $SL(n;\mathbb{R})$-structure.

*Example 9.36. Riemannian (metric) structure*: Let $E_\mathcal{G}(M)$ be a vector bundle associated with a $\mathcal{G}$-bundle $P(M,\mathcal{G})$. As each fibre of $E_\mathcal{G}(M)$ is isomorphic to $\mathbb{R}^m$, for some fixed $m > 0$, we can assign at every $p \in M$ an inner product $g_p : \pi_E^{-1}(p) \times \pi_E^{-1}(p) \to \mathbb{R}$. If for every differentiable local cross section $s : U \subset M \to E_\mathcal{G}(M)$, $g \circ s : U \to \mathbb{R}$ is a differentiable function on $U$ then $g$ is said to be a *fibre metric* on the vector bundle $E_\mathcal{G}(M)$. Let $p \in M$ and select any two elements $\xi, \eta \in \pi_E^{-1}(p)$. Set $Q(M)$ to be the collection of all $u \in P(M,\mathcal{G})$ such that at every $p \in M$ $g_p(\xi,\eta) \equiv (u(\xi), u(\eta))$, where $u \in \pi^{-1}(p)$ and $(\cdot,\cdot)$ denotes the standard inner product in $\mathbb{R}^m$. The set $Q(M)$ is a closed submanifold of $P(M,\mathcal{G})$. In fact, it is a reduction of $P(M,\mathcal{G})$ to the subgroup $\mathcal{O} \equiv \{a \in \mathcal{G} : \varrho(a) \in \mathcal{O}(m;\mathbb{R})\}$ of $\mathcal{G}$ where $\varrho$ is the representation of $\mathcal{G}$ in $GL(m;\mathbb{R})$ and where $\mathcal{O}(m;\mathbb{R})$ denotes the orthogonal subgroup of $GL(m;\mathbb{R})$.

Using the concept of partition of unity (cf. Theorem 8.20) and the fact that the bundle $E_\mathcal{G}(M)$ is locally trivial, one can show [56] that:

**Proposition 9.37.** *If $M$ is a paracompact manifold, then every vector bundle $E_\mathcal{G}(M)$ admits a fibre metric.*

Given a fibre metric $g$ on the vector bundle $E_\mathcal{G}(M)$, a connection $\omega$ on $P(M,\mathcal{G})$ is called a *metric connection* if the parallel displacement of fibres of $E_\mathcal{G}(M)$ preserves the metric. That is, the connection $\omega$ in $P(M,\mathcal{G})$ is metric if $g$ is covariantly constant with respect to this connection.

**Proposition 9.38.** *Let $g$ be a metric on a vector bundle $E_\mathcal{G}(M)$ and let $Q(M)$ be the reduction of $P(M,\mathcal{G})$ defined by the metric $g$. A connection $\omega$ on $P(M,\mathcal{G})$ is reducible to a connection on $Q(M)$ if, and only if, $\omega$ is a metric connection. Conversely, any metric connection on $Q(M)$ can be extended to a metric connection on $P(M,\mathcal{G})$.*

A *Riemannian metric* on an $n$-dimensional manifold $M$ is a fibre metric on the tangent bundle $TM$. As indicated by Proposition 9.37, $M$ admits a Riemannian metric if it is paracompact. Also, as the orthogonal subgroup

---
[8] See also [21].

$\mathcal{O}(n;\mathbb{R})$ of $GL(n;\mathbb{R})$ is closed, there is a one-to-one correspondence between the set of Riemannian metrics and the reductions $Q(M) \equiv B_\mathcal{O}(M)$ of the bundle of linear frames of $M$ (cf. Proposition 8.22). A manifold $M$ with a metric connection is called a *Riemannian manifold*. In fact [56]:

**Theorem 9.39.** *Every Riemannian manifold admits a unique symmetric metric connection, that is, a torsion-free metric connection.*

Finally, it can be shown that, although not every Riemannian structure $Q(M)$ is integrable:

**Proposition 9.40.** *A Riemannian structure $Q(M)$ is integrable if, and only if, the corresponding Riemannian metric $g$ is flat, i.e., the Riemannian connection of $g$ is curvature-free.*

*Example 9.41. G-structure induced by an object*: Consider a vector bundle $E_{GL(n;\mathbb{R})}(M)$ associated with the bundle of linear frames of $M$. A section $s: M \to E_{GL(n;\mathbb{R})}(M)$ is called a *(geometric) object* on $M$. Select as $\mathcal{G}$ the isotropy group of $s$, that is, the set of all $a \in GL(n;\mathbb{R})$ such that $R_a(s(p)) = s(p)$ at all $p \in M$. $\mathcal{G}$ is obviously a closed subgroup of $GL(n;\mathbb{R})$. Take as $B_\mathcal{G}(M)$ all frames $u \in L(M)$ such that $u(s(p))$ is constant at every $p \in M$. By Proposition 8.22 $B_\mathcal{G}(M)$ is a reduction of $L(M)$ to the group $\mathcal{G}$ called a *$\mathcal{G}$-structure induced by the geometric object $s$*. Note that any $SL(n;\mathbb{R})$-structure (cf. Example 9.35) can be viewed as an $n$-form induced $\mathcal{G}$-structure.

# 10
# Bundles of linear frames

The bundle of linear frames of a manifold was introduced in Section 8.3.1 as an example of a principal $GL(n;\mathbb{R})$-bundle (Example 8.12). In this chapter we shall look again at linear frames on a manifold but from a completely different perspective. This will enable us to extend the concept of a frame to orders higher than one. We start our presentation by reviewing some basic facts about prolongations of fibre bundles. For more detail and more in-depth presentation of these concepts and related issues the reader is referred to [9] and [80].

## 10.1 Jet prolongations of fibre bundles

Let $E(M)$ be a fibre bundle[1] over a manifold $M$ with a standard projection $\pi : E \to M$. We say that two local sections $s : U \to E$ and $r : V \to E$, where $U$ and $V$ are open subsets of $M$, are *first-order equivalent* at the point $p \in U \cap V$ if the images $s(p)$ and $r(p)$ are equal and the tangent maps $s_*(p)$ and $r_*(p)$ are identical. In other words, sections $s$ and $r$ are first-order equivalent at $p$ if $s(p) = r(p)$ and $\frac{\partial s^\alpha}{\partial x^i}(p) = \frac{\partial r^\alpha}{\partial x^i}(p)$ in some (and therefore all) local coordinate systems $\{x^i\}$ and $\{x^i, y^\alpha\}$ on $M$ and $E$, respectively. The first-order equivalence class of a section $s : U \to E$ at $p \in U$ is called the *first jet* (or 1-jet) of $s$ at $p$ and it will be denoted by $j_p^1 s$. The point $p$ is called the *source* of the first jet $j_p^1 s$ while $s(p)$ is its *target*.

The space $J^1(E)$ of all first-order equivalence classes of local sections of $E(M)$ has the canonical structure of a fibre bundle endowed with a surjective submersion $\pi^1 : J^1(E) \to M$ projecting a 1-jet of a section onto its source [80]. One can show also that the canonical projection $\pi_0^1 : J^1(E) \to E$ of a 1-jet

---
[1] This notation is a short-hand for the more standard notation $P(M, F)$. We purposely suppress indicating explicitly the standard fibre $F$ as its form is immaterial in this part of our presentation.

## 10 Bundles of linear frames

of a section onto its target, is a surjective submersion making $J^1(E)$ into yet another fibre bundle with $E$ as its base manifold. The latter structure of the space $J^1(E)$ will be discussed later.

The bundle $\pi^1 : J^1(E) \to M$ is called the *first prolongation* of the fibre bundle $E(M)$. A local section $j^1 : V \subset M \to J^1(E)$ is said to be the (first) *prolongation* of the section $s : U \to E$ if $U \cap V \neq \emptyset$ and $j^1(p) = j_p^1 s$ for every $p \in U \cap V$.

Higher-order jets of sections and the corresponding prolongations of the fibre bundle $E(M)$ can now be defined recursively. Indeed, we say that two local sections of $E$, say $s$ and $r$, are $k$-*order equivalent* at $p \in M$ if

$$s(p) = r(p) \quad \text{and} \quad \frac{\partial^l s^\alpha}{\partial x^{i_1} \cdots \partial x^{i_l}}(p) = \frac{\partial^l r^\alpha}{\partial x^{i_1} \cdots \partial x^{i_l}}(p), \tag{10.1}$$

for every $l = 1 \ldots k$, in some local coordinate systems $\{x^i\}$ and $\{x^i, y^\alpha\}$ on $M$ and $E$, respectively. Thus, the $k$-*th jet of a (local) section* of $E$ at a point $p$ is the k-order equivalence class of this section.

As indicated above, the manifold $J^1(E)$ is a fibre bundle over $M$ in its own right. Hence, its sections can also be prolonged. Indeed, generalizing the definition of the first prolongation, we say that the second order *non-holonomic prolongation* of the fibre bundle $E(M)$ is the space $\tilde{J}^2(E) \equiv J^1(J^1(E))$. Generalizing this definition further we say that the $k$-*th order non-holonomic prolongation*, $k \geq 3$, of the fibre bundle $E(M)$ is the space $\tilde{J}^k(E) \equiv J^1(\tilde{J}^{k-1}(E))$.

Note first that $\tilde{J}^k(E)$ is a fibre bundle over any of its underlying prolongations $\tilde{J}^l(E)$, $l < k \leq 0$, with $\tilde{J}^1(E) \equiv J^1(E)$, and $\tilde{\pi}^k_l$ as the standard projection onto $\tilde{J}^l(E)$. Note also that the use of the adjective "non-holonomic" implies that only some elements of the prolongation $\tilde{J}^k(E)$ can be viewed as k-order equivalence classes of sections of the original fibre bundle $E(M)$. Indeed, consider for example an element $j^2 \in \tilde{J}^2(E)$ and let $j^1 = \tilde{\pi}_1^2(j^2) \in J^1(E)$ where $\tilde{\pi}_1^2$ denotes the projection of $\tilde{J}^2(E)$ onto $J^1(E)$. By the definition of the first prolongation, there exists a local section $r$ of $\pi^1 : J^1(E) \to M$ such that $j^2 = j_{j^1}^1 r$. However, there may not exist a local section $s$ of $E(M)$ such that $j^2$ is the second-order equivalence class of $s$ at $\tilde{\pi}^2(j^2)$, where $\tilde{\pi}^2 : \tilde{J}^2(E) \to M$ is a projection satisfying $\tilde{\pi}^2 \equiv \pi^1 \circ \tilde{\pi}_1^2$. This very observation motivates (cf. [16]) the following definitions:

1. A local section $j^1 : U \subset M \to J^1(E)$ is called *admissible* at $p \in U$ if $j^1(p) = j_p^1(\pi_0^1 \circ j^1)$. In other words, $j^1$ is admissible at a point $p$ if it is a first jet of a local section of $E$ at that point.
2. An element $j^2 \in \tilde{J}^2(E)$ is called a *semi-holonomic 2-jet* if there exists an admissible local section $r : U \subset M \to J^1(E)$ at $p = \tilde{\pi}^2(j^2)$ such that $j^2 = j_p^1 r$. Note that although a semi-holonomic 2-jet is a prolongation of a section of the bundle of prolongations it may not necessarily be a second-jet of a section of the original fibre bundle $E(M)$.

3. Generalizing (2) we say that a $k$-jet $j^k \in J^1(\tilde{J}^{k-1}(E))$ is *semi-holonomic* if its projection $\tilde{\pi}_{k-1}^k(j^k)$ is a semi-holonomic jet, and there exists an admissible local section $r : U \to J^1(\tilde{J}^{k-2}(E))$ at $p = \tilde{\pi}^k(j^k)$ such that $j^k = j_p^1 r$.
4. An element $j^k \in \tilde{J}_k(E)$ is called a *holonomic $k$-jet* if there exists a local section $s : U \subset M \to E$ such that $j_p^k s = j^k$, where $p = \tilde{\pi}^k(j^k) \in U$.

Note first that a holonomic jet is always semi-holonomic. Note also that the set $\bar{J}^2(E)$ of all semi-holonomic 2-jets of $\tilde{J}^2(E)$ can be viewed as a smooth subbundle of $\tilde{\pi}_1^2 : \tilde{J}^2(E) \to J^1(E)$ as well as a subset of $\tilde{\pi}^2 : \tilde{J}^2(E) \to M$ (cf. [80]). $\bar{J}^k(E)$ will denote the set of all semi-holonomic jets of $J^1(\bar{J}^{k-1}(E))$ while $J^k(E)$ will indicate the set of all holonomic jets of $\tilde{J}^k(E)$. The spaces $\bar{J}^k(E)$ and $J^k(E)$ are called the *semi-holonomic prolongations* and *holonomic prolongations* of the fibre bundle $E$, respectively. In conclusion, one can see that

$$J^k(E) \subset \bar{J}^k(E) \subset \tilde{J}^k(E) \tag{10.2}$$

for any positive integer $k$. Realize, however, that there are still other intermediate prolongations (between fully non-holonomic and holonomic) whereby a jet may be partially semi-holonomic and/or holonomic. For example, $\bar{J}^k(E) \subset \bar{J}^i(\bar{J}^{k-i}(E)) \subset \tilde{J}^k(E)$.

## 10.2 Local coordinates on prolongations

The next step in our development of the theory of prolongations is to introduce a local coordinate system on a bundle of prolongations of sections. To this end, let $p \in M$, and let $x^1, \ldots, x^n$ be a coordinate system on an open coordinate neighbourhood $U \subset M$ containing $p$. Assume that $y^1, \ldots, y^m$ are the coordinates in the fibre of the bundle $\pi : E \to M$. Hence, the pair $(x^1, \ldots, x^n, y^1, \ldots, y^m)$ forms a local coordinate system on the trivial bundle $\pi^{-1}(U)$. A (local) section $s : U \to E$ can therefore be represented as a set of functions $\{y^\alpha(x^1, \ldots, x^n)\}$, $\alpha = 1, \ldots, m$. Its 1-jet at $p$ is given by the collection $(x^1, \ldots, x^n, y^1(p), \ldots, y^m(p), y_1^1(p), \ldots, y_n^m(p))$ where $y_l^\alpha(p) = \frac{\partial y^\alpha}{\partial x^l}(p)$, $\alpha = 1, \ldots, m$, $l = 1, \ldots, n$. This shows that the triple $\{x^i, y^\alpha, y_l^\alpha\}$ defines local coordinates on the first prolongation of $\pi^{-1}(U)$, i.e., on $J^1(\pi^{-1}(U))$. Consequently, a local section $j^1 : U \subset M \to J^1(E)$ can be represented by a collection of mappings $\{y^\alpha(x^1, \ldots, x^n), y_j^\alpha(x^1, \ldots, x^n)\}$. Henceforth, its first jet at $p$ is represented by the following set of coordinates $(x^i(p), y^\alpha(p), y_j^\alpha(p), \bar{y}_l^\alpha(p), y_{jl}^\alpha(p))$ where $\bar{y}_l^\alpha(p) = \frac{\partial y^\alpha}{\partial x^l}(p)$ and $y_{jl}^\alpha(p) = \frac{\partial y_j^\alpha}{\partial x^l}(p)$. Thus, the local coordinates of the non-holonomic second-order prolongation $\tilde{J}^2(E)$ are $\{x^j, y^\alpha, y_j^\alpha, \bar{y}_l^\alpha, y_{jl}^\alpha\}$. Conversely, let the section $j^1 : U \to J^1(E)$ be given by the collection of mappings $\{y^\alpha(x^1, \cdots, x^n), y_j^\alpha(x^1, \cdots, x^n)\}$. The section $j^1$ is admissible at $p \in M$ (as per the earlier definition) if

$$dy^\alpha(p) = \sum_{j=1}^{n} y_j^\alpha(p) dx^j, \quad \alpha = 1, \ldots, m. \tag{10.3}$$

This implies that the non-holonomic 2-jet $j_p^2 \in \tilde{J}^2(E)$ given by $(x^j(p), y^\alpha(p), y_j^\alpha(p), \bar{y}_i^\alpha(p), y_{jl}^\alpha(p))$ is semi-holonomic provided $y_j^\alpha(p) = \bar{y}_j^\alpha(p)$, $\alpha = 1, \ldots, m$, $j = 1, \ldots, n$. Thus, the semi-holonomic prolongations $\bar{J}^2(E)$ have local coordinates $\{x^i, y^\alpha, y_i^\alpha, y_{ij}^\alpha\}$. Proceeding by induction, let $\{x^i, y^\alpha, y_{j^1}^\alpha, \cdots, y_{j^1, \ldots, j^{k-1}}^\alpha\}$ be local coordinates in the neighbourhood of $j^{k-1} \in \bar{J}^{k-1}(E)$. A local section $j^{k-1}: U \subset M \to \bar{J}^{k-1}(E)$ represented by a collection of mappings $\{y^\alpha(x^1, \ldots, x^n), y_j^\alpha(x^1, \ldots, x^n), \ldots, y_{j^1, \ldots, j^{k-1}}^\alpha(x^1 \ldots, x^n)\}$ is admissible at $p = \bar{\pi}^{k-1}(j^{k-1})$ if

$$dy^\alpha(p) = \sum_j y_j^\alpha(p) dx^j, \tag{10.4}$$

$$dy_{j^1}^\alpha(p) = \sum_{j^2} y_{j^1 j^2}^\alpha(p) dx^{j^2}, \tag{10.5}$$

$$\vdots$$

$$dy_{j^1, \ldots, j^{k-2}}^\alpha(p) = \sum_{j^{k-1}} y_{j^1, \ldots, j^{k-1}}^\alpha(p) dx^{j^{k-1}}, \tag{10.6}$$

where $\alpha = 1, \ldots, m$. Hence, $\{x^i, y^\alpha, y_{j^1}^\alpha, \ldots, y_{j^1, \ldots, j^k}^\alpha\}$ defines a local coordinate system on $\bar{J}^k(E)$ in a neighbourhood of $j_{j^{k-1}}^1 j^{k-1}$, where $y_{j^1, \ldots, j^k}^\alpha(p) = \frac{\partial y_{j^1, \ldots, j^{k-1}}^\alpha}{\partial x^{j^k}}(p)$. The local coordinates $\{x^i, y^\alpha, y_{j^1}^\alpha, \ldots, y_{j^1, \ldots, j^k}^\alpha\}$ of the holonomic subbundle $J^k(E)$ are, on the other hand, completely symmetric. Namely, for any index $\alpha = 1, \ldots, m$, any $1 < q \le k$, and any permutation $\sigma$ we have that $y_{\sigma(j^1), \ldots, \sigma(j^q)}^\alpha = y_{j^1, \ldots, j^q}^\alpha$.

As our main object of interest is the bundle of second-order jets the reader may find the following example useful.

*Example 10.1.* Consider the cotangent bundle of the manifold $M$. That is, let $E(M) = T^*(M)$. A local section $\omega: U \to T^*(M)$ is a locally defined differentiable 1-form. Given such a form $\omega$ and a point $p \in U$, there exists always a smooth function $f: V \to \mathbb{R}$ such that if $U \cap V \ne \emptyset$ then $f(p) = 0$ and $df(p) = \omega(p)$. This fact enables one to identify the cotangent bundle of a manifold with the space of 1-jets of real-valued functions on $M$ with target at 0, i.e., a subset of the space of 1-jets of sections of the trivial bundle $\pi: M \times \mathbb{R} \to M$. In fact, one can show that $J^1(M \times \mathbb{R})$ is canonically diffeomorphic with $T^*(M) \times \mathbb{R}$. As per our identification, any local section $\omega$ of $T^*(M)$ is admissible as a section of a jet bundle. Therefore, for any 1-from $\omega$ its 1-jet $j_p^1 \omega$ belongs to $\bar{J}^2(M \times \mathbb{R})$, the space of semi-holonomic 2-jets of real-valued functions on $M$ with the target at 0. In other words, $J^1(T^*(M))$ is diffeomorphic to a submanifold of $\bar{J}^2(M \times \mathbb{R})$. Moreover, a 1-jet $j_p^1 \omega$ is a holonomic 2-jet, i.e., $j_p^1 \omega \in J^2(M \times \mathbb{R})$, if, and only if, $\omega$ is *closed*, that is, if there exists a real-valued function $f$ such that $\omega = df$ in some open neighbourhood of the point $p$. In a local coordinate system $x^1, \ldots, x^n$ on $U \subset M$, $\omega = y^1 dx^1 + \cdots + y^n dx^n$, while its 1-jet $j_p^1 \omega$ has coordinates $(x^j, y^i, \frac{\partial y^k}{\partial x^s})$. Accordingly, $j_p^1 \omega$ is a holonomic 2-jet if, and only if, $\frac{\partial y^k}{\partial x^s}(p) = \frac{\partial y^s}{\partial x^k}(p)$ for every $1 \le s, k \le n$.

## 10.3 Lie groups of jets of diffeomorphisms

For the remainder of this chapter we shall limit our presentation to bundles of jets of real-valued functions only. Namely, let $J^1(\mathbb{R}^n)$ denote the space of 1-jets of all local diffeomorphism of $\mathbb{R}^n$ with source at the origin. As any local diffeomorphism of $\mathbb{R}^n$ at the origin can be viewed as a local (about the origin) section of the trivial bundle $\pi : \mathbb{R}^n \times \mathbb{R}^n \to \mathbb{R}^n$, the space $J^1(\mathbb{R}^n)$ is simply the origin-based fibre of the prolongation $J^1(\mathbb{R}^n \times \mathbb{R}^n)$. It is also, as remarked earlier, a fibre bundle $\pi_0^1 : J^1(\mathbb{R}^n) \to \mathbb{R}^n$ where a 1-jet is projected onto its target. In fact, $\pi_0^1 : J^1(\mathbb{R}^n) \to \mathbb{R}^n$ is a principal fibre bundle. Its structure group $G^1$ is a fibre of $J^1(\mathbb{R}^n)$ at $0 \in \mathbb{R}^n$. In other words, $G^1$ is the space of all invertible 1-jets of (local) differentiable mappings $f : \mathbb{R}^n \to \mathbb{R}^n$ with target and source at the origin. The group $G^1$ acts on $J^1(\mathbb{R}^n)$ by composition on the right. Indeed, let $a \in G^1$ and let $j^1 \in J^1(\mathbb{R}^n)$. As $a \in (\pi_0^1)^{-1}(0)$, there exists a local diffeomorphism $g : \mathbb{R}^n \to \mathbb{R}^n$ such that $g(0) = 0$ and $a = j_0 g$. Also, by the definition of a jet of a function, there exists a local diffeomorphism $f : \mathbb{R}^n \to \mathbb{R}^n$ such that $f(0) = \pi_0^1(j^1)$ and $j_0 f = j^1$. The right action of $a$ on $j^1$ provides a jet $\hat{j}^1$ such that $j_0(f \circ g) = \hat{j}^1$. Note that the action of $G^1$ on $J^1(\mathbb{R}^n)$ is, as expected, fibre preserving due to the fact that $(f \circ g)(0) = f(0)$. Recalling the representation of a jet as a collection of first order derivatives of a diffeomorphism, we see that $G^1$ is isomorphic to the general linear group $GL(n; \mathbb{R})$.

To construct a prolongation (either holonomic or non-holonomic) of the fibre bundle $J^1(\mathbb{R}^n)$ we shall now follow the procedure developed in Section 10.1. That is, the non-holonomic $(k-1)$-prolongations $\tilde{J}^k(\mathbb{R}^n) \equiv J^1(\tilde{J}^{k-1}(\mathbb{R}^n))$, $k \ge 2$, (respectively the semi-holonomic prolongations $\bar{J}^k(\mathbb{R}^n) \subset J^1(\bar{J}^{k-1}(\mathbb{R}^n))$) are 1-jets of sections of the lower order bundles of prolongations. On the other hand, the holonomic $(k-1)$-prolongation is the space $J^k(\mathbb{R}^n)$ of k-jets of all (local) diffeomorphisms of $\mathbb{R}^n$ with the source at the origin[2]. All these prolongations are principal fibre bundles whose structure groups $\tilde{G}^k$, $\bar{G}^k$, and $G^k$, respectively, are the corresponding origin based fibres [80].

The prolongations of jets of local diffeomorphism of $\mathbb{R}^n$ are not only principal fibre bundles over $\mathbb{R}^n$ but are also, as indicated earlier, fibre bundles over

---
[2] Note that $J^k(\mathbb{R}^n)$ is only the $(k-1)$-holonomic prolongation of $J^1(\mathbb{R}^n)$ but it is not the $k$-holonomic prolongation of the fibre bundle $\pi : \mathbb{R}^n \times \mathbb{R}^n \to \mathbb{R}^n$.

the lower-order bundles of prolongations. As they project one onto another, the structure groups, that is, the origin-based fibres, project correspondingly. Denote by $\bar{\nu}_m^k : \bar{G}^k \to \bar{G}^m$, respectively $\nu_m^k : G^k \to G^m$, these induced projections. They are simple restrictions of the standard projections $\bar{\pi}_m^k$ and $\pi_m^k$ to the origin-based fibres. For any pair of positive integers, say $k > m$, the projections $\bar{\nu}_m^k$ and $\nu_m^k$ are morphisms of the corresponding Lie groups. The corresponding kernels[3] $\bar{N}_m^k$ and $N_m^k$ are (normal)[4] vector subgroups of $\bar{G}^k$ and $G^k$, respectively. In fact, all these kernels are abelian vector groups. Moreover, one can show that $\bar{N}_{k-1}^k$ is canonically isomorphic to the vector group $L^k(\mathbb{R}^n, \mathbb{R}^n)$ of multilinear $\mathbb{R}^n$-valued k-forms on $\mathbb{R}^n$. On the other hand, $N_{k-1}^k \subset \bar{N}_{k-1}^k$ may be identified with $S^k(\mathbb{R}^n, \mathbb{R}^n)$, the abelian vector group of multilinear symmetric $\mathbb{R}^n$-valued k-forms on $\mathbb{R}^n$, (cf. [56]).

Let $\bar{N}_1^k$ be the kernel of the canonical projection $\bar{\nu}_1^k : \bar{G}^k \to G^1$. Using $\bar{G}^2$ as an example, we will show below that the Lie group $\bar{G}^k$ can be seen as the semidirect product of the normal subgroup $\bar{N}_1^k$ and a subgroup (canonically) isomorphic to $GL(n; \mathbb{R})$.

*Example 10.2.* Consider the set of all semi-holonomic jets $\bar{J}^2(\mathbb{R}^n)$. Let, $\{x^i, y^i, y_j^i, y_{jk}^i\}$ be a local coordinate system in a neighbourhood of the origin. Any element, say $g$, of the group $\bar{G}^2$, that is, the zero fibre of $\bar{J}^2(\mathbb{R}^n)$, can be represented as $(g_j^i, n_{jk}^i)$, $i, j, k = 1, \ldots, n$, where $\bar{\nu}_1^2((g_j^i, n_{jk}^i)) = g_j^i$ and where $(\delta_j^i, n_{jk}^i) \in \bar{N}_1^2$. Given another element $h = (h_r^j, m_{rp}^j) \in \bar{G}^2$, and noting that $g_j^i$ and $h_r^j$ represent first derivatives of the underlying functions, while $n_{jk}^i$ and $m_{rp}^j$ are the respective second derivatives, the chain rule provides the following representation of the product $gh$:

$$(g_j^i, n_{jk}^i)(h_r^j, m_{rp}^j) = (\sum_{j=1}^n g_j^i h_r^j, \sum_{j,k=1}^n n_{jk}^i h_r^j h_p^k + \sum_{j=1}^n g_j^i m_{rp}^j), \quad (10.7)$$

(cf. [9, 20]).

## 10.4 Higher-order frame bundles

We proceed now to generalize the concept of a linear frame. To this end, let $M$ be a real, connected, (smooth) $n$-dimensional manifold. Denote by $H^k(M)$ the bundle of all $k$-jets of local diffeomorphisms $f : U \subset \mathbb{R}^n \to M$ with source at the origin. Assume that $\pi^k : H^k(M) \to M$ is the projection onto the target

---

[3] The kernel of a morphism is the inverse image of the identity of a group, i.e., given, for example, $\nu_m^k : G^k \to G^m$, $N_m^k \equiv (\nu_m^k)^{-1}(e_m)$, where $e_m$ is the identity element of $G^m$.

[4] A subgroup of a group is called a *normal subgroup* if for every element of the group the right and left subgroup based cosets are identical [78].

of a jet. If one views our diffeomorphisms as local (about the origin) sections of the product bundle $M \times \mathbb{R}^n \to \mathbb{R}^n$, the space $H^k(M)$ becomes simply the origin-based fibre of the holonomic prolongations $J^k(M \times \mathbb{R}^n)$ of the product bundle $M \times \mathbb{R}^n$. It can be shown [80] that $\pi^k : H^k(M) \to M$ is a principal fibre bundle over the manifold $M$ with structural group $G^k$. An element of $H^k(M)$ will be called a *holonomic frame of order $k$* (or a *$k$-frame*) of the manifold $M$. Note that, as any differentiable manifold is locally diffeomorphic to $\mathbb{R}^n$, the above construction is essentially identical to the construction of the space $J^k(\mathbb{R})$ of jets of real-valued functions.

The notion of a (holonomic) $k$-frame can be extended to encompass other than holonomic situations. We shall proceed by induction as in [96], but first let us note that $H^1(\mathbb{R}^n) \equiv J^1(\mathbb{R}^n)$. Next, suppose that $b : H^1(\mathbb{R}^n) \to H^1(M)$ is a local isomorphism. It induces in a natural way a local diffeomorphism $f : U \subset \mathbb{R}^n \to M$ such that $f \circ \pi^1 = \pi^1 \circ b$. The isomorphism $b$ is said to be *admissible* if the domain of $b$ contains the identity $e_1$ of the group $G^1$ and $b(e_1) = j_0^1 f$. The first jet $j_{e_1}^1 b$ of an admissible local isomorphism $b$ will be called a *semi-holonomic 2-frame*. The space $\bar{H}^2(M)$ of all semi-holonomic 2-frames is a fibre bundle over $H^1(M)$ with the projection $\bar{\pi}_1^2(j_{e_1}^1 b) = b(e_1) = j_0^1 f$. It is also a principal fibre bundle over $M$ with the projection $\bar{\pi}^2 = \pi^1 \circ \bar{\pi}_1^2$ and $\bar{G}^2$ as its structure group.

Suppose now that the fibre bundle $\bar{H}^{k-1}(M)$ of all semi-holonomic $(k-1)$-frames is already available. A local isomorphism $b^{k-1} : \bar{H}^{k-1}(\mathbb{R}^n) \to \bar{H}^{k-1}(M)$, where $\bar{H}^{k-1}(\mathbb{R}^n) \equiv \bar{J}^{k-1}(\mathbb{R}^n)$, is called $(k-1)$-*admissible* if the induced isomorphism $b^{k-2} : \bar{H}^{k-2}(\mathbb{R}^n) \to \bar{H}^{k-2}(M)$ is $(k-2)$-admissible, and $b^{k-1}(e_{k-1}) = j_{e_{k-2}}^1 b^{k-2}$, where $e_{k-1}$ and $e_{k-2}$ are the distinguished elements of $\bar{H}^{k-1}(\mathbb{R}^n)$ and $\bar{H}^{k-2}(\mathbb{R}^n)$, respectively. The 1-jet $j_{e_{k-1}}^1 b^{k-1}$ of a (k-1)-admissible local isomorphism $b^{k-1}$ is called a *semi-holonomic $k$-frame*. The space $\bar{H}^k(M)$ of all such 1-jets is a principal fibre bundle over $M$ with structural group $\bar{G}^k$.

*Remark 10.3.* It should be easy to see from the construction presented above that any holonomic frame is also semi-holonomic. Specifically, let $f : U \subset \mathbb{R}^n \to M$ be a local diffeomorphism about the origin. It induces, via its first jet at the origin, an admissible local isomorphism $f^\sharp : H^1(\mathbb{R}^n) \to H^1(M)$. Indeed, given $j^1 \in H^1(M)$ there exists a local, about the origin, diffeomorphism $g$ of $\mathbb{R}^n$ into $\mathbb{R}^n$ such that $j^1 = j_0^1 g$. Then, set $f^\sharp(j_0 g) = j_0^1(f \circ g)$ and $f^\sharp(e_1) \equiv f^\sharp(j_0^1 \mathrm{id}) = j_0^1 f$ implying that $j_{e_1}^1 f^\sharp = j_0^2 f \in \bar{H}^2(M)$. The recursive application of this argument shows, in fact, that there is a natural injection of the space of all holonomic $k$-frames into the space of all semi-holonomic $k$-frames and the corresponding natural injection $\nu^k : G^k \to \bar{G}^k$ of the respective structure groups.

The concept of a frame can be extended further yet to include the strictly non-holonomic case. Although for the precise definition of a non-holonomic

$k$-frame we refer the reader to [80] and [96], one may find the following construction instructive. Consider a differentiable mapping $f : U \to H^1(M)$ of a neighbourhood of the origin of $\mathbb{R}^n$ into $H^1(M)$ such that $\pi^1 \circ f : U \to M$ is a local diffeomorphism. The first jet of $f$ at the origin 0 can be taken as a *non-holonomic 2-frame* of $M$ at $\pi^1(f(0))$. The recursive application of this construction will then lead to the realization that the space $\tilde{H}^k(M)$ of non-holonomic $k$-frames of $M$ can be thought of as the space of first jets of all local sections of the bundle $\tilde{H}^{k-1}(M)$ of all non-holonomic $(k-1)$-frames on $M$.

The development of the notion of a frame and that of the frame bundle, as presented in this section, whether holonomic or semi-holonomic, parallels the introduction of the concept of a prolongation of a section of an arbitrary fibre bundle (see Section 10.1). Thus, one may wounder if these two concepts are "isomorphic", that is, whether or not a prolongation of a section of the bundle of linear frames $H^1(M)$ is a frame. The answer is provided by the following statement[5]:

**Theorem 10.4.** *There exists a family of canonical diffeomorphisms $\mu_k$ of the space of all semi-holonomic $k$-frames $\bar{H}^k(M)$ and the $k$-semi-holonomic prolongation $\bar{J}^{k-1}(H^1(M))$ such that*

1. *$\mu_1$ is the identity on $H^1(M)$,*
2. *For any $k \geq 1$, $\mu_k$ is a fibre bundle morphism over $M$,*
3. *For any $1 \leq m \leq k$, the following diagram commutes*

$$\begin{array}{ccc} \bar{H}^k(M) & \xrightarrow{\mu_k} & \bar{J}^{k-1}(H^1(M)) \\ {\scriptstyle \bar{\pi}^k_m} \downarrow & & \downarrow {\scriptstyle \bar{\pi}^k_m} \\ \bar{H}^m(M) & \xrightarrow{\mu_m} & \bar{J}^{m-1}(H^1(M)). \end{array}$$

In the remaining chapter we will concentrate, with a few necessary exceptions, on the analysis of bundles of linear and second-order frames only. As we have learned so far, these are principal fibre bundles. In particular, the structure group of $H^1(M)$ is the general linear group $GL(n;\mathbb{R})$ while the structure group of $\bar{H}^2(M)$ (respectively, $H^2(M)$) is the semi-direct product of the general linear group $GL(n;\mathbb{R})$ and the vector group $L(\mathbb{R}^n, \mathbb{R}^n) \equiv \bar{N}_1^2$ of bilinear $\mathbb{R}^n$-valued (respectively, the vector group $Sym(\mathbb{R}^n, \mathbb{R}^n) \equiv N_1^2$ of symmetric bilinear $\mathbb{R}^n$-valued) forms on $\mathbb{R}^n$ (see Example 10.2). Moreover, $\bar{H}^2(M)$ and $H^2(M)$ are vector bundles over $H^1(M)$ with fibres isomorphic to $\bar{N}_1^2$ and $N_1^2$, respectively.

---

[5] Yuen [96], Theorem I.7.

# 11
# Connections of higher order

We showed in the previous chapter how to generalize the concept of a linear frame to an arbitrary finite order. In this final chapter of our mathematical part, we will illustrate how to extend the idea of a (linear) connection on the bundle of linear frames to higher-order frame bundles. Despite the fact that, in the interest of clarity, we focus our presentation on connections on holonomic frame bundles, we start our exposition from the slightly more general vantage point of semi-holonomic connections so as to present the interested reader with a broader perspective of the subject[1]. For the more general case of a non-holonomic connection the reader is referred to [9, 80] and [23, 38].

## 11.1 Fundamental form

In our presentation of a linear connection in Section 9.2 we used extensively the concept of canonical form on a frame bundle (9.22) assigning to a tangent vector its coordinates in a given frame. We will now extend this important concept to bundles of frames of an arbitrary finite order. The notion of the fundamental form, which is so essential for the development of connections on bundles of linear frames of higher order, was first introduced by Ehresmann in [16]. In this exposition we will follow Kobayashi's definition in [55][2].

Consider the bundle $\bar{H}^k(M)$ of all $k$-order semi-holonomic frames of an $n$-dimensional manifold $M$. Any $k$-frame $u \in \bar{H}^k(M)$ can be represented, as shown already in the previous chapter, as $u = j^1_{e_{k-1}} b^{k-1}$ for some admissible isomorphism $b^{k-1} : \bar{H}^{k-1}(\mathbb{R}^n) \to \bar{H}^{k-1}(M)$ where $e_{k-1}$ denotes the identity of the group $\bar{G}^{k-1}$, and where $b^{k-1}(e_{k-1}) = \bar{\pi}^k_{k-1}(u)$. The tangent map of $b^{k-1}$ at $e_{k-1}$ defines a linear isomorphism $b^{k-1}_* : E^{k-1} \equiv T_{e_{k-1}} \bar{H}^{k-1}(\mathbb{R}^n) \to$

---
[1] As illustrated in Chapter 3, only holonomic and non-holonomic frames appear to have a clear physical meaning in the Continuum Mechanics context.
[2] The presentation of the material in this section is partially based on [23].

$T_{b^{k-1}(e_{k-1})}\bar{H}^{k-1}(M)$. Moreover, since $\bar{H}^{k-1}(\mathbb{R}^n) \equiv \mathbb{R}^n \times \bar{G}^{k-1}$ the tangent space at $e_{k-1}$ decomposes canonically into $\mathbb{R}^n \oplus \bar{\mathfrak{g}}^{k-1}$, where $\bar{\mathfrak{g}}^{k-1}$ denotes the Lie algebra of the group $\bar{G}^{k-1}$. In particular, $\bar{G}^1 = Gl(n;\mathbb{R})$ and $\bar{\mathfrak{g}}^1 = gl(n;\mathbb{R})$.

A *fundamental form* on $\bar{H}^k(M)$ is the $\mathbb{R}^n \oplus \bar{\mathfrak{g}}^{k-1}$-valued 1-form $\vartheta^k$ defined for any $X \in T\bar{H}^k(M)$ by the identity

$$b_*^{k-1}(\vartheta^k(X)) \equiv \bar{\pi}_{k-1*}^k(X). \tag{11.1}$$

Note that the fundamental form $\vartheta^k$ is the natural generalization of the canonical form $\vartheta$ on $L(M)$ (given by Equation (9.22)) and that $\vartheta^1 = \vartheta$. Note also, that the form $\vartheta^k$ is a pseudotensorial 1-form on $\bar{H}^k(M)$. That is, given $g \in \bar{G}^k$ and $X \in T\bar{H}^k(M)$

$$(R_g^* \vartheta^k)(X) = \varrho^k(g^{-1})\vartheta^k(X) \tag{11.2}$$

where $\varrho^k(g^{-1})$ denotes the right action of $\bar{G}^k$ on $\mathbb{R}^n \oplus \bar{\mathfrak{g}}^{k-1}$ since it is the extension of the standard action of $GL(n;\mathbb{R})$ on $\mathbb{R}^n$. Indeed, comparing the fundamental form $\vartheta^k$ on $\bar{H}^k(M)$ with the canonical form $\vartheta$ on $L(M)$ one can show (cf. [96]) that the following diagram commutes:

$$\begin{array}{ccc} T\bar{H}^k(M) & \xrightarrow{\vartheta^k} & \mathbb{R}^n \oplus \bar{\mathfrak{g}}^{k-1} \\ {\scriptstyle \bar{\pi}_{1*}^k} \downarrow & & \downarrow {\scriptstyle \bar{\pi}_*^{k-1}|_{e_{k-1}}} \\ TL(M) & \xrightarrow{\vartheta} & \mathbb{R}^n \end{array} \tag{11.3}$$

Before proceeding any further we would like to take a closer look at the action of the group $\bar{G}^k$ on its own Lie algebra and, in particular, on the tangent space to $\bar{H}^{k-1}(\mathbb{R}^n)$ at the identity, i.e., on $\mathbb{R}^n \oplus \bar{\mathfrak{g}}^{k-1}$. Consider first the space $H^1(\mathbb{R}^n)$. Its tangent space at the identity $e_1$ is $\mathbb{R}^n \oplus gl(n;\mathbb{R})$. Let $g \in G^2$ be represented by $j_0^2 f$ and let $j_0^1 \psi \in H^1(\mathbb{R}^n)$, where $f, \psi : \mathbb{R} \to \mathbb{R}$ and $f(0) = 0$. Then, the mapping $j_0^1 \psi \mapsto j_0^1(f \circ \psi \circ f^{-1})$ defines a linear isomorphism of the tangent space $\mathbb{R}^n \oplus gl(n;\mathbb{R})$ with itself, which we shall denote by $\mathrm{adj}_g$ [76]. Generalizing this to the semi-holonomic case and an arbitrary finite dimension, and taking into account the fact that $H^1(\mathbb{R}^n)$ can be viewed as a subbundle of $\bar{H}^k(\mathbb{R}^n)$, it can be shown (cf. Yuen [96]) that the adjoint action of $\bar{G}^k$ on $\mathbb{R}^n \oplus \bar{\mathfrak{g}}^{k-1}$ has the following characteristics. First, for any $\mathcal{X} \in \bar{\mathfrak{g}}^{k-1}$ and any element $g \in \bar{G}^k$

$$\varrho^k(g)\mathcal{X} = \mathrm{ad}_{(\bar{\nu}_{k-1}^k(g))}\mathcal{X}. \tag{11.4}$$

Second, given $v \in \mathbb{R}^n$

$$\varrho^k(g)(v, 0) = (\bar{\nu}_1^k(g)v, \lambda^k(g, v)) \tag{11.5}$$

where the mapping $\lambda^k : \bar{G}^2 \times \mathbb{R}^n \to \bar{\mathfrak{g}}^{k-1}$ is linear in the second argument and it vanishes if, and only if, $g \in GL(n;\mathbb{R})$. Moreover,

$$\lambda^k(gh,v) = \lambda^k(g, \bar{\nu}_1^k(h)v) + \mathrm{ad}_{(\bar{\nu}_{k-1}^k(g))}\lambda^k(h,v) \qquad (11.6)$$

for any $g, h \in \bar{G}^k$ and any $v \in \mathbb{R}^n$. Understanding the adjoint action of $\bar{G}^k$ on $\mathbb{R}^n \oplus \bar{\mathfrak{g}}^{k-1}$ allows us to decompose the fundamental form $\vartheta^k$ canonically into a sum of two 1-forms taking values in $\mathbb{R}^n$ and $\bar{\mathfrak{g}}^{k-1}$, respectively. Indeed, we set

$$\vartheta^k = \vartheta + \vartheta_{k-1} \qquad (11.7)$$

where $\vartheta$ is a projection of $\vartheta^k$ onto $\mathbb{R}^n$ while $\vartheta_{k-1}$ takes values in the algebra $\bar{\mathfrak{g}}^{k-1}$. Also, for any $X \in T\bar{H}^k(M)$ and any $g \in \bar{G}^k$

$$R_g^* \vartheta(X) = \bar{\nu}_1^k(g^{-1})\vartheta(X) \qquad (11.8)$$

while

$$R_g^* \vartheta_{k-1}(X) = \mathrm{ad}_{\bar{\nu}_{k-1}^k(g^{-1})}\vartheta_{k-1}(X) + \lambda^k(g^{-1}, \vartheta(X)). \qquad (11.9)$$

Given a $k$-frame $u \in \bar{H}^k(M)$, there exists, as pointed out earlier, an admissible isomorphism $b^{k-1} : \bar{H}^{k-1}(\mathbb{R}^n) \to \bar{H}^{k-1}(M)$ about the identity $e_{k-1}$ such that $u = j^1_{e_{k-1}} b^{k-1}$. The isomorphism $b^{k-1}$ induces, in turn, a linear isomorphism $b^{k-1}_* : \mathbb{R}^n \oplus \bar{\mathfrak{g}}^{k-1} \to T_{\bar{\pi}_{k-1}^k(u)}\bar{H}^{k-1}(M)$. Generalizing the concept of the standard horizontal vector field (Equation (9.25)) we say that the *standard horizontal space of a frame* $u \in \bar{H}^k(M)$ is the $n$-dimensional subspace $\mathcal{SH}(u)$ of $T_{\bar{\pi}_{k-1}^k(u)}\bar{H}^{k-1}(M)$ defined as

$$\mathcal{SH}(u) \equiv b^{k-1}_*(\mathbb{R}^n, 0). \qquad (11.10)$$

Consider now a local section $q : \bar{H}^{k-1}(U) \to \bar{H}^k(M)$ and assume that a $k$-frame $u$ is in the image of $q$. For every element $X \in \mathcal{SH}(u)$ the pull-back by the section $q$ of the fundamental form $\vartheta^k$ gives an element of $\mathbb{R}^n \oplus \{0\}$ as $b^{k-1}_*(\vartheta^k(q_*(X))) = \bar{\pi}_{k-1}^k(q_*(X)) = X$. This very fact provides the following convenient characterization of the standard horizontal space of a frame [26]:

**Proposition 11.1.** *Let $u \in \bar{H}^k(M)$ and let $q : \bar{H}^{k-1}(U) \to \bar{H}^k(M)$ be a local section such that $u$ is in its image. Then, $X \in \mathcal{SH}(u)$ if, and only if,*

$$q^*\vartheta_{k-1}(X) \equiv 0. \qquad (11.11)$$

## 11.2 $\mathcal{E}$-connection

In the remaining parts of this chapter we will be discussing only connections on frame bundles of second-order. But first, to get the true insight into the structure of connections of this type we need to recall the concept of the *Ehresmann connection* ($\mathcal{E}$-connection) [55, 96]. We start by introducing the notion of an equivariant section. Namely, a local section $q : \bar{H}^r(U) \to \bar{H}^k(M)$, where $r < k$ and $U \subset M$, is said to be $\bar{G}^r$-*equivariant* (or simply *equivariant*) if for any frame $u \in \bar{H}^r(M)$ and any element $g \in \bar{G}^r$

$$q(R_g(u)) = R_g(q(u)) \tag{11.12}$$

where the action of $\bar{G}^r$ on $\bar{H}^k(M)$ should be viewed as an embedded action. Note that, except for $r = 1$, there is no canonical embedding of $\bar{G}^r$ into $\bar{G}^k$ and that the image of $\bar{G}^r$ under any embedding is, in general, a submanifold but not necessarily a subgroup of $\bar{G}^k$.

An $\mathcal{E}$-*connection* of order $k+1$ is a $GL(n;\mathbb{R})$-equivariant (local) section $\mathcal{E} : H^1(U) \to \bar{H}^{k+1}(M)$. It defines a reduction of the bundle $H^{k+1}(M)$ to the general linear group $GL(n;\mathbb{R})$ which we shall denote by $M_\mathcal{E}$. The projection $N_\mathcal{E} \equiv \bar{\pi}_k^{k+1}(\mathcal{E}(H^1(U)))$ of the reduced bundle $M_\mathcal{E}$ to $\bar{H}^k(M)$ is also a reduction of $\bar{H}^k(M)$ to $GL(n;\mathbb{R})$. In addition, the submanifolds $M_\mathcal{E}$ and $N_\mathcal{E}$ are related by a $GL(n;\mathbb{R})$-equivariant local partial section $q_\mathcal{E} : N_\mathcal{E} \to M_\mathcal{E}$. The $GL(n;\mathbb{R})$-equivariant submanifold $N_\mathcal{E}$ of $\bar{H}^k(M)$, called the *characteristic manifold* of a connection, is a fundamental object, as we will see soon, for the construction of a connection on $\bar{H}^k(M)$.

Having the $\mathcal{E}$-connection defined already we proceed now to define a connection on $\bar{H}^2(M)$, the bundle of second-order semi-holonomic frames[3]. But first, let us recall that, $\bar{G}^2$ is the semi-direct product of $GL(n;\mathbb{R})$ and $\bar{N}_1^2$. Therefore, given a frame $u \in \bar{H}^2(M)$, either $u \in N_\mathcal{E}$ or there exists a unique element $n \in \bar{N}_1^2$ such that $R_n(u) \in N_\mathcal{E}$. We define a connection on $\bar{H}^2(M)$ by selecting its horizontal space at each and every frame $u$. Namely, given $u \in \bar{H}^2(M)$, if $u \in N_\mathcal{E}$ we select as the horizontal space at $u$ the standard horizontal space $\mathcal{SH}(q_\mathcal{E}(u))$. However, if $u$ is not in $N_\mathcal{E}$, and $R_n(u) \in N_\mathcal{E}$ for some $n \in \bar{N}_1^2$, then the horizontal space at $u$ is simply $R_{n^{-1}}(\mathcal{SH}(q_\mathcal{E}(R_n(u))))$, that is, an equivariant extension of the horizontal distribution on $N_\mathcal{E}$. Conversely, we say that the connection $\omega^2$ on $\bar{H}^2(M)$ is generated by an $\mathcal{E}$-connection if for every $u \in N_\mathcal{E}$ the horizontal space of $\omega^2$ at $u$ is the corresponding standard horizontal space $\mathcal{SH}(q_\mathcal{E}(u))$.

We are now in a position to represent a connection on the bundle of semi-holonomic 2-frames through the fundamental form [96, 26]:

**Proposition 11.2.** *Let $\omega^2$ be a connection on the bundle of semi-holonomic 2-frames $\bar{H}^2(M)$ and let $\mathcal{E} : \bar{H}^2(M) \to \bar{H}^3(M)$ be its generating $\mathcal{E}$-connection. Then, for any $u \in N_\mathcal{E}$ and any $g \in \bar{G}^2$*

$$\omega^2(R_g(u))(R_{g*}\xi) = \widetilde{q}_\mathcal{E}^* \vartheta_2(R_{g*}\xi) - \lambda^2(g^{-1}, \widetilde{q}_\mathcal{E}^* \vartheta(\xi)) \tag{11.13}$$

---

[3] Although the construction of an $\mathcal{E}$-connection is local only, we are able to extend it to the entire bundle $\bar{H}^2(M)$ by assuming that the base manifold $M$ is paracompact and by following the "gluing" process presented in Theorem 9.3. Note also that the construction presented here can be carried out for frames of any order, both holonomic and non-holonomic.

where $\xi \in T_u N_{\mathcal{E}}$ and where $\widetilde{q}_{\mathcal{E}}$ denotes the $\bar{G}_2$-equivariant extension of the $GL(n;\mathbb{R})$-equivariant mapping $q_{\mathcal{E}} : N_{\mathcal{E}} \to M_{\mathcal{E}}$ induced by the $\mathcal{E}$-connection.

To better understand the relation (11.13) note that if $g \in GL(n;\mathbb{R})$ then $\omega^2(R_g(u))(R_{g*}\xi) = q_{\mathcal{E}}^*\vartheta_2(R_{g*}\xi)$ as $\lambda^2$ vanishes on $GL(n;\mathbb{R})$. In addition, if $R_{g*}\xi \in \mathcal{SH}(q(R_g(u)))$ then the right-hand side vanishes identically.

Moving one step further and utilizing the fact that the standard horizontal spaces of two, in general different, $(k+1)$-frames over the same $k$-frame, are just $\bar{G}^k$ translates of each other, one can show [53] (see also [26]) the following:

**Theorem 11.3.** *Consider a connection $\omega^2$ on $\bar{H}^2(M)$ and let $N_{\mathcal{E}}$ be its characteristic manifold. Denote by $l_{\mathcal{E}} : \bar{H}^2(M) \to \bar{N}_1^2$ the $\bar{N}_1^2$-invariant mapping assigning to a frame its $\bar{N}_1^2$-translate away from the characteristics manifold $N_{\mathcal{E}}$ and such that $l_{\mathcal{E}}(ug) = g^{-1}l_{\mathcal{E}}(u)g$ for every $u \in \bar{H}^2(M)$ and $g \in GL(n;\mathbb{R})$. Also, let $\widetilde{q}_{\mathcal{E}} : \bar{H}^2(M) \to \bar{H}^3(M)$ be the $\bar{G}^2$-equivariant extension of the $q_{\mathcal{E}}$ mapping of the $\mathcal{E}$-connection of $\omega^2$. Then,*

$$\omega(u)(\xi) = \widetilde{q}_{\mathcal{E}}^*\vartheta_2(\xi) - \lambda^2(l_{\mathcal{E}}(u)^{-1}, \vartheta(\widetilde{q}_{\mathcal{E}*}(\xi))) \tag{11.14}$$

*for every $u \in \bar{H}^2(M)$ and $\xi \in T_u\bar{H}^2(M)$. Moreover, there is a one-to-one correspondence between linear connections on $\bar{H}^2(M)$ and pairs of mappings $(q_{\mathcal{E}}, l_{\mathcal{E}})$.*

The above statement confirms, in fact, as was shown in [65] using a completely different approach, that there is a one-to-one correspondence between $\mathcal{E}$-connections on $\bar{H}^3(M)$ and linear connections on $\bar{H}^2(M)$.

A connection $\omega^2$ on the bundle of semi-holonomic 2-frames $\bar{H}^2(M)$ induces through the projection $\bar{\pi}_1^2 : \bar{H}^2(M) \to H^1(M)$ a linear connection $\operatorname{proj}\omega^2$ on the bundle of linear frames $H^1(M)$, called the *projected connection* of $\omega^2$. Namely, for any $\xi \in T\bar{H}^2(M)$ the projected connection $\operatorname{proj}\omega^2$ is such that

$$\bar{\nu}_{1*}^2\omega^2(\xi) = \operatorname{proj}\omega^2(\bar{\pi}_{1*}^2(\xi)). \tag{11.15}$$

If $N_{\mathcal{E}}$ is the characteristic manifold of the connection $\omega^2$ then the characteristic manifold of the projected connection $\operatorname{proj}\omega^2$ is the projection of $N_{\mathcal{E}}$. Indeed, as implied by Proposition 11.1, we have that:

**Lemma 11.4.** *The standard horizontal space of the projection of a frame is the projection of the standard horizontal space of that frame. That is, if $u \in \bar{H}^3(M)$, then $\bar{\pi}_{1*}^2\mathcal{SH}(u) = \mathcal{SH}(\bar{\pi}_2^3(u))$.*

In what follows we will concentrate on connections which are defined locally by sections of frame bundles, that is, connections whose horizontal distributions are locally integrable. A connection $\omega$ on a principal $\mathcal{G}$-bundle $P(M, \mathcal{G})$ with the horizontal distribution $\mathcal{H}$ is called *locally integrable* if given a point $p \in M$ there exists an open neighbourhood $U$ containing $p$ and a local section $s : U \to P(M, \mathcal{G})$ such that $s(U)$ is the integral manifold of the horizontal distribution $\mathcal{H}$ restricted to the image $s(U)$. Note that the section $s$ is defined modulo the differentiable action of the structure group $\mathcal{G}$. Suppose now that $\omega^2$ is a (locally) integrable connection on $\bar{H}^2(M)$ and let $\mathcal{E}$ be the corresponding Ehresmann-connection on $H^3(V)$ where $V$ is an open subset of $M$. Assume that $s : U \to \bar{H}^2(M)$ is the defining section of the connection $\omega^2$ (the horizontal distribution of $\omega^2$ is tangent to s(U)) and that $U \cap V \neq \emptyset$. Then, there exists a local section $r : U \cap V \to H^1(M)$ and a mapping $\varepsilon : r(U \cap V) \to \bar{H}^2(M)$ such that $s(p) = \varepsilon(r(p))$ for any $p \in U \cap V$. Indeed, $r|_{U \cap V} \equiv \bar{\pi}_1^2 \circ s|_{U \cap V}$. We can extend the mapping $\varepsilon$ to a $GL(n;\mathbb{R})$-equivariant section $\varepsilon : H^1(U \cap V) \to \bar{H}^2(M)$. It defines, as shown in Proposition 11.2, a linear connection $\mathrm{i}\,\omega^2$ on $H^1(M)$, called the *induced connection of* $\omega^2$, with $\bar{\pi}_1^2(s(U \cap V)GL(n;\mathbb{R}))$ as its characteristic manifold, where $s(U \cap V)GL(n;\mathbb{R})$ denotes the extension of the image $s(U \cap V)$ by the action of the group $GL(n;\mathbb{R})$.

Thus, given a locally integrable connection $\omega^2$ on the bundle of semi-holonomic second-order frames $\bar{H}^2(M)$, it induces on $H^1(M)$ two, in general different, linear connections $\mathrm{proj}\,\omega^2$ and $\mathrm{i}\,\omega^2$. The projected connection $\mathrm{proj}\,\omega^2$ and the induced connection $\mathrm{i}\,\omega^2$ are indeed, in general, different due to the fact that the characteristic manifold $N_{\mathrm{i}\,\omega^2}$ is in most cases different from the characteristic manifold $N_{\mathrm{proj}\,\omega^2}$; one being a projection of the characteristic manifold of $\omega^2$, the other generated by the image of the section defining $\omega^2$. Conversely, however, the definition of the induced connection and the construction of a connection from its $\mathcal{E}$-connection imply the following uniqueness statement [17]:

**Proposition 11.5.** *Suppose that $\omega$ is a locally integrable linear connection on $H^1(M)$. Let $\widetilde{\omega}$ be another linear connection on $H^1(M)$ such that its characteristic manifold $N_{\widetilde{\omega}}$ is the integral manifold of the horizontal distribution $\mathcal{H}$ of the connection $\omega$. Then, there exists a unique connection $\omega^2$ on $\bar{H}^2(M)$ such that $\omega = \mathrm{proj}\,\omega^2$ and $\widetilde{\omega} = \mathrm{i}\,\omega^2$.*

## 11.3 Second-order (holonomic) connection

After discussing the role of the Ehresmann connection ($\mathcal{E}$-connection) in the construction of a semi-holonomic connection, we derive in this section the coordinate representation of a second-order connection as well as the representations of the the two linear connections induced by it. For the simplicity and clarity of our presentation we consider only connections defined on a holonomic second-order frame bundle $H^2(M)$. To make the presentation even simpler, we assume also that the base manifold $M$ can be equipped with a single (global) coordinate system.

Let $M$ be a simply connected, smooth, $n$-dimensional manifold equipped with a global coordinate system $x^1, \ldots, x^n$. Consider a linear frame $u \in$

## 11.3 Second-order (holonomic) connection

$H^1(M)$ and a holonomic 2-frame $u^2 \in H^2(M)$ such that $\pi_1^2(u^2) = u$ and $\pi^1(u) = (y^1, \ldots, y^n) \in M$. As shown in Section 10.2, the 2-frame $u^2$ has coordinates $(y^i, y_k^i, y_{kl}^i)$, $i, k, l = 1, \ldots, n$, such that $\det\{y_k^i\} \neq 0$ and $y_{kl}^i = y_{lk}^i$ for any choice $i, k$ and $l$. The corresponding linear frame $u$ is simply given by $(y^i, y_k^i)$, $i, k = 1, \ldots, n$. In contrast, an arbitrary non-holonomic second-order frame over $u$ can be represented as $(y^i, y_k^i, \overline{y}_k^i, y_{kl}^i)$ where $y_{kl}^i$ is not necessarily symmetric. On the other hand, a semi-holonomic second-order frame is characterized by the condition that $y_k^i = \overline{y}_k^i$, $i, k = 1, \ldots, n$.

Viewing frames as bases of the corresponding tangent spaces[4], we can also represent $u$ and $u^2$ in the coordinate system $x^1, \ldots, x^n$ induced standard bases $\frac{\partial}{\partial x^i}, \ldots, \frac{\partial}{\partial x^n}$ as

$$u = (y^1, \ldots, y^n; \sum_{i=1}^n y_1^i \frac{\partial}{\partial x^i}, \ldots, \sum_{i=1}^n y_n^i \frac{\partial}{\partial x^i}) \tag{11.16}$$

and

$$u^2 = (y^1, \ldots, y^n; \sum_{i=1}^n y_1^i \frac{\partial}{\partial x^i}, \ldots, \sum_{i=1}^n y_n^i \frac{\partial}{\partial x^i}; \sum_{i,s=1}^n y_{1s}^i \frac{\partial}{\partial x_s^i}, \ldots, \sum_{i,s=1}^n y_{ns}^i \frac{\partial}{\partial x_s^i}), \tag{11.17}$$

respectively.

The structure group $G^2$ of $H^2(M)$, that is, the semidirect product of the general linear group $GL(n; \mathbb{R})$ and the space of all $\mathbb{R}^n$-valued symmetric bilinear forms on $\mathbb{R}^n$ (see Section 10.3), acts on $H^2(M)$ on the right as shown in the Example 10.2. Namely, given an element $(g_k^i, n_{kl}^i) \in G^2$, $i, k, l = 1, \ldots, n$, where $n_{kl}^i = n_{lk}^i$, and a holonomic 2-frame $(y^i, y_k^i, y_{kl}^i)$

$$(y^i, y_k^i, y_{kl}^i)(g_r^k, n_{rp}^k) = (y^i, \sum_{k=1}^n y_k^i g_r^k, \sum_{k,l=1}^n y_{kl}^i g_r^k g_p^l + \sum_{k=1}^n y_k^i n_{rp}^k). \tag{11.18}$$

Suppose now that the connection $\omega^2$ on $H^2(M)$ is (locally) integrable and that its horizontal distribution is given locally by the section $s : U \to H^2(M)$ where

$$s(y^1, \ldots, y^n) = (y^1, \ldots, y^n; s_j^i(y^1, \ldots, y^n); s_{jk}^i(y^1, \ldots, y^n)), \quad s_{jk}^i = s_{kj}^i, \tag{11.19}$$

for $i, j, k = 1, \ldots, n$. The section $s$ induces a $GL(n; \mathbb{R}^n)$-equivariant section $\varepsilon : H^1(U) \to H^2(M)$ defined, as evident from Equation (11.18), by

---

[4] It is quite obvious that a linear frame $u$ can be viewed as a basis of the tangent space to the base manifold at the point $\pi^1(u)$. The same way, a 2-frame $u^2$ can be viewed as a basis of the double tangent space understood as the set of the second-order jets of curves $x : (-\epsilon, \epsilon) \to M$ at $0 \in \mathbb{R}$.

$$\varepsilon(y^i, y^i_k) = (y^i, y^i_k, \sum_{j,r,p,t=1}^{n} s^i_{rp}(s^{-1})^r_t(s^{-1})^p_j y^t_k y^j_l). \tag{11.20}$$

Note that the mapping $\varepsilon$ is nothing but the $\mathcal{E}$-connection of the induced on $H^1(M)$ connection $i\omega^2$. The section $s$ induces also a local section $r = \pi_1^2 \circ s : U \to H^1(M)$:

$$r(y^1, \ldots, y^n) = (y^1, \ldots, y^n; s^i_j(y^1, \ldots, y^n)). \tag{11.21}$$

Choosing a basis in the Lie algebra $gl(n; \mathbb{R})$, a linear connection on the bundle of linear frames $H^1(M)$ is given be a collection of real-valued $GL(n; \mathbb{R})$ equivariant forms $\omega^i_j$ (see Equation (9.35)) with the corresponding horizontal distribution spanned by the vectors

$$h_i \equiv \frac{\partial}{\partial y^i} + \sum_{j,r=1}^{n} \Gamma^k_{ij} y^j_r \frac{\partial}{\partial y^k_r}, \quad i = 1, \ldots, n \tag{11.22}$$

where $\Gamma^k_{ij}$ are the corresponding Christoffel symbols. If the horizontal distribution of a linear connection is a (local) lift by the section $r$ (or rather its tangent map) of the tangent space $TM$ to $H^1(M)$, then the horizontal distribution is spanned by

$$h_i \equiv r_*(\frac{\partial}{\partial y^i}) = \frac{\partial}{\partial y^i} + \sum_{j,r=1}^{n} \frac{\partial s^k_j}{\partial y^i} (s^{-1})^j_l y^l_r \frac{\partial}{\partial y^k_r}, \quad i = 1, \ldots, n. \tag{11.23}$$

Indeed, on the one hand, the tangent space at the image $r(U)$ is spanned by

$$r_*(\frac{\partial}{\partial y^i}) = \frac{\partial}{\partial y^i} + \sum_{j,r=1}^{n} \frac{\partial s^k_j}{\partial y^i} \frac{\partial}{\partial y^k_r}, \quad i = 1, \ldots, n, \tag{11.24}$$

while, on the other hand, any $GL(n; \mathbb{R})$-invariant vector field on $H^1(U)$ has the generic form

$$\sum_{k=1}^{n} \alpha^k \frac{\partial}{\partial y^k} + \sum_{i,j,k=1}^{n} \beta^k_j y^j_i \frac{\partial}{\partial y^k_i}. \tag{11.25}$$

A simple comparison of these two representations provides the required expression for the horizontal space of the connection induced by the section $r$.

Moreover, comparing Equations (11.22) and (11.23) implies that the Christoffel symbols of the linear connection induced by the section $r$ are

$$\Gamma^i_{jk} = \sum_{l=1}^{n} \frac{\partial s^i_l}{\partial y^j} (s^{-1})^l_k. \tag{11.26}$$

## 11.3 Second-order (holonomic) connection

The fundamental form $\vartheta^2$ on $H^2(M)$ is represented in the coordinate system $x^1, \ldots, x^n$ on $M$ by the following collection of forms [9]:

$$\vartheta^i = \sum_{k=1}^n (y^{-1})^i_k dy^k, \quad i = 1, \ldots, n \tag{11.27}$$

and

$$(\vartheta_1)^i_j = \sum_{k=1}^n (y^{-1})^i_k (dy^k_j - \sum_{l,r=1}^n y^k_{rj}(y^{-1})^r_l dy^l), \quad i,j = 1, \ldots, n \tag{11.28}$$

where $\vartheta^i$ forms represent the $\mathbb{R}^n$-valued portion of the fundamental form (see Equation (9.32)) while the forms $(\vartheta_1)^i_j$ take value in the Lie algebra $gl(n;\mathbb{R})$. Invoking the representation of the $\mathcal{E}$-connection (Equation (11.20)) as well as Proposition 11.2, a straightforward calculation shows that the Christoffel symbols of the induced on $H^1(M)$ connection $i\omega^2$ take the form

$$\widetilde{\Gamma}^i_{kl} = -\sum_{p,r=1}^n s^i_{pr}(s^{-1})^p_k(s^{-1})^r_l, \quad i,k,l = 1, \ldots, n. \tag{11.29}$$

Finally, we note that the horizontal distribution of the second-order connection $\omega^2$ on $H^2(M)$ generated by the section $s$ is spanned by the vectors

$$s_*(\frac{\partial}{\partial y^i}) = \frac{\partial}{\partial y^i} + \frac{\partial s^j_k}{\partial y^i}\frac{\partial}{\partial y^j_k} + \frac{\partial s^l_{rp}}{\partial y^i}\frac{\partial}{\partial y^l_{rp}}, \quad i = 1, \ldots, n. \tag{11.30}$$

On the other hand, as shown in [9], any invariant vector field on $H^2(M)$ is spanned by

$$\frac{\partial}{\partial y^j} - \sum_{k,l,r=1}^n \Gamma^k_{il}\frac{\partial}{\partial y^k_r} - (\Gamma^s_{im}y^m_{rk} + \Gamma^s_{iml}y^m_r y^l_k)\frac{\partial}{\partial y^s_{rk}}, \quad i = 1, \ldots, n \tag{11.31}$$

where the coefficients $\Gamma^i_{jk}$ and $\Gamma^i_{jkl}$, called the *generalized Christoffel symbols* of a second order connection, are functions of position. All of the above confirms that the generalized Christoffel symbols of the integrable holonomic second-order connection $\omega^2$ on $H^2(M)$ generated by the section $s$ can be expressed as

$$\Gamma^i_{jk} = \sum_{l=1}^n \frac{\partial s^i_l}{\partial y^j}(s^{-1})^l_k \tag{11.32}$$

and

$$\Gamma^i_{jkl} = \sum_{m,p,r,s=1}^n \frac{\partial s^i_r}{\partial y^j}(s^{-1})^r_s(s^{-1})^m_k(s^{-1})^p_l s^s_{mp} - \sum_{r,p=1}^n \frac{\partial s^i_{rp}}{\partial y^j}(s^{-1})^r_k(s^{-1})^p_l. \tag{11.33}$$

Note that the first generalized Christoffel symbols $\Gamma^i_{jk}$ are identical, as expected, to the Christoffel symbols (11.26) of the projected connection $\operatorname{proj}\omega^2$.

## 11.4 Simple connections

We finish our discussion of connections on bundles of (linear) frames by addressing the following question: under what conditions is the horizontal distribution of a connection on a second-order frame bundle locally generated by a coordinate chart of the base manifold? In preparation for the discussion of this important issue we shall first introduce the concept of the prolongation of a connection and the notion of a simple connection. Thus, let us consider once again a connection $\omega^2$ on a semi-holonomic frame bundle $\bar{H}^2(M)$. Assume that $\mathcal{E} : H^1(U) \to \bar{H}^3(M)$ is the corresponding (local) $\mathcal{E}$-connection generating the characteristic manifold $N_\mathcal{E}$ and the induced $GL(n; \mathbb{R})$-equivariant section $q_\mathcal{E} : \bar{H}^2(U) \to \bar{H}^3(M)$, where $U$ is an open neighbourhood on $M$. The *prolongation of a connection* $\omega^2$ is a connection, say $\mathcal{P}(\omega^2)$, on $\bar{H}^3(M)$ such that its horizontal distribution is the $q_\mathcal{E}$-lift of the horizontal distribution of $\omega^2$. That is, if $\mathcal{H}$ denotes the horizontal distribution of $\omega^2$ while $\mathcal{P}(\mathcal{H})$ is the horizontal distribution of the connection $\mathcal{P}(\omega^2)$, the latter is the prolongation of the former provided

$$\mathcal{P}(\mathcal{H})_{q_\mathcal{E}(u)} = q_{\mathcal{E}*}(\mathcal{H}_u) \tag{11.34}$$

for every $u \in N_\mathcal{E}$.

The following facts are straightforward consequences of the definition of the prolongation of a connection [17]:

**Proposition 11.6.** *Let $\omega^2$ be a connection on $\bar{H}^2(M)$. Then,*

1. *Its prolongation $\mathcal{P}(\omega^2)$ is unique,*
2. *The projection of the prolongation $\mathcal{P}(\omega^2)$ onto $\bar{H}^2(M)$ coincides with the original connection $\omega^2$,*
3. *Suppose that $\omega^2 = \text{proj}\,\omega^3$ for some third-order connection $\omega^3$. Then, $\omega^3$ is a prolongation of the second-order connection $\omega^2$ if, and only if, $q_\mathcal{E}(N_\mathcal{E})$ is its characteristic manifold.*

We are now ready to introduce the concept of a simple second-order connection, an intermediate step from a flat connection. We say that a connection $\omega^2$ on $\bar{H}^2(M)$ is *simple* if it is a prolongation of some linear connection, say $\omega$. In other words, $\omega^2$ is simple if there exists a linear connection $\omega$ such that $\omega^2 = \mathcal{P}(\omega)$. As implied by the definition of the prolongation, the horizontal distribution of a simple connection $\omega^2$ is a $GL(n; \mathbb{R})$-equivariant lift of a horizontal distribution of its own projection, thus providing, according to Proposition 11.6, the following convenient characterization of a simple connection.

**Proposition 11.7.** *If a second-order connection $\omega^2$ is simple then its horizontal distribution is tangent to its characteristics manifold at all points.*

Furthermore, if $\omega^2$ is a locally integrable second-order connection its simplicity has even greater consequences, as implied by Propositions 11.5 and 11.7:

**Corollary 11.8.** *A locally integrable second-order connection $\omega^2$ is simple, i.e., $\omega^2 = \mathcal{P}(\mathrm{proj}\,\omega^2)$, if, and only if, the projected linear connection $\mathrm{proj}\,\omega^2$ and the induced linear connection $\mathrm{i}\,\omega^2$ are identical.*

Indeed, when $\omega^2$ is simple and integrable its induced connection $\mathrm{i}\,\omega^2$ has the same characteristic manifold as the the projected connection $\mathrm{proj}\,\omega^2$. This property plus the fact that both connections have the same generating $q_\mathcal{E}$-section imply that they are, in fact, identical. The converse is almost automatic [17]. Moreover, using the same arguments as above, one can show that the definitions of the induced linear connection and simplicity imply that the local integrability of a connection is preserved by the prolongation. That is:

**Corollary 11.9.** *A simple second-order connection $\omega^2$ is locally integrable if, and only if, its projected connection $\mathrm{proj}\,\omega^2$ is locally integrable.*

We are now ready to state explicit conditions under which a holonomic second-order connection (a connection defined on a holonomic frame bundle) is locally flat, i.e., is locally equivalent to the standard flat connection on the trivial bundle $\mathbb{R}^n \times G^2$. But first, generalizing the concept of curvature of a linear connection (see Section 9.2) we say that the *curvature* $\Omega_{\omega^2}$ of a second-order connection $\omega^2$ is the $\mathfrak{g}^2$-valued 2-form $d\omega^2$ restricted to the horizontal distribution $\mathcal{H}_{\omega^2}$. Similarly, by the *torsion form* $\Theta_{\omega^2}$ of a connection $\omega^2$ we understand the $\mathbb{R}^n \times gl(n;\mathbb{R})$-valued 2-form $d\vartheta^2$ restricted again to the horizontal distribution $\mathcal{H}_{\omega^2}$. Note [9] that the curvature and torsion of a connection $\omega^2$ and its projection $\mathrm{proj}\,\omega^2$ are related by

$$\pi_1^{2*}\Theta_{\mathrm{proj}\,\omega^2} = d\vartheta\big|_{\mathcal{H}_{\omega^2}}, \tag{11.35}$$

$$\pi_1^{2*}\Omega_{\mathrm{proj}\,\omega^2} = \nu_{1*}^2 \Omega_{\omega^2}. \tag{11.36}$$

Hence, if a connection $\omega^2$ has a vanishing torsion and/or a curvature its projection $\mathrm{proj}\,\omega^2$ exhibits exactly the same properties.

This simple observation, as it was shown in [96] and independently in [25] and [62], leads to the following important conclusion:

**Theorem 11.10.** *A second-order connection is locally flat if, and only if, it is simple and its curvature and torsion vanish.*

As we are especially interested in integrable connections, and as the above stated conditions of local flatness are often difficult to verify, we shall try to investigate if there are simple and more practical ways of identifying flat connections. In particular, let us note that as shown in [53] and [96] (see also [58]):

**Proposition 11.11.** *Let $\mathcal{E}$ be an Ehresmann connection on the holonomic frame bundle $H^3(M)$ inducing a holonomic second-order connection $\omega^2$. Then, the connection $\omega^2$ has a vanishing torsion.*[5]

---

[5] This property can be proved directly from Theorem 11.3 using the definition of the torsion form. Note also that, in general, the Ehresmann connection defining a

Consequently, the definition of the induced connection implies that if $\omega^2$ is a holonomic connection then the Ehresmann connection of $\mathrm{i}\,\omega^2$ is a section of the holonomic frame bundle $H^2(M)$. Therefore:

**Corollary 11.12.** *If a holonomic second-order connection $\omega^2$ is curvature free its induced connection $\mathrm{i}\,\omega^2$ has a vanishing torsion.*

Thus, the definition of the prolongation of a connection implies independently that:

**Corollary 11.13.** *A second-order (locally intergrable or not) connection cannot be prolonged into the holonomic frame bundle $H^3(M)$ unless it is torsion-free.*

Finally, based on all the above partial statements, we easily conclude that:

**Proposition 11.14.** *A simple holonomic second-order connection $\omega^2$ is locally flat if, and only if, its projected connection $\mathrm{proj}\,\omega^2$ is locally flat as well.*

What remains to be shown is that if the torison of the projection $\mathrm{proj}\,\omega^2$ vanishes then its prolongation has a vanishing torsion too. In fact, the definition of the prolongation of a connection and Corollary 11.13 imply that:

**Proposition 11.15.** *A simple, locally-integrable second-order holonomic connection is locally flat.*

The above statement conveys that in the case of a locally integrable holonomic second-order connection being locally flat is equivalent to being simple, that is, a prolongation of a linear connection. Combining this fact with the property that a simplicity of a locally integrable connection is equivalent to having its projected linear connection identical to its induced linear connection (Corollary 11.8), we obtain:

**Theorem 11.16.** *A curvature-free holonomic second-order connection $\omega^2$ is locally flat if, and only if, its projected counterpart $\mathrm{proj}\,\omega^2$ is identical to its induced connection $\mathrm{i}\,\omega^2$.*

A curvature-free linear connection is locally flat, that is, is locally generated by a coordinate chart of the base manifold, if it is symmetric, i.e., its torsion vanishes identically. In somewhat similar way a curvature-free holonomic second-order connection $\omega^2$ is locally flat provided, as stated in Theorem 11.16, its tensorial $\mathfrak{g}^2$-valued two-form

$$\mathcal{D}_{\omega^2} \equiv \mathrm{proj}\,\omega^2 - \mathrm{i}\,\omega^2 \tag{11.37}$$

vanishes. We summarize these findings in the following form:

---

holonomic second-order connection does not need to be a section of a holonomic frame bundle. On the other hand, if it is, the connection it induces has very special properties.

## 11.4 Simple connections

**Proposition 11.17.** *Let $\omega^2$ be a curvature-free holonomic second-order connection. Then,*

1. *$\omega^2$ is locally flat if, and only if, $\mathcal{D}_{\omega^2} = 0$,*
2. *If the 2-form $\mathcal{D}_{\omega^2}$ vanishes, then the projection $\operatorname{proj}\omega^2$ is a symmetric linear connection.*

*However,*

3. *Given a locally flat linear connection, its holonomic and/or semi-holonomic prolongations are by definition simple but, in general, non-flat. That is, although the vanishing of $\mathcal{D}_{\omega^2}$ implies the vanishing of the torsion $\Omega_{\operatorname{proj}\omega^2}$ the converse is not necessarily true.*

We feel that it is appropriate to add here that a similar, but not identical, statement can be made regarding the local flatness of a semi-holonomic second-order connection. Indeed, given a locally integrable semi-holonomic second-order connection $\overline{\omega}^2$, the vanishing of the corresponding 2-form $\mathcal{D}_{\overline{\omega}^2}$ is still a necessary condition for the local flatness of $\overline{\omega}^2$. However, it is no longer a sufficient condition. To make it sufficient, one needs to require also that the projected linear connection $\operatorname{proj}\overline{\omega}^2$ be torsion-free[6]. The difference between the holonomic and the semi-holonomic case comes from the fact that the semi-holonomicity of $\overline{\omega}^2$ does not, in general, guarantee the vanishing of the torsion of its induced connection $\mathrm{i}\overline{\omega}^2$. That is, although the vanishing of the form $\mathcal{D}_{\overline{\omega}^2}$ forces $\operatorname{proj}\overline{\omega}^2$ to be identical to $\mathrm{i}\overline{\omega}^2$ it does not imply that $\mathrm{i}\overline{\omega}^2$ is symmetric. However, if in addition to $\mathcal{D}_{\overline{\omega}^2} = 0$ one requires that the torsion of $\operatorname{proj}\overline{\omega}^2$ is zero, then not only $\operatorname{proj}\overline{\omega}^2$ is locally flat but its prolongation is simple and reducible to a holonomic connection, thus locally flat [26].

Going now back to our holonomic second-order example considered in Section 11.3 we point out that as stated in Proposition 11.17 a connection $\omega^2$ induced locally be a section $s(y^1, \ldots, y^n) = (y^1, \ldots, y^n; s^i_j(y^1, \ldots, y^n);$ $s^i_{jk}(y^1, \ldots, y^n))$, $i, j, k = 1, \ldots, n$, is simple and so locally flat if

$$(\mathcal{D}_{\omega^2})^i_{jk} = \Gamma^i_{jk} - \widetilde{\Gamma}^i_{jk} = 0 \tag{11.38}$$

where the Christoffel symbols $\Gamma^i_{jk}$ and $\widetilde{\Gamma}^i_{jk}$ of the projected linear connection $\operatorname{proj}\omega^2$ and the induced counterpart $\mathrm{i}\omega^2$ are defined by Equations (11.26)

---

[6] We would like to point out that the local flatness of a non-holonomic, locally integrable, connection is characterized by a similar set of conditions but applied to different connections. Indeed, as presented in the first part of this monograph, and as probably evident for the observant reader from the definition of a non-holonomic frame and its coordinate representation (see Section 11.3), a local section of a non-holonomic frame bundle induces a linear connection (borrowed from the first order structure) and a second-order connection somewhat independent from the first one. The local flatness of such a non-holonomic connection is then characterized by the vanishing of the torsion of the first-order connection and the vanishing of the $\mathcal{D}$ tensor of the second-order connection [26, 38].

and (11.29), respectively. Note that, the above condition is coordinate independent as the form $\mathcal{D}_{\omega^2}$ is a tensorial 2-form (see also Proposition 9.23).

Recalling the forms of the Christoffel symbols of the projected and induced connections of our second-order holonomic example from Section 11.3, we see that the vanishing of the form $\mathcal{D}_{\omega^2}$ is equivalent to

$$\sum_{k=1}^{n} \frac{\partial s_j^i}{\partial y^k} s_l^k = s_{jl}^i \tag{11.39}$$

for any $i, j, l = 1, \ldots, n$. As $s$ is a local section into the second-order holonomic frame bundle $H^2(M)$ its coefficients $s_{jk}^i$ are symmetric in lower indices. Therefore, the above relation is simply an integrability condition for the section $s$ proving that there exists a local coordinate system on the base manifold $M$, say $x^1, \ldots, x^n$, such that

$$s_j^i = \frac{\partial y^i}{\partial x^j} \quad \text{and} \quad s_{jk}^i = \frac{\partial^2 y^i}{\partial x^j \partial x^k}, \quad i, j, k = 1, \ldots, n. \tag{11.40}$$

# A
# Groupoids

## A.1 Introduction

In general terms, one may say that, in much the same way as a group represents the *internal*, or local, symmetries of a geometric entity, a groupoid[1] encompasses also the *external*, or distant, symmetries of the same object. A classical example is the tiling of a bathroom floor. Each square tile has a symmetry group consisting of certain rotations and reflections. But it is also intuitively recognized that the floor as a whole has a repetitive pattern and, therefore, some extra symmetry. Because the floor is not infinite, however, we cannot describe all of these extra symmetries by means of a group of global transformations of the plane, such as translations. We will see how the notion of groupoid circumvents this problem. In the case of a material body, the fact that two distant points are made of the same material should be understood as an extra degree of symmetry that the body possesses, just as in the case of the bathroom floor, where distant tiles happen to have the same shape. This analogy should not be pushed too far, but it serves to trigger a useful picture and to understand the unifying role that the concept of groupoid plays in terms of encompassing all types of symmetries.

## A.2 Groupoids

The abstract notion of a *groupoid* emerges as the common structure underlying many constructions that arise naturally in a variety of apparently disconnected applications in algebra, topology, geometry, differential equations, and practically every branch of mathematics. In a restricted way, it can be seen as a generalization of the notion of group, but it is better to understand it as an important mathematical concept in its own right.

---

[1] A thorough treatment of groupoids can be found in [66]. A more informal, but illuminating, explanation is given in [93].

# A  Groupoids

A groupoid consists of a total set $\mathcal{Z}$, a base set $\mathcal{B}$, two ("projection") submersions:
$$\alpha : \mathcal{Z} \longrightarrow \mathcal{B} \quad \text{and} \quad \beta . \mathcal{Z} \longrightarrow \mathcal{B} \tag{A.1}$$
called, respectively, the *source* and the *target* maps, and a binary operation ("composition" or "product") defined only for those ordered pairs $(y, z) \in \mathcal{Z} \times \mathcal{Z}$ such that:
$$\alpha(z) = \beta(y). \tag{A.2}$$
This operation (usually indicated just by reverse apposition of the operands) must satisfy the following properties:

1. **Associativity**:
$$(xy)z = x(yz), \tag{A.3}$$
   whenever the products are defined;
2. **Existence of identities**: for each $b \in \mathcal{B}$ there exists an element $id_b \in \mathcal{Z}$, called the *identity at* $b$, such that $z\, id_b = z$ whenever $\alpha(z) = b$, and $id_b\, z = z$ whenever $\beta(z) = b$;
3. **Existence of inverse**: for each $z \in \mathcal{Z}$ there exists a (unique) *inverse* $z^{-1}$ such that
$$zz^{-1} = id_{\beta(z)} \quad \text{and} \quad z^{-1}z = id_{\alpha(z)}. \tag{A.4}$$

It follows from this definition that to each ordered pair $(a, b)$ of elements of $\mathcal{B}$ one can associate a definite subset $\mathcal{Z}_{ab}$ of $\mathcal{Z}$, namely the subset: $\{z \in \mathcal{Z} \mid \beta(z) = b,\ \alpha(z) = a\}$. It is clear that these sets (some of which may be empty) are disjoint and that their union is equal to $\mathcal{Z}$. It is also clear that the various identities are elements of subsets of the form $\mathcal{Z}_{bb}$.

**Remark A.1. The groups inside the groupoid:** It is not difficult to verify that each set of the form $\mathcal{Z}_{bb}$ is a group.

A useful way to think of a groupoid is just as a collection of symbols $(a, b, c, ...)$ and arrows $(x, y, z, ...)$ connecting some of them. The symbols correspond to the elements of the base set $\mathcal{B}$, while the arrows correspond to the elements of the total set $\mathcal{Z}$. The tail and tip of an arrow $z$ correspond to the source $\alpha(z)$ and the target $\beta(z)$, respectively. Two arrows $z$ and $y$ can be composed if, and only if, the tip of the first ends where the tail of the second begins. The result is an arrow $yz$ whose tail is the tail of $z$ and whose tip is the tip of $y$:

$$\beta(y) = \beta(yz) = c \quad \overset{y}{\longleftarrow} \quad \alpha(y) = b = \beta(z) \quad \overset{z}{\longleftarrow} \quad a = \alpha(z) = \alpha(yz)$$

(with $yz$ the composite arrow spanning from $a$ to $c$)

$$\tag{A.5}$$

For this picture to correspond more or less exactly to the more formal definition of a groupoid, however, we have to add the proviso that for each arrow $z$ connecting point $a$ to point $b$, there exists an "inverse" arrow $z^{-1}$ connecting point $b$ with point $a$. It is also very important to bear in mind that there is no need for a given pair of points to be connected by one or more arrows. Some may be connected and some may not. In fact, an extreme case can occur whereby no two (different) points are thus connected. In this extreme case, the set $\mathcal{Z}$ becomes simply the disjoint union of the groups $\mathcal{Z}_{bb}$. In our tiled-floor analogy, we have a case of a floor patched up with tiles all of which are of different sizes or shapes.

---

**Box A.1 Some examples of groupoids**
To show the versatility of the concept of groupoid, we list a few examples drawn from different areas of Mathematics.

1. **The product groupoid:** Given a set $\mathcal{B}$, the Cartesian product $\mathcal{B} \times \mathcal{B}$ is a groupoid with $\alpha = pr_1$ and $\beta = pr_2$. The reader should be able to complete this trivial, but important, example.
2. **The general linear groupoid $GL(\mathbb{R})$:** Take as the total set the collection of all non-singular square matrices of all orders. The base space will be taken as the set $\mathbb{N}$ of natural numbers. The binary operation is matrix multiplication. We can see that this groupoid is nothing but the disjoint union of all the general linear groups $GL(n, \mathbb{R})$.
3. **The fundamental groupoid:** Let $\mathcal{T}$ be a topological space. For each pair of points $a, b \in \mathcal{T}$ we consider the collection of all continuous curves starting at $a$ and ending at $b$. Two curves starting and ending at the same points are said to be *homotopic* if, keeping these ends fixed, it is possible to transform continuously one curve into the other. Accordingly, we partition this set into equivalence classes, two curves being considered equivalent if they are homotopic, and we define $\mathcal{Z}_{ab}$ as the quotient set (namely, the set of these equivalence classes). The composition of curves is done just like with the arrows of our pictorial description. [Question: why is the partition into equivalence classes needed?]
4. **The tangent groupoid:** Given a differentiable manifold $\mathcal{B}$, we form its tangent groupoid $\mathcal{T}(\mathcal{B})$ by considering, for each pair of points $a, b \in \mathcal{B}$ the collection of all the non-singular linear maps between the tangent spaces $T_a\mathcal{B}$ and $T_b\mathcal{B}$.

Question: Which of the groupoids above are transitive and which aren't?

---

*Remark A.2.* **Conjugation:** The reader should be able to prove that if $\mathcal{Z}_{ab} \neq \emptyset$, then the groups $\mathcal{Z}_{aa}$ and $\mathcal{Z}_{bb}$ are *conjugate*[2], and the conjugation between them is achieved by any element of $\mathcal{Z}_{ab}$. Moreover, the set $\mathcal{Z}_{ab}$ is spanned completely by composing any one of its elements with $\mathcal{Z}_{aa}$ or with $\mathcal{Z}_{bb}$ (to the right or to the left, of course).

---

[2] Strictly speaking, conjugation is a concept that applies to two subgroups of a given group. We will, however, abuse the terminology and still call two groups, $\mathcal{Z}_{aa}$ and $\mathcal{Z}_{bb}$, conjugate if they are related by an equation of the form $\mathcal{Z}_{bb} = z\, \mathcal{Z}_{aa}\, z_{12}^{-1}$, for some $z \in \mathcal{Z}$.

A groupoid is said to be *transitive* if for each pair of points $a, b \in \mathcal{B}$ there exists at least one element of the total set with $a$ and $b$ as the source and target points, respectively. In other words, a groupoid is transitive if, and only if, $\mathcal{Z}_{ab} \neq \emptyset \ \forall (a, b) \in \mathcal{B} \times \mathcal{B}$. In a transitive groupoid all the local groups $\mathcal{Z}_{bb}$ are mutually conjugate (recall Remark A.2).

A groupoid is a *Lie groupoid* if the total set $\mathcal{Z}$ and the base set $\mathcal{B}$ are differentiable manifolds, the projections $\alpha$ and $\beta$ are smooth, and so are the operations of composition and of inverse. It follows from the definition that each of the sets $\mathcal{Z}_{bb}$ is a Lie group.

## A.3 Transitive Lie groupoids and principal bundles

Let $b \in \mathcal{B}$ be a fixed point in the base manifold of a transitive Lie groupoid $\mathcal{Z}$. Consider the subset of the total set $\mathcal{Z}$ formed by the disjoint union $\tilde{\mathcal{Z}}_b$ of all the sets $\mathcal{Z}_{bx}, \forall x \in \mathcal{B}$. The elements $\tilde{z}$ of this set have the property $\alpha(\tilde{z}) = b$. The group $\mathcal{Z}_{bb}$ has a natural effective right action on $\tilde{\mathcal{Z}}_b$, as can be verified directly by composition. Moreover, two elements of $\tilde{\mathcal{Z}}_b$ that differ by the right action of an element of this group must have the same target. In other words, the equivalence classes corresponding to this action consist precisely of the sets $\mathcal{Z}_{bx}$ and, therefore, the quotient set is precisely the manifold $\mathcal{B}$. We are thus led to a principal bundle with total space $\tilde{\mathcal{Z}}_b$, structural group $\mathcal{Z}_{bb}$ and projection $\beta$ (or, rather, the restriction of $\beta$ to $\tilde{\mathcal{Z}}_b$).

If we were to start from a different point, $c$ say, of $\mathcal{B}$, the previous construction would lead to a principal bundle whose structural group $\mathcal{Z}_{cc}$ is conjugate to $\mathcal{Z}_{bb}$, and it is not difficult to show that the two principal bundles are isomorphic. We see, therefore, that giving a transitive Lie groupoid is tantamount to giving an equivalence class of isomorphic principal bundles, each one conveying the same information as the groupoid. The choice of the reference point of departure is somewhat analogous to the choice of a basis in a vector space. No information is lost, but there is a certain loss of objectivity, in the sense that one is no longer working with the actual objects but rather with their representation in the chosen reference.

### Box A.2 From principal bundles to groupoids

Somewhat more surprising than the previous passage from a groupoid to any one of its representative principal bundles is the fact that, given an arbitrary principal bundle $(\mathcal{P}, \pi_P, \mathcal{B}, \mathcal{G}, \mathcal{G})$, one can construct a groupoid of which it is a representative. Indeed, define the quotient space $\mathcal{Z} = (\mathcal{P} \times \mathcal{P})/\mathcal{G}$ as the total set of the intended groupoid. The reason for this choice is clear: we want to assign an arrow between two points of the base manifold to each diffeomorphism between their fibres which is consistent with the right action of the structural group. More precisely, let $a, b \in \mathcal{B}$. The diffeomorphisms $z : \pi_P^{-1}(a) \longrightarrow \pi_P^{-1}(b)$ we are referring to are those satisfying:

$$z(pg) = z(p)g, \quad \forall p \in \pi_P^{-1}(a), \ g \in \mathcal{G}. \tag{A.6}$$

These are exactly the diffeomorphisms generated by pairing any two points $p \in \pi_P^{-1}(a)$, $q \in \pi_P^{-1}(b)$ (i.e., one from each fibre) and then assigning to $pg$ the point $qg$. As $g$ runs through the group the diffeomorphism is generated. In other words, we assign the same arrow to the pair $(p, q) \in \mathcal{P} \times \mathcal{P}$ as we assign to $(pg, qg)$. Hence the quotient $\mathcal{Z} = (\mathcal{P} \times \mathcal{P})/\mathcal{G}$.

The remainder of the construction is straightforward. The inverse of the arrow with the pair $(p, q)$ as representative, is the arrow represented by the pair $(q, p)$. The identity at $b \in \mathcal{B}$ is represented by any pair $(p, p)$ with $\pi_P(p) = b$. The source and target maps are simply: $\alpha(p, q) = \pi_P(p)$ and $\beta(p, q) = \pi_P(q)$. Finally, the composition of arrows is effected by composition of maps. More carefully, let $(p, q)$ be an arrow between $a$ and $b$, and let $(r, s)$ be an arrow between $b$ and $c$, with $a, b, c \in \mathcal{B}$. We proceed to change the representative of the equivalence class $(r, s)$ by applying the right action of the unique $g \in \mathcal{G}$ such that $rg = q$. We obtain the new representative of the *same* arrow as $(q, sg)$. The successive application of the arrows is the arrow from $a$ to $c$ whose representative is the pair $(p, sg)$ (the common intermediate element being cancelled out, as it were).

# References

1. Arnold V (1978) Mathematical methods of classical mechanics, Springer-Verlag, New York
2. Binz E, Elżanowski M (2002) Another look at the evolution of material structures, *Mathematics and Mechanics of Solids* 7/2: 203–214
3. Brauer R (1965) On the relation between the orthogonal group and the unimodular group, *Arch. Rational Mech. Anal.* 18: 97–99
4. Capriz G (1989) Continua with microstructure. *Springer tracts of natural philosophy* 35, Springer, Berlin
5. Cohen H, Epstein M (1983) Homogeneity conditions for elastic membranes, *Acta Mechanica* 47: 207–220
6. Cohen H, Epstein M (1984) Remarks on hyperelastic uniformity, *International Journal of Solids and Structures* 20/3: 233–243
7. Coleman B D, Gurtin M E (1967) Thermodynamics with internal state variables, *The Journal of Chemical Physics* 47/2: 597–613
8. Coleman B D, Noll W (1964) Material symmetry and thermostatic inequalities in finite elastic deformations, *Arch. Rational Mech. Anal.* 15: 87–111
9. Cordero L A, Dodson C T J, de León M (1989) Differential geometry of frame bundles. Kluwer, Dordrecht
10. Cosserat E, Cosserat F (1909), Théorie des corps déformables, Hermann, Paris
11. Cowin S C, Hegedus D H (1976) Bone remodeling I: theory of adaptive elasticity, *Journal of Elasticity* 6: 313–326
12. Di Carlo A, Quiligotti S (2002) Growth and balance, *Mechanics Research Communications* 29: 449–456
13. DiCarlo A (2005) Surface and bulk growth unified. In: *Mechanics of Material Forces* (P. Steinmann and G.A. Maugin, eds.), Advances in Mechanics and Mathematics 11: 53–64, Springer-Verlag, New York
14. DiCarlo A, Naili S, Quiligotti S, Teresi L (2005) Modeling bone remodeling. In: *Proceedings of the COMSOL Multiphysics Conference* (J M Petit, J Daluz eds): 31–36. Paris, France.
15. Duff G F D (1956) Partial differential equations, University of Toronto Press, Toronto
16. Ehresmann Ch (1953) Introduction à la theéorie des structures infinitesimales et des pseudogroupes. In: *Colloq Intern Geom Diff*, Strassbourg: 97–110

17. Elżanowski M (1995) Mathematical theory of uniform material structures, Kielce Technical University Press, Kielce
18. Elżanowski M, Epstein M (1985) Geometric characterization of hyperelastic uniformity, Arch Rational Mech. Anal. 88: 347–357
19. Elżanowski M, Epstein M (1991) The interplay of uniformity with constrained elasticity, Stability and Applied Analysis of Continua 1/1: 95–107
20. Elżanowski M, Epstein M (1992) The symmetry group of second-grade materials, Int J Non-linear Mech 27(4): 635–638
21. Elżanowski M, Epstein M, Śniatycki J (1990) G-structures and material homogeneity, Journal of Elasticity 23: 167–180
22. Elżanowski M, Preston S (1996) On non-holonomic second-order connections with applications to continua with microstructure, Extracta Mathematicae 11(1): 51-58
23. Elżanowski M, Preston S (2004) Linear connections of higher order (unpublished manuscript)
24. Elżanowski M, Preston S (in press) A model of a self-driven evolution of a defective continuum, Mathematics and Mechanics of Solids
25. Elżanowski M, Prishepionok S (1993) Connections on holonomic frame bundles of higher order contact and uniform material structures. Research Reports, Department of Mathematical Sciences, Portland State University, Portland
26. Elżanowski M, Prishepionok S (1994) Connection on higher order frame bundles. In: Tamassy L, Szenthe J (eds) Proc Colloq on Differential Geometry. University of Debrecen. Kluwer Academic Publishers, Dordrecht Boston London
27. Epstein M (1987) A question of constant strain, Journal of Elasticity 17: 23–34
28. Epstein M (1992) Eshelby-like tensors in thermoelasticity. In: Nonlinear Thermodynamic Processes in Continua, TUB-Dokumentation, Berlin, 61: 147–159
29. Epstein M (1999) Toward a complete second-order evolution law, Mathematics and Mechanics of Solids 4: 251–266
30. Epstein M (2002) From saturated elasticity to finite plasticity and growth, Mathematics and Mechanics of Solids 7/3: 252–283
31. Epstein M (2002) The Eshelby tensor and the theory of continuous distributions of inhomogeneities, Mechanics Research Communications 29/6: 501–506
32. Epstein M (2005) Self-driven continuous dislocations and growth. In: Mechanics of material forces, (P Steinmann, G A Maugin eds.): 129–139, Springer
33. Epstein M, Bucataru I (2003) Continuous distributions of dislocations in bodies with microstructure, Journal of Elasticity 70/1: 237–254
34. Epstein M, Burton D, Tucker R (2006) Relativistic anelasticity, Classical and Quantum Gravity 23: 3545–3571
35. Epstein M, Elżanowski M (2003) A model for the evolution of a two-dimensional defective structure, Journal or Elasticity 70: 255-265
36. Epstein M, Elżanowski M, Śniatycki J (1985) Locality and uniformity in global elasticity. In: Lectures notes in Mathematical Physics 1139: 300–310, Springer-Verlag, New York
37. Epstein M, de León M (1996) Homogeneity conditions for generalized Cosserat media, Journal of Elasticity 43: 189-201
38. Epstein M, de León M (1997) Uniformidad material de vigas alabeadas, Anales de Ingeniería Mecánica 3: 539-545
39. Epstein M, de León M (1998) Geometrical theory of uniform Cosserat media, Journal of Geometry and Physics 26: 127–170 (1998).

40. Epstein M, de León M (1998) On uniformity of shells, *International Journal of Solids and Structures* 35/17: 2173–2182
41. Epstein M, de León M (2000) Homogeneity without uniformity: toward a mathematical theory of functionally graded materials, *International Journal of Solids and Structures* 37: 7577–7591
42. Epstein M, de León M (2001) Continuous distributions of inhomogeneities in liquid-crystal-like bodies *Proc Royal Society London* A/457: 2507-2520
43. Epstein M, Maugin G A (1990) The energy-momentum tensor and material uniformity in finite elasticity, *Acta Mechanica* 83: 127–133
44. Epstein M, Maugin G A (1990) Sur le tenseur de moment matériel d'Eshelby en elasticité non-linéaire *Comptes Rendus de l'Académie des Sciences* II-310: 675–678
45. Epstein M, Maugin G A (1991) Energy-momentum tensor and J-integral in electrodeformable bodies, *International Journal of Applied Electromagnetics in Materials* 2: 141–145
46. Epstein M, Maugin G A (1995) On the geometrical material structure of anelasticity, *Acta Mechanica* 115: 119–131
47. Epstein M, Maugin G A (1995) Thermoelastic material forces: definition and geometric aspects, *Comptes Rendus de l'Académie des Sciences* IIb-320: 673–68
48. Epstein M, Maugin G A (2000), Thermomechanics of volumetric growth in uniform bodies, *International Journal of Plasticity* 16: 951-978
49. Epstein M, Śniatycki J (2005), Non-local inhomogeneity and Eshelby entities, *Phil. Mag.* 85/33-35: 3939-3955
50. Eringen A C, Kafadar C B (1976), Polar field theories, *Continuum Physics* (ed A.C. Eringen) IV/1, Academic Press, New York
51. Eshelby J D (1951) The force on an elastic singularity, *Phil. Trans. Roy. Soc. London* A244: 87–112
52. Fujimoto A (1972) Theory of G-structures. Publications of the Study Group of Geometry I, Tokyo
53. Garcia P (1972) Connections and 1-jet fibre bundles *Rendiconti del Seminario Matematico della Università di Padova* 47: 227–242
54. Gurtin M E (2000), Configurational forces as basic concepts of continuum physics, Springer-Verlag, New York
55. Kobayashi S (1961) Canonical forms on frame bundles of higher-order contact *Proc Symp Pure Math* 3: 186–193
56. Kobayashi S (1995) Transformation groups in Differential Geometry. Springer-Verlag, Berlin, Heidelberg, New York
57. Kobayashi S, Nomizu K (1996) Foundations of Differential Geometry. Vol. I, John Wiley & Sons, Inc., New York
58. Kolar I (1994) Torsion-free connections on higer order frame bundles. In: Tamassy L, Szenthe J (eds) Proc Colloq on Differential Geometry. University of Debrecen. Kluwer Academic Publishers, Dordrecht Boston London
59. Kondo K (1955) Geometry of elastic deformation and incompatibility, *Memoirs of the unifying study of the basic problems in engineering sciences by means of geometry*, (Tokyo Gakujutsu Benken Fukyu-Kai, IC)
60. Kroener E (1958) Kontinuumstheorie der Versetzungen und Eigenspannungen. Springer-Verlag, Berlin
61. Lang S (1972) Differential manifolds. Springer-Verlag, New York, Heidelberg

62. de León M, Epstein M (1993) On the integrability of second-order G-structures with application to continuous theories of dislocations, *Reports on Mathematical Physics* 33: 419–436
63. de León M, Epstein M (1994) Material bodies of higher grade *Comptes Rendus de l'Académie des Sciences* I-319: 615–620
64. de León M, Epstein M (1996) The geometry of uniformity in second-grade elasticity, *Acta Mechanica* 114: 217–224
65. Libermann P (1997) Introduction to the theory of semi-holonomic jets, *Arch Mathematicum (Brno)* 33: 173–189
66. MacKenzie K (1987) Lie groupoids and Lie algebroids in Differential Geometry. London Mathematical Society Lecture Note Series 124, Cambridge University Press, Cambridge
67. Marín D, de León M (2004) Classification of material G-structures, *Mediterranean Journal of Mathematics* 1: 375–416.
68. Maugin G A (1993) Material inhomogeneities in elasticity, Chapman and Hall
69. Maugin G A, Epstein M (1991) The electroelastic energy-momentum tensor *Proceedings of the Royal Society London* A433: 299–312
70. Maugin G A, Epstein M, Trimarco C (1992) Pseudomomentum and material forces in inhomogeneous materials, *International Journal of Solids and Structures* 29/14-15: 1889–1900
71. Morgan A J A (1975) Inhomogeneous materially uniform higher order gross bodies, *Arch. Rational Mech. Anal.* 57/3: 189–253
72. Morris S A (1977) Pontryagin duality and the structure of locally compact Abelian groups. In *London Mathematical Society Lecture Notes Series* 29. Cambridge University Press, Cambridge
73. Moser J (1965) On the volume elements of a manifold, *Transactions of the American Mathematica Society* 120: 286–294
74. Noll W (1965) Proof of the maximality of the orthogonal group in the unimodular group, *Arch. Rational Mech. Anal.* 18: 100–102
75. Noll W (1967) Materially uniform bodies with inhomogeneities, *Arch. Rational Mech. Anal.* 27: 1–32
76. Okubo T (1987) Differential Geometry. Marcel Dekker, New York and Basel
77. Podio-Guidugli P (2002) Configurational forces: are they needed? *Mechanics Research Communications* 29: 513–519
78. Pontryagin L.S (1966) Topological groups. Gordon and Breach, New York
79. Poor W A (1981) Differential geometric structures. MacGraw-Hill, New York
80. Saunders D J (1989) The Geometry of jet bundles. Cambridge University Press, Cambridge
81. Segev R (1986) Forces and the existence of stresses in invariant Continuum Mechanics, *Journal of Mathematical Physics* 27/1: 163–170
82. Segev R, Epstein M (1996) On theories of growing bodies. In: *Contemporary research in the mechanics and mathematics of materials (CIMME)* (R C Batra, M F Beatty, eds): 119–130, Barcelona
83. Steenrod N (1951) The topology of fibre bundles. Princeton University Press, Princeton, New Jersey
84. Sternberg S (1983) Lectures on Differential Geometry. Chelsea, New York
85. Synge J L, Schild A (1969) Tensor calculus. Dover, New York
86. Toupin R A (1964), Theories of elasticity with couple-stress, *Arch. Rational Mech. Anal.* 17: 85-112

87. Truesdell C, Noll W (1965), The non-linear field theories of mechanics, *Handbuch der Physik* III/3, Springer Verlag, Berlin
88. Truesdell C and Toupin R A (1960), Principles of classical mechanics and field theories, *Handbuch der Physik* III/1, Springer Verlag, Berlin
89. Wang C-C (1967) On the geometric structure of simple bodies, a mathematical foundation for the theory of continuous distributions of dislocations, *Arch. Rational Mech. Anal.* 27: 33–94
90. Wang C-C (1972) Material uniformity and homogeneity in shells, *Arch. Rational Mech. Anal.* 47: 343–368
91. Wang C-C, Truesdell C (1973), Introduction to Rational Elasticity, Noordhoff, Leyden
92. Warner F W (1983) Foundations of differentiable manifolds and Lie groups. Springer-Verlag, New York, Heidelberg
93. Weinstein A (2000) Groupoids: unifying internal and external symmetry. A tour through examples, *Notices of the American Mathematical Society* 43/7: 744–752
94. Wolff J (1892) Das Gesetz der Transformation der Knochen. A. Hirschwald, Berlin
95. Yavari A, Marsden J E, Ortiz M (2006) On spatial and material covariant balance laws in elasticity, *Journal of Mathematical Physics* 47/4: 042903(1–53)
96. Yuen P-C (1971) Cahiers de Topologie et Géométrie Diff 13(3): 333–370

# Index

1-form, 198
   components of, 198

action of a group on a manifold, 204
actual evolution, 142, 186
adjoint representation, 203
admissible
   coordinate system, 231
   isomorphism, 241
   section, 236
admissible isomorphism, 241
algebra of differentiable functions, 195
ambient space, 195
Ambrose-Singer theorem, 221
archetype
   change of, 15, 62, 78, 143
   Cosserat body, 75
   second-grade body, 42, 48
   simple body, 13
associated fibre bundle, 209
atlas
   compatible, 194
   complete, 194

balance equations
   extra, 156, 159
   in growth and remodelling, 145
base manifold of a fibre bundle, 205
Bianchi identity, 220
   second, 224
bracket operation, 197
bundle chart, 205

canonical

1-form on a Lie group, 215
   flat connection, 222
   form on a frame bundle, 222
Cauchy stress, 113
centralizer of a group, 99
characteristic manifold of a connection, 246
characteristic object, 38
chart, 194
Christoffel symbols, 18, 225, 251
   generalized, 251
closed form, 239
co-vectors, 198
compatible atlas, 194
complete atlas, 194
complete parallelism, 29
configuration
   Cosserat body, 69, 70
      change of, 73, 80
      homogeneous, 78
   functionally graded body
      homosymmetric, 100
      relaxed, 101
   second-grade body
      homogeneous, 59, 61
      stress free, 61
   simple body, 3, 6
      change of, 35
      homogeneous, 30
      natural (stress free), 31, 36
      reference, 4, 6
connection, 213
   $f$-induced, 217

canonical flat, 222
characteristic manifold of, 246
Christoffel symbols of, 225
curvature form of, 220
curvature of, 253
Ehresmann, 246
flat, 222
form, 214
generalized Christoffel symbols of, 251
holonomy group of, 221
induced, 248
linear, 223
locally integrable, 247
material, 19
  Cosserat body, 79
  second-grade body, 61
metric, 233
projected, 247
prolongation of, 252
reducible, 217
reduction of, 217
simple, 252
structure equation of, 220
torsion of, 223, 253
constant strain state, 37
constraints, 105
contorted aelotropy, 37
coordinates of prolongations, 237
cotangent bundle, 210
covariant derivative
  material, 19
  of a section, 226
creep, 165
cross section, 22
cross section of a fibre bundle, 211
curvature, 253
  field, 228
  form, 220
  of a second-order connection, 253
curve
  differentiable, 195
  graph of, 195
  horizontal, 216
  horizontal lift of, 216, 218
  parametrization of, 195

deformation, 4
  gradient, 4

diffeomorphism, 200
differentiable
  curve, 195
  distribution, 201, 232
  functions
    algebra of, 195
  manifold, 195
  mapping, 195
  r-form, 7, 198
  vector field, 196
differential form
  equivariant, 214
  left invariant, 203
distant parallelism, 29
distribution, 201
  differentiable, 201, 232
  horizontal, 213
  integral manifold of, 201
  involutive, 201
  maximal integral manifold of, 201

E-connection, 246
elastic material, 5
embedded submanifold, 200
embedding, 200
equivariant 1-form, 214
equivariant section, 246
Eshelby stress, 112, 114, 117
  and hyperstress
    in a Cosserat body, 132
    in a second-grade body, 129
evolution law
  second-grade material, 183
  self-driven, 175
  simple material, 139
    bulk growth, 169
    viscoelasticity, 162
exponential mapping, 203
exterior differential, 199

f-invariant vector field, 202
FGB, 97
fibre bundle, 21, 204
  base manifold of, 205
  cross section of, 211
  first prolongation of, 236
  homomorphism of, 205
  induced by a mapping, 206
  lift of a function to, 206

local chart of, 205
local section of, 211
local trivialization of, 205
material, 21
principal, 206
projection of, 204
standard fibre of, 204
structure group of, 205
tangent space
  vertical subspace of, 218
total space of, 205
transition functions of, 205
trivializable, 205
fibre bundle tangent space
  horizontal subspace of, 218
fibre metric, 233
field of a quantity, 227
first jet, 235
first prolongation of a fibre bundle, 236
first-order equivalent sections, 235
flat connection, 222
fluid crystal point, 39
fluid point, 38
form
  closed, 239
  exterior covariant derivative of, 219
  horizontal, 219
  pseudotensorial, 219
frame
  holonomic, 49, 90, 241
  non-holonomic, 86, 242
  second-order, 49
  semi-holonomic, 89, 241
frame bundle
  canonical form on, 222
  fundamental form on, 244
  of a manifold, 23
frame bundle of a manifold, 208
Frobenius theorem, 201
function
  support of, 211
  total differential of, 198
fundamental form, 244
fundamental vector field, 206

G-structure, 208, 230
  admissible coordinate system of, 231
  induced by an object, 234
  integrable, 231

locally equivalent, 63
locally flat, 35, 231
material, 25, 53, 90, 100
standard flat, 231
G-structures
  conjugate, 26, 230
  isomorphic, 230
  locally equivalent, 34, 230
general linear group, 208
geometric object, 234
  on a manifold, 210
group
  conjugate, 12
  effective action of, 204
  free action of, 204
  of jets of diffeomorphisms, 239
  representation of, 218
groupoid, 258
  Lie, 260
  locally flat, 36
  tangent, 36, 54, 259
  transitive, 260
growth, 145

Hausdorff space, 195
heat-conduction tensor
  material or Eshelby-like, 127
holonomic
  frame of order k, 241
  jet, 237
  prolongation, 237
holonomy
  bundle, 221
  group, 221
holonomy group, 221
homeomorphism, 194
homogeneity
  local and global, 30
homogeneity, local and global, 60, 79
homosymmetry
  local and global, 100
horizontal
  curve, 216
  distribution, 213
  form, 219
  lift of a curve, 216
  lift of a vector, 213
  space, 213
  vector, 213

272    Index

hyperelastic material, 5

immersion, 200
implant, 12
induced
   connection, 248
   fibre bundle, 206
inhomogeneity
   localized (with compact support), 120
   velocity gradient, 140
injective mapping, 200
integrable G-structure, 231
integral
   curve of a vector field, 201
   manifold, 201
invariant section, 93
inverse motion, 114
involutive distribution, 201
isomorphic G-structures, 230

J-integral, 122
Jacoby identity, 197
jet
   holonomic, 237
   of a map, 44
   of a section, 235
   semi-holonomic, 236
   source of, 44, 235
   target of, 44, 235
jets
   composition of, 45

k-frame, 241
k-order
   equivalent sections, 236
   non-holonomic prolongation, 236
k-th jet, 236

left invariant differential form, 203
left translation, 202
Lie algebra, 202
   constants, 220
Lie group, 202
   canonical 1-form on, 215
   left translation of, 202
   right translation of, 202
Lie subgroup, 203
lift of a function to a fibre bundle, 206
linear connection, 223

Christoffel symbols of, 225
linear frame, 208
local coordinate system, 195
local section of a fibre bundle, 211
locally equivalent G-structures, 63, 230
locally finite covering of a manifold, 211
locally flat G-structure, 231
locally flat groupoid, 36
locally integrable connection, 247

Mandel stress, 117
manifold
   differentiable, 195
   differential action of a group on, 204
   frame bundle of, 208
   geometric object on, 210
   locally finite covering of, 211
   parallelizable, 232
   product of, 197
   Riemannian, 234
   smooth, 195
   tangent bundle of, 209
mapping
   differentiable, 195
   exponential, 203
   injective, 200
   pull-back of, 200
   rank of, 200
   surjective, 200
mass diffusion, 188
mass flux, possibility of, 151
material groupoid, 27, 54, 91
material isomorphism
   Cosserat body, 75
   mass-consistency condition, 11
   second-grade body, 48
   simple body, 8
material parallelism, 17
material symmetry consistency, 144, 187
material symmetry group
   Cosserat body, 76
   second-grade body, 49
   simple body, 10
material uniformity
   Cosserat body, 75
   in thermoelasticity, 123
   second-grade body, 48
   simple body, 10

Maurer-Cartan equation, 203
maximal integral manifold, 201
metric connection, 233

N-structure of a FGB, 100
natural associated bundle of a G-bundle, 212
natural state, 101
non-holonomic
   frame, 242
   prolongation, 236
normal subgroup, 240
normalizer of a group, 98

object induced structure, 234
one-parameter group of diffeomorphisms, 201
open submanifold, 200

paracompact space, 211
parallel
   displacement, 216
   field, 228
   section, 218
   transport, 216
parallel transport, 19
parallelism
   complete, 232
   material, 17
parallelizable manifold, 232
partition of unity, 212
Piola stress, 112, 116
principal fibre bundle, 24, 206
   connection on, 213
   fibre bundle associated with, 209
   homomorphism of, 207
   reduction of, 207
   structure group of, 206
   subbundle of, 207
principal G-bundle, 206
principal-bundle isomorphism, 70
product manifold, 197
projected connection, 247
projection of a fibre bundle, 204
prolongation, 95
   coordinates of, 237
   holonomic, 237
   integrable, 96
   of a connection, 252

   of a fibre bundle, 236
   of a section, 236
   semi-holonomic, 237
pseudo-group of transformations, 194
pseudotensorial form, 219
pull-back, 7, 200

r-form, 198
rank of a mapping, 200
reduced bundle, 207
reducible connection, 217
reduction
   of a group to a subgroup, 207
   theorem, 221
   to the archetype, 141
reduction to the archetype, 185
remodelling, 145
representation of a group, 218
rheological models, 167
Riemannian
   manifold, 234
   metric, 233
   structure, 233
right translation, 202

second Bianchi identity, 224
second structure equation, 224
second-grade body, 47
second-grade inhomogeneity tensor, 61
section
   admissible, 236
   covariant derivative of, 226
   equivariant, 246
   firts jet of, 235
   jet of, 236
   k-th jet of, 236
   non-holonomic prolongation of, 236
   of a fibre bundle, 211
   parallel, 228
   prolongation of, 236
sections
   first-order equivalent, 235
   k-order equivalent, 236
semi-holonomic
   frame, 241
   prolongation, 237
   second-jet, 236
semi-holonomic frame, 241
simple connection, 252

simple material, 5
smooth manifold, 195
solid point, 36
spherical dilatation, 99
stabilizer, 105
standard fibre of a bundle, 204
standard flat G-structure, 34, 231
standard frame bundle, 34
standard horizontal space of a frame, 245
standard horizontal vector field, 223
structure constants, 220
structure equation, 220, 223, 224
structure group
   of a fibre bundle, 22, 205
   of a principal fibre bundle, 206
subbundle of a principal fibre bundle, 207
submanifold
   embedded, 200
   open, 200
submersion, 205
support of a function, 211
surjective mapping, 200
symmetry-automorphism of a FGB, 98
symmetry-isomorphism of a FGB, 98

tangent
   bundle of a manifold, 209
   groupoid, 36, 54, 259
   map, 4, 200
   space, 6, 196
   vector, 195
thermal stresses, 124
thermoelasticity, 122
torsion, 33
   field, 228
   form, 253
   form of a linear connection, 223

   of a second-order connection, 253
total differential, 198
total space of a fibre bundle, 205
Toupin subgroup, 57
   generalized, 57
transition functions, 195, 205
trivializable fibre bundle, 205
trivialization of a fibre bundle, 205

undistorted state, 101
uniformity field, 12
unisymmetric homogeneity of a FGB, 101
unisymmetry of a FGB, 98

vector
   horizontal, 213
   horizontal component of, 213
   tangent, 195
   vertical, 213
   vertical component of, 213
vector bundle, 226
vector field
   components of, 196
   differentiable, 196
   f-invariant, 202
   fundamental, 206
   horizontal lift of, 213
   integral curve of, 201
   left invariant, 202
   right invariant, 202
   standard horizontal, 223
vertical vector space, 213
virtual power
   of configurational forces, 158
   standard, 157
volume form, 232

Wolff's law, 171